PRAISE FOR
Don't Know Much Abou

"Kenneth Davis is the high school teacher we all wish we'd had—smart, funny, and irreverent. *Don't Know Much About® Mythology* is a crystalized reminder of what's enduring about the past, and why it continues to matter today. It's a perfect companion to Harry Potter, *The Da Vinci Code*, and the Bible—and the best excuse I know to get deserted on a Greek island. Between Odysseus and Icarus, it might even give you a few clues for how to find your way home—and how not to."

—Bruce Feiler, author of *Walking the Bible*

"Because Davis ranges widely and with such sparkling wit through a broad sweep of myths, his survey provides a superb starting point for entering the world of mythology."

—*Publishers Weekly* (starred review)

"An engaging handbook on gods, goddesses, and the civilizations they have inspired. . . . [Davis's] goal as an author is to infect readers with his own intellectual eagerness, and he succeeds admirably with this idiosyncratic tour of world mythology. . . . Even professors will have to concede that Davis has done his research—his annotated bibliography is excellent—and that he's a laudably conscientious scholar. An accessible and informed guide to an always-fascinating subject, and an ideal reference for the general reader."

—*Kirkus Reviews*

"A massive overview of every myth under the sun. Davis shatters commonly held myths about myths, differentiating them from allegories and legends, and explores the history of such tales in societies and religions around the globe, from Mesopotamia's Gilgamesh to Genesis' Noah. You can read here, too, about Native Americans' use of peyote, a tempestuous Nordic god of thunder, and a debate over the meaning of evil."

—*Daily News*

"Who are we? In his thoughtful and entertaining *Don't Know Much About® Mythology*, Kenneth C. Davis suggests that, in large part, we are a product of our own creation, our best instincts and worst prejudices—reflected in the stories we tell. We have become our myths, Davis suggests, though they are not necessarily true. In the Americas, myths justified slavery and the destruction of native societies. Yet myth can empower, pulling us upward toward greater creativity and humanity. For all who choose to know just who we are, you must read this book."

—Richard M. Cohen, author of *Blindsided: Lifting a Life Above Illness*

"Over the long development of human culture, the stories of mythology are like a chronology of human evolution. They tell us who we are, and hint at the answer to the growing spiritual intolerance we see today: at the level of the soul, we all want the same things. In *Don't Know Much About® Mythology*, Kenneth C. Davis illuminates these ideas in a popular and entertaining way. I highly recommend this book."

—James Redfield, author of *The Celestine Prophecy*

"With his trademark wit and fiercely entertaining style, Kenneth Davis draws us into mythological worlds, preserving ancient mysteries and enchantments even as he clarifies, orders, and makes sure we have the stories straight. *Don't Know Much About® Mythology* frames questions that arouse curiosity and produces answers that lead to astonishment. Whether you want a crash course on North American Native myths or a refresher course on Gilgamesh, this book will provide a great read and remain a permanent reference manual."

—Dr. Maria Tatar, Department of Folklore and Mythology, Harvard University, and author of *The Annotated Brothers Grimm*

"In each of his Don't Know Much About® books, Kenneth C. Davis has brought the forgotten child to the front row, reminding those of us who hated school that one size doesn't fit all in education—that the desire to learn is far better served by the pursuit of individual passion than by classroom conformity. In *Don't Know Much About® Mythology*, Davis uses the intense passion that stirred in his own soul as a fifth-grade boy reading *The Odyssey* to take us to a place of magic, imagination, and transcendence. Davis not only presents an entertaining exploration of humanity's most sacred stories across many civilizations, he brings us face to face with our most distant ancestors, who were driven by innate curiosity to explain life's mysteries. Davis's book is a masterpiece. I couldn't stop turning the pages."

—Albert Clayton Gaulden, founder and director of the Sedona Intensive and author of *Signs and Wonders*

About the Author

KENNETH C. DAVIS, the *New York Times* bestselling author of *Don't Know Much About® History*, was recently dubbed the "King of Knowing" by Amazon.com. He often appears on national television and radio, and has served as a commentator on NPR's *All Things Considered*. His *USA Weekend* column is read by millions. In addition to his adult titles, he writes the Don't Know Much About® Children's series, published by HarperCollins. He and his wife live in New York City and Vermont.

Also by Kenneth C. Davis

Two-Bit Culture: The Paperbacking of America

Don't Know Much About the Bible

Don't Know Much About the Civil War

Don't Know Much About Geography

Don't Know Much About History

Don't Know Much About the Universe

Mythology

EVERYTHING
YOU NEED TO KNOW
ABOUT THE GREATEST STORIES
IN HUMAN HISTORY
BUT NEVER LEARNED

KENNETH C. DAVIS

HARPER

NEW YORK • LONDON • TORONTO • SYDNEY

HARPER

A hardcover edition of this book was published in 2005 by HarperCollins Publishers.

FIRST HARPER PAPERBACK PUBLISHED 2006.

Designed by Elliott Beard

The Library of Congress has catalogued the hardcover edition as follows:
Davis, Kenneth C.

Don't know much about mythology : everything you need to know about the greatest stories in human history but never learned / Kenneth C. Davis.—1st ed.

p. cm.

Includes bibliographical references (p.507) and index.

ISBN 978-0-06-019460-4

1. Mythology. I. Title.

BL312.D37 2005

201'.3—dc22 2005043341

ISBN 978-0-06-093257-2 (pbk.)

15 QK/RRD 20 19 18 17 16 15 14 13 12

For my Muse,
Joann

CONTENTS

I want to know what were the steps by which men passed from barbarism to civilization.

—VOLTAIRE

Throughout the inhabited world, in all times, and under every circumstance, the myths of man have flourished; and they have been the living inspiration of whatever else may have appeared out of the activities of the human body and mind. . . .

—JOSEPH CAMPBELL,
The Hero with a Thousand Faces

We have not met our forgotten ancestors, but we begin to sense their presence in the dark. We recognize their shadows here and there. They were once as real as we are. We would not be here if not for them. Our natures and theirs are indissolubly linked despite the aeons that may separate us. The key to who we are is waiting in those shadows.

—CARL SAGAN and ANN DRUYAN,
Shadows of Forgotten Ancestors

INTRODUCTION

"In the olden days"—that seems like a good opening for a book about myths—when I was about eleven years old, I could not sit still at my fifth-grade desk. I squirmed. I fidgeted. My mind wandered. Oh, I tried, but I didn't remember much of anything I was supposed to learn. Except that at the end of each day, as the clock on the wall ticked slowly toward three o'clock and freedom, I would sit like a stone in anticipation of those few minutes before dismissal when our teacher set aside the math and science to read aloud from the *Odyssey*.

Magically connected over the vastness of centuries to the people who heard these tales once sung around campfires, I was captivated. Instead of fighting fractions and verbs, I was aboard a ship, sailing mythical oceans, battling witches, demons, and one-eyed monsters—trying to find my way home with brave Odysseus, the wily hero of Homer's epic.

Those daily doses of this great Greek story made my day, gave me a taste for literature and poetry, and certainly whetted my appetite for more mythology. When I had a chance, I would spend hours in the school library, devouring books about the myths—and not just the classics of Greece and Rome. I read about Norse gods such as Thor and the trickster Loki, and the Egyptian gods who inspired the pyramids. There was Sigurd slaying the mighty dragon Fafnir, and the fearsome Celtic hero Cuchulainn single-handedly battling hundreds of enemies in showers of gore that might make Arnold Schwarzenegger wince. I had discovered a whole new world. It was a world of gods, heroes, monsters, and legends—and it was a lot more interesting to me than school!

A few years later, my first job was delivering the *Daily Argus*, the local newspaper in my hometown of Mount Vernon, New York. By all accounts, I was a curious boy, so I wanted to know what "argus" meant. I soon discovered that in Greek mythology, Argus was a monster whose body was covered with eyes—exactly how many eyes he had depends on the source; some say four, some say a hundred—but only two of his eyes ever closed at any one time.

Argus played a supporting role in a tale about Zeus, the randy lord of the gods, and Io, the daughter of a local river god. She was just one of the many women—mortal and divine—desired by the seemingly insatiable Zeus. To conceal his dalliance with Io from his jealous wife, Hera, Zeus transformed the young maiden into a snow-white heifer. But Hera was no dummy when it came to Zeus and his philandering ways with nubile young women. Like some Olympian Alice Kramden of *The Honeymooners* always foiling Ralph's best-laid plans, Hera saw through Zeus's attempted ruse. To get back at her seemingly sex-addicted, cheating husband, Hera claimed Zeus's pet heifer for herself. Hera had Io placed in chains and then set the ever-watchful Argus to guard over her youthful rival.

Zeus didn't give up so easily. He struck back by sending the god Hermes to lull Argus to sleep and free Io. In one version of the story (many Greek myths have variations), Hermes tried to put Argus to sleep by playing on his magical pipe, but that didn't work. So he bored Argus to sleep with a long, tedious story—then cut off his head. To honor Argus, the grieving Hera placed his many eyes on the tail of her favorite bird, the peacock—and that's why the peacock's tail looks the way it does. Hera, however, wasn't finished. Poor Io, still in the form of a heifer, was freed. But Hera just tormented Io with a gadfly that drove her, itching madly, on a wild gallop across Europe and Asia until she finally dove into the sea (the Ionian Sea, which is named for her). Io swam to Egypt, where Zeus returned her to human form and she bore what the tabloids call a "love child." But that's another story. With the Greeks, there's almost always another story.

For me, the link between the monstrous Argus and the newspaper I carried every day was now clear—our local daily was supposed to be the ever-watchful eyes of the community. I'm not sure how accurate that was, but I did become a newspaper junkie from about that time—and this connection between a commonplace, everyday item like a newspaper and ancient myths just made me love the subject all the more.

Myths continue to fascinate me—and millions of others. Only most of us don't call it "mythology."* We like to call it "going to the movies." For instance, on a cold Vermont night a few years ago, I went to see the second installment in *The Lord of the Rings* trilogy with my two teenagers and another friend. We were lucky to get tickets, as they sold out quickly. As we took our seats, people were scrambling futilely to find places, and I envisioned one of my worst personal fears: a raucous crowd of kids on Christmas break talking throughout the show.

But as soon as the lights went down, the extraordinary occurred. There was complete silence in this small Rutland theater. When the nearly-three-hour-long movie was over, the silence continued for a moment. And then the crowd exploded with loud and sustained applause.

There was some debate among the legions of Tolkien lovers about the faithfulness of these screen versions to their source. (Confession: I was one of those die-hard fans. When I was fourteen, I read all three books straight through while on a sick leave from school, which I extended a few days beyond the illness.) Merits of the film aside, I was struck at how *reverent* the audience was.

Chances are, a good many people in that audience were not churchgoers, and sitting in this darkened theater may have been as close to some form of collective spiritual encounter as any they might ever have experienced. And I thought further that this experience probably connected this twenty-first-century collection of strangers back to something

*This might be a good place to distinguish more precisely between myths and mythology. Many people use the words interchangeably—as the title of this book does. But to be specific, myths are the stories themselves, while mythology is actually the study of those myths. Even though the words have come to mean the same thing in common usage, there is a distinction. This book discusses the myths in great detail and, in chapter 1, offers a brief history of mythology—what people have thought about myths over the course of thousands of years.

much deeper, the act of sitting around a campfire three thousand years ago as someone recounted timeless exploits of heroes and monsters, Good versus Evil.

Looking at some of the box-office hits of the past few years merely confirms this idea. Recently, theaters have been filled with hits like *The Matrix*, *Finding Nemo*, *X-Men*, and the *Terminator* trilogy. To a large degree, all of these Hollywood blockbusters tap into ancient myths and tales of legendary heroes and epic quests. In the spring of 2004, the enduring appeal of myth got a fresh wind with *Troy*. Although moviegoers saw more of Brad Pitt's butt than Achilles' heel,* the success of the movie sparked lively new interest in a story that comes from the misty dawn of history. It is a story that still seems to speak volumes about men, women, and war. That said, the film *Troy* had about as much to do with Homer's epic *Iliad* as *Gone With the Wind* had to do with the real Civil War. We might start with the presentation of Patroclus, who is identified as the "cousin" of Achilles in the film. Homer's Achilles and Patroclus were "good friends," very possibly in the way Greek comrades-in-arms often were. But Hollywood wasn't having Brad Pitt do a gay Achilles.

Couple these recent Hollywood offerings with other successes, such as *E.T.*, a Disney—animated and very sanitized—version of *Hercules*, the Civil War romance *Cold Mountain*, and the Coen Brothers' *O Brother, Where Art Thou* (both works loosely based on the *Odyssey*), and, above all, the *Star Wars* saga, and you see even more evidence of the enduring appeal of ancient myths.

All of these films draw on mythic themes and often include very specific mythical references. (In *The Matrix* trilogy, for instance, the names Morpheus, Niobe, and Oracle are all drawn directly from mythic Greek characters.) Perhaps it is no accident that some of them are among the highest-grossing films worldwide. Throw in the extraordinary Harry Potter phenomenon—another spin on the mythic quest of an ordinary boy who learns to fly and has miraculous powers, like Luke Skywalker of *Star*

*When Achilles was born, his mother was told that he would be impervious to harm if bathed in a sacred pool. His mother dipped him in the water, but held him by the heel, which was the only place a wound would kill him. So, an Achilles' heel has come to mean a person's most vulnerable spot.

Wars and Neo of *The Matrix*—and you have yet another powerful piece of evidence that we still love myths.

And it's not just the myths of Greece and Rome. Among the popular attractions at Disney World is a ride based on the 1946 film *Song of the South*. Perhaps best known for the famed Disney song "Zip-A-Dee-Doo-Dah," this cartoon was inspired by the "Br'er Rabbit" stories, popular among African-American slaves. These stories, in turn, came from ancient tales of a mythic African Hare, a trickster god who crossed the Atlantic in the terrible Middle Passage and found new life in the American South. Tricksters, one of the most popular types of gods found in many societies, were greedy, mischievous, evil—kind of like the Joker in *Batman*—and sexually aggressive. Often, they took animal form, like the African Hare or Native American Coyote.

Hmmm. A mischievous rabbit and a greedy coyote trying to outsmart other animals. Sounds a little like Bugs Bunny and the Roadrunner's tireless nemesis Wile E. Coyote. And you thought myths were dead.

Just smartly packaged mass media, you say? I don't think so. Many of these films, cartoons, or books are well made and highly entertaining. But their broad, international popularity crosses the boundaries of age and sex, tapping into our basic human need for myth. As Homer—the poet, not the father of Bart Simpson—put it, "All men have need of the gods."

And it's not just entertainment. Do you like Halloween and its Hispanic equivalent, Día de los Muertos ("the day of the dead")? Both are modern vestiges of ancient mythical celebrations. Or perhaps you celebrate Christmas and Easter? With its candles and gift-giving, Christmas is based on old pagan Roman holidays, including the Saturnalia, a week-long festival of the winter solstice dedicated to the god of agriculture. Many of the trappings of modern Christmas, including Christmas trees, wreaths, mistletoe, holly, and ivy, are borrowed from ancient druidic traditions from northern Europe in which the evergreen symbolized the hope for new life in the dead of winter. The Easter celebration of the resurrection of Jesus Christ is closely associated with pagan festivals celebrating the coming of spring. Early Christians appropriated this familiar mythic notion to celebrate the new life Christians gain through Jesus's death and resurrection. The word "Easter" itself may have come from an early-English word *Eastre*, possibly the name of an Anglo-Saxon goddess of spring. (Other scholars believe the word "Easter" comes from the early German word *eostarun*, which means "dawn.")* The blending of Christian teachings with local myths in such places as ancient Celtic Ireland, Mexico, and Central America, and in the Caribbean and American South, where Christianity and myth fused in voodoo and Santeria—so-called "primitive," African-influenced religions still widely practiced today—is one of the most fascinating and ignored aspects of ancient mythology alive in our world today.

The intermingling of pagan myth and Christian rites and beliefs is one key element in the plot of the best-selling sensation *The Da Vinci Code*, a thriller which draws on the adaptation—or theft—of ancient pagan religions and rituals by Christians in ancient Rome and the first Church fathers. While many of its most controversial elements are historically questionable, the book's runaway international success is another tip-off that lots of people think there are deeper connections to ancient myths and mysteries than we've ever been told by mainstream religion. The fascination with *The Da Vinci Code*, as with *The Celestine Prophecy*, another novel that posits an elaborate Church conspiracy to conceal ancient truths, plays to a deep-seated skepticism about organized religion, but also taps into a level of curiosity about ancient spiritual ideas and wisdom—in other words, myths.

To get another gauge of the impact of myths, you could simply check the calendar. Is today a Thursday in March? A Saturday in June? The names of these days and months all come from Greek, Roman, and Norse mythology. From the calendar to the planets in our solar system—all except Earth are named for Roman gods—our language is loaded with words from our mythic past. Do you buy your books from *Amazon*.com? Are you wearing a pair of *Nikes*? Do you worry about a *Trojan horse* virus infecting your computer? Does the idea of a *panacea tantalize* you? Or perhaps your *arachnophobia* puts you in a *panic*? "Hypnosis," "morphine," "Golden Fleece," "Labor of Hercules," "leprechaun,"

*Easter's place on the calendar is itself probably a vestige of mythic beliefs related to the moon. It's one of the movable feasts of the Christian religion, and the date of Easter varies each year, but for most Christians it usually falls on the first Sunday after the first full moon on or after March 21.

"typhoon," and "hurricane" are just a few of the words and phrases that come from the world of mythology and color our speech. Is there an American Express card in your wallet? Then you don't leave home without Hermes (or Mercury), the Greco-Roman god of commerce whose image appears on that card.

Hell! Even *hell* comes from the name of the Norse goddess Hel, ruler of an icy underworld where oath-breakers, evildoers, and those unlucky enough not to have died in battle were sent. Unlike Christianity's fiery place of eternal torment, the Norse hell, you might say, was "frozen over."

In other words, myths have been, and remain, a powerful force in our lives, often without our even recognizing them. Myths surround us—in literature, in pop culture, in our language, and in the news. Rarely do you pick up a newspaper or magazine without finding words and phrases that contain references to ancient myths. And sometimes myths are part of the news. In Mexico, the planned construction of a Wal-Mart-owned supermarket was met with fierce resistance, because it was so close to the Pyramid of the Sun in the ancient ruins of Teotihuacán, the place where, the Aztecs believed, "men became gods." (Despite the protests and the discovery of an altar during excavation the store opened in November 2004.)

In modern India, many Hindus still make offerings of their hair to a temple deity in thanks for help in medical crises or to ask for good grades on exams. What some of these devout Hindus didn't know was that their hair was then used to manufacture high-priced wigs that accounted for a $62-million export business. But one set of beliefs ran head-on into another, because many of those wigs were purchased by Orthodox Jewish women who observe an ancient code of modesty that forbids the public display of their hair after marriage. When Orthodox Jewish rabbis in Israel declared that these wigs were made with hair offered for purposes of idolatry, their use was forbidden. Thousands of Orthodox women, according to the *New York Times*, publicly burned their human-hair wigs.

Another story from India is less benign. As recently as 2004, people have been charged with the very rare practice of ritual human sacrifice. The goddess Kali is an ancient Hindu goddess who slays evil but has always been known as a demandingly bloodthirsty deity. Millions of Hindus still travel to temples dedicated to Kali in eastern India. Most buy innocuous souvenirs of plastic swords and postcards featuring Kali's fearsome images on which she is bedecked with skulls and belts of severed feet. But several disciples of Kali have been accused of ritual murder, a chilling vestige of an ancient past—not at all exclusive to India—when human sacrifice was viewed as a necessary means to please or propitiate the gods. Recent discoveries of Peruvian and Celtic mummies, Egyptian sacrifices, and Mesopotamian mass graves are grim evidence that some of these victims went as willing sacrifices to help their people in this world or their divine leader in the next.

Myths also play a serious role in history. Perhaps the most deadly historical example of the impact of myth comes from World War II, when Adolf Hitler drew upon ancient Germanic myths to help enthrall an entire country. In the classic history of Hitler's climb to power, *The Rise and Fall of the Third Reich*, William L. Shirer wrote, "Often a people's myths are the highest and truest expressions of its spirit and culture, and nowhere is this more true than in Germany." Shirer recalled Hitler saying, "Whoever wants to understand National Socialist Germany must know Wagner." Hitler was deeply taken by Wagner's operas, which drew vividly on the world of German heroic myths, pagan gods and heroes, demons and dragons. Hitler intrinsically understood the deep emotional power of the symbols of these myths. Massive statues of ancient Germanic gods played a prominent role in the Nazi mass rallies at Nuremberg in the 1930s. Hitler grasped the visceral power, as well as the propaganda value, of a shared Teutonic myth in uniting the German people with a "master race" ideology.

One need only watch Leni Riefenstahl's famed—or notorious—documentary, *The Triumph of the Will*, to get a sense of the operatic mythology behind these mass pageants. Hitler deliberately mingled Christian and pagan elements, and when he solemnly marched up to stand before a wreath honoring Germans who had died in battle, he appeared to be playing the role of high priest in what one of his biographers has called a "pagan rite of communion." There is a scholarly debate as to whether Hitler himself was a true believer in the occult, but Nazi officials certainly set out to find historic and religious symbols and artifacts—apparently, including the Holy Grail—to embellish the Nazi cult of power.

In wartime Japan during the same period, myths were the source of the national Shinto religion, as the Japanese emperor Hirohito was supposedly descended from the Shinto sun goddess, Amaterasu. In the waning days of the war, this devotion to the emperor-god led to the use of the notorious kamikaze* pilots. With the war going against Japan in 1945, young men were recruited and given enough training to fly their dynamite-laden planes in suicide crashes aimed at American warships. While it once might have seemed difficult for modern Westerners to imagine, an ancient myth-religion was used to drive these young fighters—and an entire nation—with fanatical devotion to its emperor. That was barely half a century ago, in a thoroughly modern, industrialized, and well-educated society.

And, of course, it doesn't end there, as recent history has proved too well. In the past few years, the world has witnessed the combustible mixture of belief and fanatical devotion. "The virgins are calling you," Mohamed Atta wrote to his fellow hijackers just before 9/11. The notion of dying a martyr's death and gaining entrance to a paradise with the promise of virgins is clearly a powerful idea that continues to drive the terrorists who strap explosives to their bodies, or drive cars filled with explosives, or fly hijacked jets into buildings. They are motivated by beliefs whose roots stretch back to the most ancient of times. The idea of warriors gaining access to paradise through death is certainly not exclusive to any one mythology or faith.

When Afghanistan's Taliban was still in power and banning music, television, and kite-flying, the harsh regime destroyed several massive Buddha statues chiseled into a cliff on the ancient Silk Road. Beyond eliminating what they viewed as idolatrous images, these Islamic fundamentalists were attempting to eradicate a vestige of a 2,500-year-old belief system derived from the complex myths of India. The destruction of these irreplaceable cultural artifacts shocked the world and raised a deeper question: Can you kill beliefs and ideas by killing an image?

*The word "kamikaze" means "divine wind" and referred to a typhoon that saved Japan by preventing a Mongol invasion in 1281. In 1945, the young Japanese pilots were supposedly going to be the equivalent of that divine wind and turn away the American invading forces. While they killed many American sailors and destroyed numerous American ships, the kamikaze attacks ultimately did not affect the war's outcome.

That is not a new idea. The Spanish conquistadors and priests who followed them into Mexico in the 1500s may have leveled the temples and buildings of the Aztec capital of Tenochtitlán, but did they completely eradicate the beliefs behind them? The Spanish in the Americas, the British in Ireland and Australia, and the United States government have all attempted to "control" defeated people by taking away their language and beliefs. It doesn't always work.

So, myths can be a powerful business. And that is one reason they have been around for thousands of years. As old as humanity, the first myths belong to a time when the world was full of danger, mystery, and wonder. In the earliest of human times, every society developed its own myths, which eventually played an important part in the society's daily life and religious rituals.

One of the chief reasons that myths came into being was because people couldn't provide scientific explanations for the world around them. Natural events, as well as human behavior, all came to be understood through tales of gods, goddesses, and heroes. Thunder, earthquakes, eclipses, the seasons, rain, and the success of crops were all due to the intervention of powerful gods. Human behavior was also the work of the gods. For instance, the Greeks, like most early civilizations, had a story to explain the existence of all the bad things that happen in the world—from illness and pestilence to the idea of evil itself. The Greeks believed that, at one time, all of the world's evils and problems were trapped inside a jar (not a box!). When this jar was opened by the first woman, all of the world's misfortunes escaped before this woman—Pandora—was able to close the lid.

The Blackfoot Indians of the American Plains also blamed the woes of the human condition on a troublesome female. When Feather Woman dug up the Great Turnip after being told not to do so, she was cast out of Sky Country—or the heavenly paradise. Yet another woman is seen as the source of the world's ills in a tale told by a nomadic group from the ancient Near East—the "cradle of civilization," as they called it back in your school days. In one version, her name was Havva, and she disobeyed her god when she ate from a forbidden tree. Of course, most of us know her by the more familiar name—Eve.

Obviously, we now have many more scientific answers for most of our questions about the world and universe. We know why the sun rises and sets. Why the rain falls in some seasons and not in others. What makes crops grow. We have a much better understanding of where we came from. We understand illness and death—to a certain degree. And although the source of evil in the world—and why bad things happen to good people—is still a great mystery, we have even begun to unravel the beginnings of the universe.

But in earlier times, people invented stories to explain these beginnings. In the Creation story of the Krachi people of Togo in Africa, for instance, the creator god Wulbari and man lived close together, and Wulbari lay on top of Mother Earth. But there was so little space to move about that the smoke of the cooking fires got in Wulbari's eyes and annoyed the god. In disgust, Wulbari went away and rose up to the present place where humans can admire him but not reach him.

In another African tale, of the Kassena people, the god We also moved out of the reach of man, because an old woman, anxious to make a good soup, used to cut off a bit of him at each mealtime. Annoyed at such treatment, We went higher to escape this daily eating of his flesh.

These may seem like amusing legends of "primitive" people. A god who goes to the heavens because smoke gets in his eyes and another god who is peeved at being cut up for the day's soup. But consider these mythical stories: A god who is so angry when a woman eats a piece of fruit that he makes childbirth eternally painful for all women. In his anger, this ancient Hebrew god—who also liked to walk around the Garden of Eden in the cool of the evening—gradually removes himself, like Wulbari and We, from his creation. Or a god whose body and blood are consumed each week at a ritual of sacrifice called the Eucharist.

In other words, what we call one person's "myth" is often another person's religion. One of this book's essential goals is to explore that transformation of myth into religion. And how that transformation has changed history.

Many books about myths approach the subject from one of two perspectives: geographically—that is, simply grouping myths together by a region or particular civilization; or thematically— the broad range of typical myths, such as Creation stories or other explanatory myths. Creation myths set out to explain the origin of the world, the birth of gods and goddesses, and eventually the creation of human beings. Explanatory, or causal, myths try to give a mythic reason for natural events, such as the Norse belief that Thor made thunder and lightning by throwing his hammer.

Don't Know Much About Mythology takes a slightly different tack. It sets out to examine all the fascinating myths created by these ancient cultures and relate them to their histories and achievements. Besides Creation and explanatory myths, another fundamental type of myth is the "foundation" story, which explains the beginnings of a society—often with the distinct sense of superiority that direct descent from the gods clearly implies. For instance, it is impossible to understand Egyptian history and culture without understanding Egyptian mythology. For the Egyptians, their elaborate system of myths and beliefs was life itself—it was the critical underpinning of this amazing empire that lasted three thousand years.

Part of that mix of myth and history is the way in which myths became the means to rule and domination. Once local rulers understood that connecting themselves to the gods would cement their hold over people, myth was elevated to an institution that could prove more powerful than an army. Most of the great ancient civilizations—whether in Egypt, China, or Mesoamerica—were theocracies, in which there was no difference between religion and state. With connections to the gods and usually the cooperation of a potent priesthood, divinely anointed rulers held the power of life and death over their subjects. Even in societies that did not produce a godly king and a central government tied to the beliefs, the most respected and feared person in the society was the shaman, sometimes known as the "witch doctor"—a man whose deep connection to the gods enabled him to heal or kill. In his groundbreaking book *Guns, Germs and Steel*, Jared Diamond singled out the power of belief as one of the key means for the wealthy and powerful—what he calls the "kleptocracy"—to maintain their hold over the poor and powerless.

The history of myth, in other words, goes hand in hand with the history of civilization. Stop and think about "ancient civilization." What does it mean? The wheel. Zero. Writing. Bronze. Glass. Fireworks. Paper. Noodles. Indoor plumbing. Beer. These are only a few of the pleasant and delightful creations devised by the ancient civilizations of Egypt, Mesopotamia, China, Greece, India, Rome, Africa, Central America, and Japan. They also gave us astronomy, democracy, the calendar, God, philosophy, and a whole set of complex ideas that have driven students crazy for centuries. The scientific discoveries, practical inventions, laws, religions, art, poetry, and drama of these ancient people have driven human life and culture—civilization as we know it.

These same ancients "invented" the myths that grew hand in hand with their civilizations, making it impossible to separate one from the other. While the impact of myths may seem less obvious than that of the wheel, writing, or a mug of beer, these ancient legends are still a powerful force in our lives today. They remain alive in our art, literature, language, theater, dreams, psychology, religions, and history.

With that in the background, *Don't Know Much About Mythology* traces the story of myths through the ages and shows how myths helped make civilization. It also looks at the way myths moved from one group to another in the exchange of civilizations. The familiar mythology of the Greeks did not emerge full-blown from the sea—the way Aphrodite supposedly did. It drew upon ideas from Mesopotamia, Egypt, Crete, and other ancient neighbors. While many of us may be somewhat familiar with the stories of Adam, Eve, Noah, and the later tales of the Hebrew patriarchs which were set down in the Book of Genesis, we may not know about their connection to much older Mesopotamian stories, such as the epic poem *Gilgamesh*, a tale of a very flawed hero from the same part of the world. Myths don't just spring up from virgin ground—they are often borrowed from older sources and then molded and remade into new myths.

In telling the story of the connections between these age-old traditions and civilizations, this book is an outgrowth of *Don't Know Much About the Bible*. In writing that book, I learned about the deep, primordial connections between the civilizations of the ancient Near East and the people who emerged as the Jews of the Old Testament. Some scholars and historians believe that the idea of the one God of the Hebrews might have been inspired by an Egyptian pharaoh named Akhenaten who unsuccessfully tried to replace the vast pantheon of Egyptian deities with a single sun god. Some historians believe this concept may have been adopted by the ancient Hebrews while they were in Egypt. It is a controversial and unproved idea, but there is no question that comprehending the myths and civilizations of Egypt and ancient Mesopotamia adds to an understanding of the Judeo-Christian world, which was later, similarly, influenced by the Greek and Roman worlds, in which Christianity was born, and by the world of the "pagan" people to whom the early Christian missionaries began to preach the gospel of Jesus—all of these worlds alive with myth and ancient religions.

To accomplish this, I use the techniques I have employed in all the books of the Don't Know Much About series: questions and answers, timelines that show historical connections, "voices" of both real people and mythic sources, and stories about the "household names" of ancient myths—including Hercules, Jason, Ulysses, Romulus and Remus, as well as many more unfamiliar names from other cultures. This book also draws on a vast array of recent archaeological and scientific discoveries that have shed new light on the ancient societies that created these myths.

The chapters are organized by the various civilizations, starting with the two with the earliest known mythologies and worship systems—Egypt and Mesopotamia. The book then moves through the other major Western mythologies in rough chronological order—Greece, Rome, and Northern Europe. The major Eastern myth systems of India, China, and Japan come next, followed by chapters on the remaining areas of the world as they were opened up to Europeans: sub-Saharan Africa; the Americas; and the Pacific island areas—the last places "discovered" in the world.* That raises two important points. First, while this guided world-tour is an overview of the world's principal civilizations and their myths, it is obviously not an "encyclopedic" treatment. It would be impossible to cover every myth and every god from each civilization—large and small—in a single volume such as this. Instead, this book focuses on the "need to know" approach, as do all of the other Don't Know Much About books. This book aims to highlight, in an accessible and entertaining manner, the most significant aspects of these myths and cultures and present the "first word" on this subject, not the "last word." An extensive bibliography lists the many resources and wide range of literature available to further explore the world of myth.

*It is important to remember that we "discover" new things all the time. As this book was being written, researchers announced the discovery of a previously unknown group of three-foot-tall "dwarf" humans who lived in a remote section of Indonesia within the time span of "modern" humans. Interestingly, the existence of such "little people" was part of the local mythology.

And second. This is, admittedly, a rather "Eurocentric" organization that looks at history as it progressed from a Western perspective. The chapters proceed in a rough chronology from the dawn of Western history through its gradual contacts with the rest of the world and through the impact on the West of that growing contact with these "new worlds." Frankly, the myths of Egypt, Mesopotamia, and Greece have a lot more to do with Western history than the myths of early China or the San people of the Kalahari Desert. That does not imply that one set of myths is superior, or that one is more "right" or "wrong"—just that I have tried to organize the book to reflect the role myth has played in our history. It is also worth noting that so many of these myths—regardless of their geographic origin—are often more alike than different, a point that will be underscored many times in this book.

In this way, I hope to provide an accessible portal into the myths and the civilizations that created them. Typically, our schools may teach a little bit about one or two of these groups, but rarely discuss them in connection with each other. What did the Greeks learn from the Egyptians? How were they different? Were the Egyptians really Africans? Did the Chinese influence the Hindus or vice versa? How did a handful of Spaniards overthrow mighty empires and convert thousands of Aztecs and Incas from their old beliefs to Catholicism? These are the sort of questions that make this book a somewhat unique addition to the vast literature of mythology.

No small task! The scope of this book is much wider than simply retelling old stories from a modern—and perhaps skeptical—perspective. Unfortunately, for most of us, learning about ancient civilizations—if we learned anything at all— wasn't very interesting. One of the key objectives of the Don't Know Much About series is to revisit all those subjects we should have learned about back in school but never did, because they were dull, dreary, and boring, as well as badly taught and riddled with misinformation.

But beyond that, *Don't Know Much About Mythology* also tries to spin a thread that winds through all of my Don't Know Much About books. The history of world myths is deeply connected to such subjects as geography, biblical history, and astronomy. And one of my goals in this series has always been to show how seeing those connections makes learning about these subjects become far more compelling.

Finally, this book and the subject of mythology touch on something deeper still. In the late nineteenth century, a generation of scholars began to view myth as part of the primal human need for a spiritual life. In a classic study of myths, called *The Golden Bough*, Sir James Frazer tried to demonstrate that every ancient society was deeply invested in a ritual of sacrifice that involved a dying and reborn god whose rebirth was essential to the society's continuing existence.

A little later, Sigmund Freud argued that myths were part of the human subconscious, universally shared stories that reflected deeply rooted psychological conflicts—most of them sexual, in Freud's view. Then Swiss psychoanalyst Carl Gustav Jung, Freud's disciple who eventually split with Freud, suggested that myths were rooted in what he called the "collective unconscious," a shared common human experience as old as mankind itself. Jung believed that this collective unconscious was organized into basic patterns and symbols—which he called **archetypes**. Our dreams, art, religion, and, perhaps most important, our myths are all among the ways that man has expressed those archetypes. Jung also believed that all myths have certain common features—characters such as gods and heroes; themes such as love and revenge; and plots such as a battle between generations for control of a throne, or a hero's quest—which are fundamental to our humanity.

For more than a hundred years, scholars have debated different views of the role that myths have played in the human experience. Religion, psychology, anthropology—all of these are lenses through which we can view that role. This book takes into account such approaches to mythology and raises another set of questions: Are these timeless stories simply collections of amusing tall tales from long, long ago? Did myths begin as the ancient world's version of *The Sopranos*? Are they simply old versions of entertaining stories about sex and violence—or were they created to cement the social order with divine kings lording over the common person? Do myths reach some deeper level of human thought and experience, as many anthropologists and psychologists suggest? And finally, how do the ancient ideas found in myths speak to us today?

In his 1949 classic, *The Hero With a Thousand Faces*, Joseph Campbell wrote: "Religions, philosophies, arts, the social forms of primitive and historic man, prime discoveries in science and technology, the very dreams that blister sleep, boil up from the basic magic ring of myth."

Throughout human history, myths have provided what T. S. Eliot, a poet deeply interested in myth, called "the roots that clutch." Exploring Campbell's "magic ring of myth" takes *Don't Know Much About Mythology* into territory that has been touched upon in my previous books, especially those about the Bible and the universe—the powerful connections between belief and science, the conflicts between faith and the rational world, and a deeper sense of the mysterious in human life, which are all part of man's search for meaning.

Underpinning these books, I hope, is an idea expressed by the Irish poet William Butler Yeats, who said, "Education is not the filling of a pail, but the lighting of a fire." How *Promethean*! (See? I told you myths live in our language.)

Over the course of the more than fifteen years that I have been writing the Don't Know Much About series, I have discovered that people are not ignorant about subjects like history and religion by choice. On the contrary, I've discovered that people of all ages are eager to learn and have endless curiosity. One of the saddest things I have witnessed in these years—especially when I visit schools—is how the innate and insatiable curiosity young children have about the world gets absolutely killed by the tedium of school.

I remember so well how myths saved one little boy from that tedium. And I also believe that the story of myth is ultimately about our innate human curiosity. Like that curious newspaper boy who wanted to know what "argus" meant. Or that curious woman who wanted to know what was in the jar given to her by the gods. Or that curious pair in Eden who wanted to gain knowledge. This is what got us where we are today. The human experience is about a boy asking questions and pushing the envelope of curiosity. Across centuries of time and great distances of culture, mythology is about that common human experience and that driving curiosity about other people, the world, the heavens. Deeper than intellect alone, it's a piece of what makes us what we are—call it soul, collective unconscious, or even superstition. I hope, if nothing else, this book will help you discover that childlike curiosity that has driven us from dark caves to the outermost edges of the universe.

CHAPTER ONE

ALL MEN HAVE NEED OF THE GODS

Badness you can get easily, in quantity: the road is smooth, and it lies close by. But in front of excellence the immortal gods have put sweat, and long and steep is the way to it, and rough at first. But when you come to the top, then it is easy, even though it is hard.

—HESIOD (c. 700 BCE), *The Theogony*

Know thyself.

—Inscription at the Delphic Oracle, attributed to the Seven Sages (c. 650 BCE–550 BCE)

No science will ever replace myth, and a myth cannot be made out of any science. For it is not that "God" is a myth, but that myth is the revelation of a divine life in man. It is not we who invent myth, rather it speaks to us as a word of God.

—CARL GUSTAV JUNG

The awe and dread with which the untutored savage contemplates his mother-in-law are amongst the most familiar facts of anthropology.

—Sir James Frazer, *The Golden Bough*

The highest point a man can attain is not Knowledge, or Virtue, or Goodness, or Victory, but something even greater, more heroic and more despairing: Sacred Awe!

—Nikos Kazantzakis, *Zorba the Greek*

What are myths?

Myths, legends, fables, folktales: What are the differences?

Where does the urge to make myths come from?

Are all myths historical?

Who was the man who "found" Troy?

How did an ancient myth cast doubt on the divinity of the Bible?

When does myth become religion? And what's the difference?

Are myths all in our minds?

You're driving down the highway and you pass an accident along the road. Admit it. Without even thinking, you slow down and rubberneck, just like everybody else. Instantly, your mind seeks an explanation for what you see.

You may have had only a fleeting glimpse of the accident scene—maybe you saw sets of skid marks, a car upended, dazed people talking to the police. You hear an ambulance wail in the distance as a trooper or firefighter waves you past. You don't know what happened. But you see the effects and want to explain the cause. If you are like most people, you begin to stitch together a theory of what went wrong. Almost without consciously thinking about it, you begin to manufacture a narrative of what happened.

"That driver was probably drinking." "He had to be going too fast." "The driver must have fallen asleep and swerved across the road." "One car probably cut off the other."

In other words, without any facts or much evidence, you try to create a coherent story to explain what you have seen. Maybe it is that simple: this is what makes us truly human. The innate need to explain and understand is what has gotten us to where we are today, in the early days of the twenty-first century.

Myths may have begun, in the oldest sense, as a way for humans to explain the "car wrecks" of their world—the world they could see as well as the world they could not see. Long before science envisioned a Big Bang. Long before Greek philosophers reasoned, Siddhartha Gautama sought enlightenment, or Jesus walked the shores of the Sea of Galilee. Long before there was a Bible or a Koran. Long before Darwin proposed natural selection. Long before we could know the age of a rock and before men walked on the moon, there were myths.

Myths explained how Earth was created, where life came from, why the stars shine at night and the seasons change. Why there was sex. Why there was evil. Why people died and where they went when they did.

In short, myths were a very human way to explain everything.

MYTHIC VOICES

Look now how mortals are blaming the gods, for they say that evils come from us, but in fact they themselves have woes beyond their share because of their own follies.

—HOMER, The Odyssey *(c. 750 BCE)*

What are myths?

When people use the word "myth" today, they often have in mind something that is widely believed but untrue. Like alligators in the sewers of New York City—which is really not a myth at all but an "urban legend." In another sense, it is now common to talk about the "myth" of the cowboys of the Old American West, and there are plenty of other so-called myths in American history—old ones that die hard and new ones being created all the time. Myths about the Founding Fathers, the Civil War, slavery, the Sixties—just about any period or movement in America's past has been "mythologized" and layered with legend to some degree.

In bookstores today, you'll also find a profusion of books with titles and subtitles that underscore this notion of a myth as something that is commonly believed but is not true: *The Beauty Myth*, *The Mommy Myth*, *The Myth of Excellence*. Most of these recent books with "myth" in the title tend to treat a "myth" as an old and possibly dangerous idea that needs to be debunked.

Like most words, "myth" means different things to different people, but in its most basic sense, a myth is defined as "A traditional, typically ancient story dealing with supernatural beings, ancestors, or heroes that serves as a fundamental type in the world view of a people, as by *explaining aspects of the natural world* or delineating *the psychology, customs, or ideals of society*." (*American Heritage*, emphasis added.)

"Explaining aspects of the world"—that's another way to say "science" or "religion," the two principal ways people have used to explain the world.

"The psychology, customs, or ideals of a society." That's a large mouthful that covers just about everything else *not* covered by science and religion—but gets to the heart of what we think and believe, even if we can't "know" it.

In the ancient world, myth had a meaning that is almost completely opposite to our modern concept of myth as an "untruth." In the earliest days of humanity, myths existed to convey *essential truths*. They were, in a very real sense, what many people today might call **gospel**. Or as David Leeming put it in *A Dictionary of Creation Myths*, "A myth is a . . . projection of a . . . group's sense of its sacred past and its significant relationship with the deeper powers of the surrounding world and universe. A myth is a projection of . . . a culture's soul." Ananda Coomaraswamy, a twentieth-century Indian philosopher, put it this way: "Myth embodies the nearest approach to absolute truth that can be expressed in words."

Viewed in this very ancient and much broader sense, myths are about what makes us tick. They are as old as humanity and as current as the news.

The word **myth** is derived from the Greek word *mythos*, for "story," and when the Greek philosopher Plato coined the word "mythology" more than two thousand years ago, he was referring to stories that contained invented figures. In other words, the great Greek thinker conceived of myths as elaborate fiction, even if they expressed some larger "Truth." Plato—using the voice of Socrates as his Narrator—criticized the myths as a corrupting influence, and in his ideal state, set out in *The Republic*, banned poets and their tales.

"The first thing will be to establish a censorship of writers of fiction, and let the censors receive any tale of fiction which is good, and reject the bad; and we will desire mothers and nurses to tell their children the authorized ones only." He goes on to say about the stories of the gods: "These tales must not be admitted into our State, whether they are supposed to have an allegorical meaning or not. For a young person cannot judge what is allegorical and what is literal; . . . and therefore it is most important that the tales which the young first hear should be models of virtuous thoughts."

On the other hand, Plato himself was not above using **allegories** (a Greek word that essentially means "saying something in a different way") as a teaching device; Plato's own tale of Atlantis, a mythical idealized world, and his famed allegory of the Cave, in which most men are trapped in a world of illusion and ignorance, seeing only flickering shadows of reality, are inventions—stories that are meant to convey a greater eternal and universal Truth.

For thousands of years we have invented stories to tell one another and our children. But why? Myths clearly fulfill some basic function in human life. But what is that function? And was the philosopher Plato mistaken? Is mythology more than a set of elaborate fictions?

There is no question or doubt that the creation of myths and their role in everyday life is one of the most common of all human endeavors. As Homer said, "All men have need of the gods." So, we need myths. It is clear that myths developed in the dawn of human consciousness, from the very first evidence of human culture—cave paintings, carved pieces of bone, fertility figurines, household idols, and ancient burial practices. Even the famed Neanderthal, an early species of human that was eventually overtaken and supplanted by the Cro-Magnon more than fifty thousand years ago, had burial practices that show an interest in what comes after death.

Even if myths only started out as a way to pass the time on long nights sitting around the campfire, they clearly became far more than elaborate entertainment. The people who formed the first civilizations developed myths. Over time, as their villages became cities and cities became states, the myths grew into complex, interconnected tales that formed the basis of intricate belief systems. These stories of gods and ancestors became one of the central organizing principles in these cultures, and they dictated religious rituals, the social order and customs, everyday behavior, and even the organization of entire civilizations.

It is the essential sacred, spiritual—or religious—significance in myths that fundamentally separates them from other types of very old stories, such as legends and folktales. While the mythologies of many cultures grew to include related legends and folktales, myths were usually considered sacred and absolutely true—a notion that is completely at odds with the modern concept of myth as fallacy.

Of course, the ancient myths usually were about gods or other divine beings who possessed supernatural powers. That simply makes it all the more interesting to see that many of the gods, goddesses, and heroes of myth exhibited very recognizable human characteristics in spite of these powers. The gods of every civilization seemed to be wholly subject to the same sorts of whims and emotions—love and jealousy, anger and envy—experienced by the people who worshipped them. Zeus, the greatest of the Greek gods, had what in modern terms is called "a zipper problem." He couldn't avoid temptation in any form—goddess, mortal, or even young boy. His divine wife, Hera, was most irritated by his behavior but put up with it, a model of sort of the long-suffering, wronged but devoted wife.

Myths are also filled with sibling rivalries, one of the most basic of human emotions. In ancient Egypt, Seth kills his brother Osiris out of envy, jealousy, and a desire for power. Then Osiris' son Horus continues the fight against his uncle Seth—a cosmic family feud that was the rivalry around which their national religion was focused. Other mythologies feature "trickster" tales filled with stories of unscrupulous characters, like the Native American Coyote finding endless ways to deflower maidens. The vanity of three goddesses of Mount Olympus, each wanting to be called the most beautiful, is the force that powers the chain of events which leads to the affair between Paris, the prince of Troy, and the beautiful mortal Helen—the daughter of Leda, seduced by Zeus in the form of a swan. Those events bring the Greeks and Trojans to ten years of the Trojan War.

All of which leads to another question still—one that people have been asking for thousands of years:

Are the gods created in man's image or is it the other way around?

MYTHIC VOICES

In my opinion mortals have created their gods with the dress and voice and appearance of mortals. If cattle and horses, or lions, had hands, or were able to draw with their feet and produce the works which men do, horses would draw the forms of gods like horses, and cattle like cattle, and they would make the gods' bodies the same shape as their own. The Ethiopians say that their gods have snub noses and black skins, while the Thracians say that theirs have blue eyes and red hair.

—XENOPHANES *(c. 570–475 BCE)*

Myths, legends, fables, folktales: What are the differences?

A famous old ad campaign for luxurious fur coats illustrated with photographs of usually over-the-hill celebrities used to ask the question, "What Becomes a Legend Most?" What do the so-called legends of Hollywood have to do with stories of ancient gods and heroes? Not much.

Although the words "myth" and "legend" are often used interchangeably, there are some notable distinctions. When most people think of myths, they have in mind some pantheon (another Greek word, it is formed from *pan* for "all" and *theos* for "gods") of Greek and Roman gods. According to the myths, these divinities were believed to be supernatural beings who actually controlled events in the natural world.

Legends are really an early form of history—stories about historical figures, usually humans, not gods, that are handed down from earlier times. Most Americans, for instance, are familiar with the legend of young George Washington and the cherry tree. This story of Washington as a young boy chopping down his father's cherry tree and being incapable of lying about it was the purely fictional creation of a "biographer" named Parson Weems, who passed himself off as the rector of the parish at Mount Vernon. His stories of young Washington were tailored to fit into a neat collection of morality tales for children, well after Washington was dead. Nonetheless, stories such as the tale of the cherry tree became part of the national American legend of George Washington and his unquestioned honesty. But while Washington was certainly celebrated as a legend, both in his times and for centuries afterwards, nobody actually ever thought of him as a god.

Another familiar example of the distinction between myth and legend comes from ancient Great Britain. King Arthur is a historical figure about whom legendary stories have been created and retold for more than a thousand years, including, in recent times, in T. H. White's popular *The Once and Future King* and the musical *Camelot*, which provide most of our images of King Arthur and the Knights of the Round Table. A figure in British prehistory, Arthur was most likely based upon an actual person, possibly a tribal chieftain in Wales, around whom an elaborate cycle of heroic tales was collected. George Washington's biography and times were reasonably well documented, but the life of Arthur is more difficult to pin down with hard facts. Many of the stories about a king called Arthur first began to be collected in *History of the Kings of Britain*, written by Geoffrey of Monmouth between 1136 and 1138, anywhere from five hundred to a thousand years after the real-life model for King Arthur might have lived.

But older legends about Arthur go back even further, to an ancient Welsh collection of tales called the *Mabinogion*. These stories—which may have had their origins even earlier in Celtic Ireland, and then migrated to Wales—contain some of the earliest-known references to a character named Arthur. He is also mentioned as a military chieftain in post-Roman Britain, fighting against the Saxon and Norse invaders, and his name is derived from Artorius, a Latin name recorded in Britain in the second century. Spinning off from these ancient Celtic and Welsh myths came elaborate, Christian-influenced tales of Arthur and his wife Guinivere, as well as the collection of noble knights in their quest for the Holy Grail—the cup from which Jesus drank at the Last Supper. These tales were recycled and rewritten by many writers, working over the next several centuries, gradually transforming the legendary king and his court into the more-familiar image we hold of Arthur today—chivalrous knight in medieval armor. These medieval romances were complex fictions created by later generations looking to turn this Dark Age Welsh warlord into a Christian king of England. By the Middle Ages, when the concept of the knightly order and the ideas of chivalry developed, the legends of Arthur were draped in these medieval fashions, removed by centuries from Arthur's far more primitive historical beginnings.

Another intriguing example of a legendary figure is Christianity's St. George, famed as the slayer of dragons and best known as the patron saint of Britain (and Portugal). Like Arthur's, George's origins pose a tricky question and show how myths and legends sometimes merge. Based upon an ancient story from the Near East, the legend of George was transformed into a Christian allegory, and he was later sainted by the Roman Catholic Church. The source of the St. George story has been traced to Palestine, where European Crusaders probably first learned of it during the period of the Crusades in the eleventh through thirteenth centuries. During the First Crusade, a vision of St. George supposedly led the Christians into battle at Antioch, against the Saracens. During the Third Crusade, King Richard I placed himself and his army under the protection of George, who came to be seen as the patron of soldiers.

Little is known about the Christian St. George's actual life. He probably came from Lydda, in what is now Israel. According to religious tradition, George became a soldier in the Roman army and rose to high rank. But after he converted to Christianity, he was arrested and executed, possibly during the persecution of the Christians by the Roman emperor Diocletian around 303 CE. Before his martyrdom, George supposedly helped convert thousands of new Christians after slaying a dragon that had terrorized the countryside. This dragon had been appeased with the regular sacrifice of two sheep, but when sheep grew scarce, the dragon demanded a human victim, chosen by lot. When the king's daughter was selected as the victim, George promised to kill the dragon if the people agreed to be baptized.

But the stories of George slaying the dragon are much older than the Christian era. In one earlier version, set in Libya, in North Africa, George came to the aid of a group of local people who were obliged to sacrifice a virgin each day by feeding her to a dragon. George slew the dragon and rescued the maiden, who was chained to a rock. According to this local legend, George also had the power to fertilize barren women, who, by visiting one of his shrines in northern Syria, were said to be magically impregnated by him. While it is possible that someone resembling St. George might have once lived, he belongs to the misty era of early Christianity and even earlier pagan eras, unlike Arthur, whose living, breathing inspiration probably existed in Roman Britain. Shrouded in stories that mixed magic and Christianity, dragons and Roman persecution, George was adopted as patron of the Order of the Garter by King Edward III (1327–1377) and he was invoked as England's patron by Henry V at the famous Battle of Agincourt in 1415. (St. George is not the only Christian saint drawn from older "pagan" sources. Another example is St. Brigid, one of Ireland's patron saints, who is very much like an older Celtic goddess also named Brigid. See chapter 5.)

George's evolving story is a perfect example of how myths are sometimes shared and become layered with meaning as they are adopted and adapted over the course of time. The story of a dragon slayer is one of the most common themes in ancient myths, and there are connections between the stories of St. George and of the Greek hero Perseus, and of even older dragon slayers in both Egyptian and Mesopotamian myths.

The dragon, in fact, is one of the most universal archetypes, often related to evil and chaos, and images of these creatures have been found in Egyptian tombs, on the Ishtar gate in Babylon, on Chinese scrolls, on Aztec temples, and even in Inuit bone carvings.

Another mythical dragon slayer was the Canaanite god Baal. In one myth, Baal slays the dragon Lotan (whose named was changed to Leviathan in the Hebrew Bible), the symbol of chaos. For this act, Baal was rewarded with a beautiful palace built by the gods in his honor. Readers of the Bible are familiar with Baal in another context. He was seen as one of the chief "false gods" whom the Hebrew people of the Old Testament had to overcome in establishing the Promised Land of Israel.

Fables are simple, usually brief, fictitious stories, typically teaching a moral, or making a cautionary point, or, in some cases, satirizing human behavior. In many fables, the moral is usually told at the end, in the form of a proverb. Often, they feature animals that speak and act like human beings, as in the most famous examples—those attributed to Aesop, a Greek slave who supposedly lived about 600 BCE, but about whom little else is known. Stories like "The Tortoise and the Hare," in which slow and steady wins the race, or "The Grasshopper and the Ant," in which a fun-loving but lazy grasshopper plays while the ant dutifully stores away food for the winter, were simple morality tales.

For generations, "Aesop's Fables," which may have come from more ancient sources, were handed down orally, until around 300 BCE, when they were gathered into a collection, *Assemblies of Aesopic Tales*. Compiled by an Athenian politician named Demetrius of Phaleron, this collection was later translated into Latin by Phaedrus, a freed Greek slave. About five hundred years later, in 230 CE, another Greek writer combined Aesop's fables with similar tales from India and translated all of them into Greek verse. Among the world's oldest known fables are those from India, called the *Panchatantra*, an anonymous collection written in Sanskrit (and translated as "five treasures"). Derived from Buddhist sources, they were probably written as instructions for the children of royalty.

The fables of Aesop, which are sometimes intermingled with Greek myths, have remained an essential part of Western culture and are as familiar to children today as they may have been to Athenian children two thousand years ago. Such stories as "Androcles and the Lion," in which a slave saves his own life by removing a thorn from the paw of a lion, or "The Crow and the Pitcher," in which a thirsty crow fills a pitcher with stones to raise the water level so he can drink (moral: necessity is the mother of invention), are still widely told. And they permeate our language and literature. In "The Fox and the Grapes," for example, a fox decides that some grapes growing too high for him to reach are probably sour anyway. The moral of the fable—that people often express a dislike for what they cannot have—is the source of the common expression "sour grapes."

Related to fables are **folktales**, another type of story usually handed down orally, which often deals with common people and is primarily meant to entertain rather than instruct. Unlike legends, folktales are not supposed to have actually happened and don't usually involve national heroes. Although the term "folktale" is often used interchangeably with "fairy tale," they are two different forms. Folktales generally tell of the customs, superstitions, and beliefs of ordinary people; **fairy tales** are usually filled with elves, pixies, fairies, and other supernatural creatures with magical powers. In both of these, the central character tends to be a person of low status, frequently trapped in a case of mistaken identity, who has been victimized or persecuted, like Cinderella by her wicked stepsisters. Over time, and often with magical help, they overcome adversity and their goodness is rewarded as they are restored to their proper place in society. In other words, for average people, folk- and fairy tales are the equivalent of stories of people winning the lottery, always holding out hope that some stroke of luck or miraculous intervention will change their luck and fortune forever.

The tales of *Arabian Nights*, including "Ali Baba and the Forty Thieves," "Aladdin's Lamp," and "Sinbad the Sailor," are examples of the best-known folk- and fairy tales. Another collection of the most familiar folk- and fairy tales are *Grimm's Fairy Tales*, the famous German stories collected by the brothers Jakob and Wilhelm Grimm between 1807 and 1814. These include "Hansel and Gretel," "Little Red Riding Hood," "Snow-White," "Rumpelstiltskin," "Sleeping Beauty," "Cinderella," and "Rapunzel," many of which were drawn from much older, mythic sources.

Where does the urge to make myths come from?

Like the witness to the car accident at the beginning of this chapter, people everywhere love—or perhaps need—to create a good story. And if the details change a bit in the retelling, what's the difference? Who hasn't slightly embellished their biography or stretched the truth with a touch of dramatic flair to add some color and spice to an encounter at the supermarket or an argument with the boss? Often, these stories—just like everyday rumors and tabloid-newspaper reports—change with each retelling. It has always been that way, and in the broadest sense, the myths of every culture include all of these types of "stories"—legends, fables, folktales, and fairy tales—to form a broad worldview.

But the larger question remains: where do these stories come from? Are they all inspired—as is most likely the case of King Arthur and possibly St. George—by some actual person or events? Or are all of these mythical stories simply the work of human imagination? People have been arguing about that for more than 2,500 years.

As early as 525 BCE, a Greek named Theagenes, who lived in southern Italy, identified myths as scientific analogies or allegories—an attempt to explain natural occurrences that people could not understand. To him, for instance, the mythical stories of gods fighting among themselves were allegories representing the forces of nature that oppose each other, such as fire and water. This is clearly the source of a great many explanatory or "causal" myths, beginning with the accounts found in every society or civilization that explain the creation of the universe, the world, and humanity. These "scientific" myths attempted to explain the seasons, the rising and setting of the sun, the course of the stars. Myths like these were, in some ways, the forerunners of science. Old mythical explanations for the workings of nature began to be replaced by a rational attempt to understand the world, especially in the remarkable era of Greek science and philosophy that began about 500 BCE.

The most fundamental and universal "explanatory" myths are Creation myths, found in every culture. Quite often, there is more than one Creation myth for a particular group, whether a tribe or civilization. Sometimes these are variations on a theme; other times they represent different traditions that arose in different periods. Or some can reflect different regions or cities that generated their own Creation myths. In

Egypt, for instance, there were at least four major Creation accounts, each one from a different major religious center. These are myths that set out to explain the ordering of the universe and are very often associated with myths that explain the appearance of humans. (The major Creation myths of each civilization will be discussed in each of the following chapters.)

Are all myths historical?

Whether searching for the historical Jesus, King Arthur, Atlantis, or Troy, people for centuries have had a deep fascination with the possibility that all of these stories and mythic characters are based on identifiable historical events. This concept, called "historical allegory," is not a recent one, but goes back to a very old explanation for the source of myths—the notion that they all began with real people and actual events. With the passage of time, and the retelling of the stories, the events and the people involved became distorted and layered with legend.

One of the first people to suggest that all myths are based on real people and events was a Greek scholar named Euhemerus (a native of what was then a Greek colony on the island of Sicily), who lived during the late 300s and early 200s BCE. Like an ancient Greek *Gulliver's Travels*, his *Sacred History* described a journey that Euhemerus said he had taken to three fantastic islands in the Indian Ocean. On one of these islands called Panchaea, Euhemerus claimed he had found old inscriptions written by the great god Zeus himself. Euhemerus insisted that he discovered these inscriptions on a pillar inside a golden temple on the island. The writings proved, according to Euhemerus, that Zeus and the other gods of Greece were all based upon an early king from the island of Crete. To Euhemerus, the gods of Olympus and other characters of Greek myths were all real heroes and conquerors who had been deified, and he claimed to be able to document the entire primitive history of the world from these inscriptions.

While the tale Euhemerus told was clearly a work of fiction, his idea that all the gods were representations of people who had once lived had significant influence for centuries, even carrying over into the Christian era. The belief that all of the Greek myths were based on actual events was used by early Christians to dismiss what they called pagan mythology as a purely human invention. In other words, Christians argued that the Greek gods—who were later adapted by the Romans—were not divine at all, and everyone should acknowledge the one true Christian God.*

This line of thinking about myths—that the gods were once humans—was later called "euhemerism" in honor of Euhemerus. It has continued into modern times, as people search for the historical foundations of many mythic characters and events, whether the historical reality of the Trojan War or the existence of the biblical Abraham or Moses. Even the greatest scientist of the Enlightenment, Isaac Newton (1642–1727), once attempted to document the myths as if they were actual events that could be identified. Universally recognized as a giant of science for his laws of physics, Newton was also a very devout Christian who devoted the last years of his life to an esoteric quest to bring his astronomical calculations in line with biblical history. Although it seems an odd activity for such an eminent man of science, Newton tried to base his chronology of world events on a mythical occurence—the famed voyage of Jason and the Argonauts in search of the Golden Fleece, one of the great quest stories of ancient Greece. Like other Christians, Newton credited the doctrine of Euhemerus and, accordingly, thought that the mythical voyage of the Greek hero Jason aboard his ship, the *Argo*, must have been a fact. Using his own, carefully kept astronomical records, Newton believed he could fix this event to an actual date. Accomplishing this, Newton argued, would also lead to calculating an exact date of the fall of Troy and hence of the founding of Rome by Aeneas, a refugee from the destroyed city of Troy. Newton, who may have been going mad in his later years from mercury poisoning, was never successful in this endeavor.

But a century after Newton, the search for the history behind myths took another great leap. Major archaeological discoveries during the

*The word "pagan," which has come to broadly mean anyone who is not Christian, Jewish, or Muslim, was originally coined by early Christians in Rome and meant "country dweller" and "civilian," in the sense that pagans were not members of the so-called army of God.

nineteenth century transformed the European view of ancient civilizations. With the power of church and kings weakened after the Protestant Reformation, the Enlightenment-era quest for rational explanations of natural events began to replace the orthodox Christian view of the ancient world as simply barbarous. One of the key events spurring this quest was the discovery of the Rosetta Stone by Napoleon's army in 1799. Half-buried in the mud near Rosetta, a city not far from Alexandria, Egypt, the stone is made of black basalt. Measuring 11 inches (28 centimeters) thick, it is about 3 feet 9 inches (114 centimeters) high and 2 feet 4½ inches (72 centimeters) across. This stone had been carved to commemorate the crowning of Ptolemy V Epiphanes, king of Egypt from 203 to 181 BCE. (The Ptolemies were the rulers of Egypt who were heirs to Alexander the Great after he conquered Egypt. The line of the Ptolemies ended with Cleopatra and her royal machinations and disastrous affair with Julius Caesar and marriage to Marc Antony.) The Rosetta Stone contained three separate inscriptions: the first inscription was in ancient Egyptian hieroglyphics; the second was in demotic, the commonly spoken language of Egypt at that time; and, at the bottom, the message appeared again in Greek.

Until this time, the language of ancient Egypt had been a mystery to the world. But a French scholar named Jean François Champollion (1790–1832) was able to decipher the Egyptian hieroglyphics. With a knowledge of Coptic—a form of Egyptian that was written mainly with Greek letters—and using the Greek text as a guide, he was able to pick out the same names in the Egyptian text and learn the sounds of many of the Egyptian hieroglyphic characters, which enabled him to translate many Egyptian words in the inscription. In 1822, Champollion published a pamphlet that opened up the literature of ancient Egypt to scholars. Champollion, viewed as the "father of Egyptology," died of a stroke at the age of forty-one. (Continuing its history as a spoils of war, the Rosetta Stone was later taken to England, where it remains in the British Museum.)

Discoveries such as the Rosetta Stone were keys to opening up the past at a time when more of the world was being opened up to Europe. As the British Empire spread into the Middle East, Asia, and the Pacific, geographers, astronomers, and naturalists, including Charles Darwin, were routinely sent aboard British ships to map and study the natural world. Obviously done in the name of the empire, this unprecedented but deliberate combination of exploration and colonization, discovery and scholarship was having a profound impact on the academic world. As ancient worlds and civilizations were revealed by Great Britain's explorers and mapmakers, archaeologists, linguists, and the first generation of anthropologists followed suit. The academic world was beginning to view myth as an essential ingredient in understanding the past, not simply as the superstitious beliefs of barbarous "heathens" who needed to be Christianized. Once purely the domain of "classicists" who used the Greek myths to teach Greek, the world of mythology was now a fertile field for scholars who wanted to "prove" that these myths were based in the realities of the ancient world.

Who was the man who "found" Troy?

The most famous—and controversial—of this generation of archaeologists was Heinrich Schliemann (1822–1890), a successful German businessman who converted a boyhood fascination with Homer's Troy into a lifelong study of ancient Greece. Schliemann's life was the stuff of a wild, Dickensian novel. Born the son of a poor Protestant minister in northern Germany, Schliemann had been a cabin boy and was shipwrecked as a teenager. After returning to Europe, he taught himself to speak English, French, Dutch, Portuguese, Spanish, and Italian while working as an office boy. He translated his linguistic talents into building an import-export business that made him extremely wealthy. He came to California in the Gold Rush era, started a bank, and by the time he was in his thirties, Schliemann was a bank director and wealthy merchant who could afford to start a second life as an amateur archaeologist with a single, obsessive goal. Guided by his love of the *Iliad*, he set out to find the actual ruins of Homer's Troy. With his second wife, Sophia, a Greek girl thirty years his junior, he focused his efforts on a mound at what is now Hissarlik, in northwestern Turkey. Underwriting the costs of the digs with his considerable personal fortune, Schliemann began to excavate in September 1871.

Ignored or ridiculed by skeptical professionals, Schliemann had the last laugh. His faith paid off when he discovered the buried city of

Troy—actually, the Schliemanns had discovered the *cities* of Troy. At the site where he had predicted he would find Troy, nine cities were uncovered, each successive layer built on the ruins of the one before it. Carelessly digging through these many levels in his frenzied search for Homer's Troy, Schliemann probably destroyed many relics in which he had no interest. But near the bottom level, the Schliemanns found objects of bronze, gold, and silver, in the city they believed was the Troy of the *Iliad*.

Schliemann was clearly a bit of a P. T. Barnum, or a carnival huckster. Using his showman's instincts, he had his beautiful young wife photographed wearing jewels they had discovered, as though she were a modern incarnation of Helen of Troy, and the two became internationally famed. After the triumph of finding Troy, the Schliemanns returned to Greece, where they explored the fabled site of Mycenae, an ancient Greek city where, in 1876, they unearthed five royal graves full of jewels and other treasures. Although the Schliemanns incorrectly believed that they had discovered the burial site of Homer's King Agamemnon, their discoveries opened up the world to the possibility that the classical myths were all based on historical incidents. Even though their scholarship, methods, and personal use of the treasures they found would raise eyebrows for years to come, the Schliemanns had provoked new fascination with the "dead" world of ancient civilizations. As Daniel Boorstin wrote in *The Discoverers*, "The vast watching public came to believe that the earth held relics and messages from real people in the distant past."

How did an ancient myth cast doubt on the divinity of the Bible?

Schliemann's anstonishing finds stimulated a new popular appreciation and interest in mythology. As a flood of information about newly discovered cultures swept through Europe in the wake of nineteenth-century exploration and colonization, a wave of late nineteenth-century scholarship was revolutionizing long-held views of the ancient world. At nearly the same moment as Schliemann found Troy, another discovery had similarly dramatic and far-reaching consequences, although it did not garner the headlines that the beguiling Sophia Schliemann had in her "Helen-ish" jewels. In Nineveh, once the capital of the ancient Assyrian empire, and a city prominent in biblical history, a large number of clay tablets were discovered in the ruins of a temple. Nineveh was the "wicked city" where the Hebrew prophet Jonah of "large fish" (not a whale) fame was sent by God in the Bible. Needless to say, the discovery of writings from so important a site in biblical history attracted a significant amount of attention.

When an exhibit of these Assyrian objects—which had been taken to London's British Museum—opened in 1850, Victorian London was astonished. Assyrians had been viewed as the "bad guys" of the Bible, cruel conquerors who had enslaved the Jews. But here, on display, were carvings and statuary that fascinated Londoners. Jewelers began to make replicas of the Assyrian ornaments, and they became the fashion rage. But even more significant was the impact of the Assyrian discoveries on biblical scholarship. George Smith, a young man when the Assyrian display opened, practically became obsessed with the exhibit—much the way America was fascinated by all things Egyptian when the King Tut exhibit toured museums around the country during the 1970s. With little formal training, Smith was able to win a job at the museum, and in 1872, delivered a paper before the Society of Biblical Archaeology with translations from these ancient tablets. Smith had translated portions of *Gilgamesh*, an ancient Babylonian epic poem that tells of an imperfect hero named Gilgamesh and his quest for immortality, a poem widely considered the world's oldest known work of literature.

But its impact went far beyond exciting a few professors of literature and ancient languages. The contents of *Gilgamesh* that Smith revealed turned the accepted world of Christian biblical beliefs on its head. Smith's translations of *Gilgamesh* included episodes of a great flood that contained clear parallels with the biblical accounts of Noah's flood, along with many other elements shared with the Book of Genesis. His paper set off shockwaves, and a London newspaper commissioned Smith to head for Mesopotamia to do further research. On an expedition to Nineveh, Smith contracted a virulent strain of fever and died at the age of thirty-six.

But Smith's translations had unleashed a flood of another sort. Reaction to the material was earthshaking in the world of biblical scholarship. When a leading German scholar delivered a lecture titled "Babel und Bible" in 1902, and stated that the Bible was not the world's oldest book, as Christian and Jewish scholars had taught for centuries, there was complete outrage. Germany's kaiser Wilhelm II heard the lecture and was neither impressed nor amused. "Religion has never been the result of science," the kaiser wrote, "but the outpouring of the heart and being of man from his intercourse with God."

Smith's translations and the suggestion that the Bible was not the divine word of God came just as the foundations of religion were being shaken in the later nineteenth century by new scientific theories and discoveries. Darwin and the ideas about natural selection and evolution he introduced in *Origin of Species* in 1859 were rattling the foundations of orthodox science and religion. Archaeologists and linguists were now shaking up old ideas about the sources of the Bible and the ancient roots of Judaism and Christianity. With this extraordinary intellectual ferment in the background, a new approach to mythology began to focus on the spiritual and religious significance of myths and their connection to accepted Christian belief.

The translation of another ancient sacred text introduced still another approach to mythology. It came from German Sanskrit scholar Max Müller, who translated the Rig-Veda, the earliest Hindu scriptures, beginning in 1849. Müller believed that myths were expressions of ideas that could not be conveyed in language. According to Müller, "Where we speak of the sun following the dawn, the ancient poets could only speak and think of the Sun loving and embracing the Dawn. What is with us a sunset, was to them the Sun growing old, decaying or dying." In Müller's view, all of the gods and mythical heroes were simply representations of nature, especially the sun. While Müller's ideas have been largely dismissed by modern scholars, his work was another example of the incredible ferment that was sweeping the academic world in this time of radical new assessment of ancient myths. And it was spilling over into the world of religion and politics.

Around this time, the serious study of cultural anthropology was also invented, and one of its chief proponents was Edward Burnett Tylor (1832–1917), who would later become the first professor of anthropology at Oxford University in 1896. A young Quaker suffering from tuberculosis, in 1855 Tylor was sent to the Caribbean, where he became more interested in the ways of the newly discovered remote peoples in the Americas. As a Quaker and abolitionist, Tylor was interested in what was then called "ethnology," and his fascination was more than just academic. He had a missionary zeal, believing studies of "primitive" people could help him document "human brotherhood." Proving the connections between races, Tylor believed, would aid the antislavery cause. The goal of his journey was, as he put it, to "trace the course which the civilization of the world has actually followed." Among the areas he pursued most vigorously was religion, and he coined the word "animism" to describe the most simple belief in spiritual beings and in everything having a soul. According to Tylor's 1871 landmark book *Primitive Culture*, there appeared to be no tribes that "have no religious conceptions whatever." Myths, he believed, were born in the attempt to explain natural phenomena but were rooted in fear and ignorance. Tylor's theories transformed the field, even though his ideas have been largely dismissed, in part because of their somewhat racist overtones.

The controversial but growing connection between mythology and religion reached a new height with the late-nineteenth-century work of Sir James George Frazer (1854–1941). Born in Glasgow, Scotland, he was a classical-scholar-turned-anthropologist, who believed that myths began in the great cycle of nature—birth, growth, decay, death, and rebirth. Frazer's theory, which formed the basis for his twelve-volume masterwork *The Golden Bough* (it appeared between 1890 and 1915), developed from his attempt to explain an ancient Italian ritual called the king of the woods at Nemi, a place near Rome. According to the legend, the king of the woods held an uneasy grasp on the throne, because he was always threatened by challengers who wanted to kill him and take his place. The challenger had to break off a golden bough—hence the title of his study—from a sacred tree in the grove. The death of the king and his replacement by a younger, more virile successor ensured the fertility of the crops.

Frazer's central idea was that all myths were part of primitive religions that centered on fertility rites, and he collected and coordinated hundreds of myths and folktales from all over the world. Frazer was most intrigued by what he called the "Great Mother," and her relationship with a younger male consort who was often sacrificed as a sacred king. According to Frazer, this theme of the dying and reborn god appears in almost every ancient mythology, either directly or symbolically. Some of the most significant examples he singled out were Ishtar and Tammuz from Mesopotamia, and Isis and Osiris in Egypt, but he also made a connection between this idea and Jesus and Mary.

Much of Frazer's work is now dismissed by modern scholars, and aspects of *The Golden Bough* have been discredited—including most of the story of the king of the woods at Nemi. But at the time, Frazer's work was revolutionary. He gave credibility to mythology as a serious study that explained the primitive roots of religion. He also profoundly influenced a whole generation of anthropologists and, perhaps just as important, a generation of writers. James Joyce, T. S. Eliot, and William Butler Yeats were among the twentieth-century writers whose work was shaped in part by Frazer's ideas. (Eliot's famed poem *The Wasteland* makes reference to *The Golden Bough*, though Frazer reportedly said he couldn't make sense of it.)

Closely related to Frazer's ideas were those of the so-called **ritualists**, who believed that all myths are derived from rituals or ceremonies. One of the first scholars to develop this theory was Jane Ellen Harrison (1850–1928), a British classicist of the late 1800s and early 1900s, who argued that people create myths in order to justify already-established magical or religious rituals. "Gods and religious ideas generally reflect the social activities of the worshipper," she wrote in a 1912 book about Greek religion, *Themis: A Study of the Social Origins of Greek Religion*.

A colleague of Frazer, Harrison disagreed with many of his ideas. One of her key contributions was to emphasize the importance of female divinities. "The Great Mother is prior to the masculine divinities," she argued, introducing an idea that is being revived with the so-called Goddess worship, which has enjoyed renewed popularity in recent years. Harrison's theories, while clearly influential, have also been diminished, because it is difficult to say what came first—the ritual or the myth.

MYTHIC VOICES

One of the reasons why religion seems irrelevant today is that many of us no longer have the sense that we are surrounded by the unseen. Our scientific culture educates us to focus our attention on the physical and material world in front of us. This method of looking at the world has achieved great results. One of its consequences, however, is that we have . . . edited out the sense of the "spiritual" or "holy" which pervades the lives of people in more traditional societies at every level and which was once an essential component of our human experience of the world.

—KAREN ARMSTRONG, A History of God *(1993)*

When does myth become religion? And what's the difference?

For most people, the answer to this question might be this simple formulation: "If I believe it, it is religion. If you believe it, it is a myth."

For most of the past two centuries, people educated in the world of science and the rational explanation for the workings of the universe readily dismissed myths as the primitive beliefs of backward people who didn't know better. But to the ancient people of Mesopotamia, Greece, Egypt, India, or China, myths were not myths at all, but religion. They dictated life and formed the basis of the social structure.

As best-selling religious historian Karen Armstrong writes in *A History of God*, "It seems that creating gods is something that human beings have always done. When one religious idea ceases to work for them, it is simply replaced. These ideas disappear, like the Sky God, with no great fanfare."

Consider the Lord's Prayer, or the "Our Father," familiar to millions of Christians the world over as the prayer taught by Jesus in the Gospel of Matthew:

"*Our Father, who art in heaven, hallowed be thy name.*"

Some might consider this idea heretical, but if you know that prayer, stop for a moment and substitute "Zeus" or "Re" for "Our Father." The entreaties that Christians make in this familiar prayer to their divinity are simple yet universal desires that have been part of ritualized prayers across many diverse religions for thousands of years.

"Thy will be done on earth as it is in heaven"—make earth a paradise

"Give us this day our daily bread"—help the crops to grow

"Forgive us our trespasses (or debts)"—we all make mistakes, but have mercy on us

"Lead us not into temptation"—we are weak and do things we know are wrong

"Deliver us from evil"—keep us safe from all the bad things that are out there in the dark and dangerous world

Similar prayers of entreaty can be found in almost every religion and culture.

For instance, this is a traditional African prayer from Sudan:

Our Father it is thy universe, it is thy will.
Let us be at peace, let the souls of the people be cool.
Thou art our Father, remove all evil from our path.

Prayers are—in a most fundamental and ancient way—the essence of belief in the supernatural world. Prayers found in Egyptian tombs are more than five thousand years old. In modern America, a majority of people claim that they believe in prayer and say they pray regularly. The essential underlying question is: If myths were once created to answer fundamental questions and solve problems outside the control of mortals, when did these myths morph into religions?

Setting aside Karl Marx's famous statement that "Religion . . . is the opium of the people," **religion** is an organized system of beliefs, ceremonies, practices, and worship that may center on one supreme God or Deity, or a number of gods or deities. The earliest recorded evidence of religious activity dates from about 60,000 BCE, and today there are thousands of religions in the world. The eight major ones are Buddhism, Christianity, Confucianism, Hinduism, Islam, Judaism, Shinto, and Taoism. But throughout history, mythologies and religions have shared some basic traits and characteristics:

- *Religious rituals* are central to both mythic belief systems and religion. Every tradition has some basic practices that include acts and ceremonies by which believers appeal to and serve God or other sacred powers. The Hebrew Bible, or Old Testament, is filled with elaborate instructions for the ritual sacrifice of animals, practices not very different from the animal sacrifices that were common in ancient Mesopotamia or Greece. Christians around the world participate in a ritual in which bread and wine are believed to be transformed into the body and blood of Jesus, which was shed in a rite of sacrifice, an act that also has very ancient roots.

  Prayer is probably the most common ritual. When praying, a believer or someone on behalf of believers addresses words and thoughts to an object of worship. Most major religions have a daily schedule of prayer.

  Many religions also have rituals intended to purify the body. For example, Hindus consider the waters of the Ganges River in India to be sacred, and every year, millions of Hindus purify their bodies by bathing in the river, especially at the holy city of Varanasi. The common practice of Christian baptism—either infant or adult—is another widely practiced ritual with deep roots in mythical practices. The mother of Achilles dipping her infant son into sacred waters is not so different from the Christian priest anointing an infant's head with holy water to consecrate and protect the newborn, or the Hindu pilgrims who travel to the Ganges River for a dip.

- *Belief in a deity.* "Who knows this truly, and who will now declare it, what paths lead together to the gods?" the ancient Rig-Veda of Hinduism asks. "Only their lowest aspects of existence are seen, who exist on supreme mystical planes."

  Like ancient myths, most religions believe in one or more deities who govern or greatly influence the actions of human beings as well as events in nature. While Judaism, Christianity, and Islam are monotheistic—believe in one god—Hinduism teaches that a world spirit called Brahman is the supreme power but that there are numerous other gods and goddesses. (Confucianism is one of the few significant atheistic religions.)

- *Sacred stories.* The Bible, Koran, Bhagavad-Gita, Popol Vuh. Every religion has its collection of sacred or divine stories, which, in essence, are myths. After all, myths were first told to describe how the sacred powers directly influenced the world.

  As Ninian Smart writes in *The World's Religions,* "Experience is channeled and expressed not only by ritual but also by sacred narrative or myth. . . . It is the story side of religion. It is typical of all faiths to hand down vital stories: some historical; some about that mysterious primordial time when the world was in its timeless dawn; some about great founders, such as Moses, the Buddha, Jesus and Muhammad; some about assaults by the Evil One. . . . These stories are called myths. The term may be a bit misleading, for in the modern study of religion, there is no implication that a myth is false."

This is the essence of one problem between believers of different faiths and traditions. Jews and Christians who hold on to the Bible as the divinely inspired word of God don't necessarily allow that the Mayan Popol Vuh, a collection of sacred stories, is anything more than an invention, superstition that is far beneath their own "holy writ." Arguments over these sacred stories even split people with shared religious traditions. For instance, Catholics and Protestants don't even agree on what should be in the Bible. The portions of the Bible known as the Apocrypha are considered sacred by Catholics, but are less than divine according to Protestant belief. Many Jews and Christians believe that the Creation accounts in Genesis provide the literal and historical explanation for the beginning of the universe and life on earth. Other believers accept the scientific explanations for the creation of the universe and view the biblical account as a metaphor, accepting the message contained in the stories without treating the specific details as literal truth.

In other words, the difference between myth and religion may exist only in the eye of the believer—or the nonbeliever. And Ninian Smart's conclusion brings this chapter nearly full circle. Myths are the sacred stories that may convey essential truths, even if they come in the guise of tales about ancient gods behaving badly.

Mythic Voices

I believe that a large portion of the mythological conception of the world which reaches far into the most modern religions, is nothing but psychology projected to the outer world.

—Sigmund Freud *(1856–1939)*

Myths go back to the primitive storyteller and his dreams, to men moved by the stirring of their fantasies. These people were not very different from those whom later generations called poets or philosophers.

—Carl Gustav Jung *(1875–1961)*

Are myths all in our minds?

Beyond exploring the sacred, spiritual, or religious component at the heart of myths, the twentieth century brought one more significant explanation for the source of mythology. With the beginnings of modern psychology, pioneers of psychological thought, such as Sigmund Freud, maintained that myths originate in the unconscious mind. Freud called myths "the dreams of early mankind," and he frequently made reference to mythic characters, most famously in the term "Oedipus complex," which drew on the story of the Greek king Oedipus who killed his father and then married his mother, and which was set forth in Freud's landmark *The Interpretation of Dreams* (1900).

To Freud, myths were a product of personal psychology and dreams were the source of these myths. Specifically, he believed that most myths were sexual in nature. Heroes slaying dragons and gods killing other gods were all really forms of every male's desire to kill his father and have sex with his mother. Or, as Barry Powell summarizes Freud's thinking in *Classical Myth*, "Mythical kings and queens represent parents, sharp weapons are the male sexual organ, and caves, rooms and houses symbolize the mother's containing womb. The imagery of myths can therefore be translated into that of sex. . . ."

In the early 1900s, Swiss psychoanalyst Carl Gustav Jung, a disciple of Freud, took this concept—that all myths are generated in the unconscious mind—in a different direction. Born in Basel, Switzerland, Carl Gustav Jung was the son of a minister. As a boy, he was fascinated with superstition, mythology, and the occult, and he first planned to study archaeology at the University of Basel. Instead he graduated as a physician from the University of Zurich in 1902 and soon became a student of Freud's psychiatry, using Freud's psychoanalytical theories.

But Jung broke with his mentor over Freud's emphasis on sexuality. Rejecting Freud's belief that the symbolism of the unconscious was primarily sexual, Jung said dreams came both from what he called the personal and the collective unconscious. While the personal unconscious reflects an individual's experiences, the collective unconscious is inherited, shared by all humankind. According to Jung, art, religion, dreams, and myths are the means in which the unconscious finds expression. Jung believed that the entire psychological development of humanity could be traced by studying myths, fairy tales, and fables.

The collective unconscious, Jung stated, was organized into basic patterns and symbols, which he called **archetypes**, and which all mythologies shared. Gods and heroes, mythic places—such as the home of the gods or the underworld—and battles between different generations for control of a throne were all found in every system of myths. Jung asserted these mythical archetypes were so fundamental to humanity that "If all the world's traditions were cut off at a single blow, the whole of mythology and the whole history of religion would start all over again with the next generation."

Where Freud had taken a limited view of the importance of religion or a sense of the sacred in the realm of psychology, Jung saw mythology as a powerful connection to the sacred and he regretted the modern loss of faith in this mysterious part of the human experience. "From time immemorial, men have had ideas about a Supreme Being (one or several) and about the Land of the Hereafter," Jung wrote in *Man and His Symbols*. "Only today do they think they can do without such ideas."

Acknowledging that humanity had progressed into a complex, rational, scientifically ordered world, which rejected those things that cannot be proved, Jung argued for a spiritual component in life that mythology—and, later, organized religion—has traditionally provided throughout human history.

"There is, however, a strong empirical reason why we should cultivate thoughts that can never be proved. It is that they are known to be useful," Jung wrote in *Man and His Symbols*. "A sense of wider meaning to one's existence is what raises a man beyond mere getting and spending. If he lacks this sense, he is lost and miserable."

Echoing Jung's theories, late in his life, Albert Einstein wrote, "The most beautiful thing we can experience is the mysterious. It is the source of all true art and science. He to whom this emotion is a stranger, who can no longer pause to wonder and stand rapt in awe, is as good as dead: his eyes are closed. This insight into the mystery of life, coupled though it be with fear, has also given rise to religion. To know what is impenetrable to us really exists, manifesting itself as the highest wisdom and the most radiant beauty which our dull faculties can comprehend only in their most primitive forms—this knowledge, this feeling, is at the center of true religiousness. In this sense, and in this sense only, I belong in the ranks of devoutly religious men."

Science, history, anthropology, language, psychology, rituals, religion, and spirituality. All of these frameworks help to explain how myth has operated since the dawn of human time. Yet none of them alone does it completely. As classicist Barry B. Powell has observed, "Myth taken together is too complex, too many-faceted, to be explained by a single theory."

That myths reflect so many aspects of the human condition—our history, our innermost thoughts, our best and worst behavior, an acceptable code of conduct—makes trying to fit them into one neat theoretical framework impossible. It is like trying to make many different people wear a single suit of clothes. There are just too many sizes and shapes for that to work.

Needless to say, for thousands of years, the myths that have organized human civilizations and given faith to worshippers across all time are clearly something greater than a collection of compelling stories about dysfunctional gods, flawed heroes, sex-crazed tricksters, or primeval monsters lurking in the closets of our minds.

Heady stuff. It may be wise to remember the words of American humorist James Thurber, who once wrote: "It is better to know some of the questions than all of the answers."

CHAPTER TWO

GIFT OF THE NILE

The Myths of Egypt

Hail to you gods . . .
On that day of the great reckoning.
Behold me, I have come to you,
Without sin, without evil,
Without a witness against me,
Without one whom I have wronged. . . .
Rescue me, protect me,
Do not accuse me before the great god!
I am one of pure mouth, pure of hands.

—*The Book of the Dead*
(c. 1700–1000 BCE)

Creator uncreated
Sole one, unique one, who traverses eternity,
Remote one, with millions under his care;
Your splendor is like heaven's splendor.

—*First Hymn to the Sun God*
(c. 1411–1375 BCE)

Egypt was old, older than any culture known at the time. It was already old when the political policy of the future Roman Empire was being formed in the first meetings on the Capitoline Hill. It was already old and blighted when the Germans and Celts of the north European forest were still hunting bears. When the First Dynasty came into power about five thousand years ago . . . marvelous cultural forms had already been evolved in the land of the Nile. And when the Twenty-sixth Dynasty died out, still five hundred years separated European history from our era. The Libyans ruled the land, then the Ethiopians, the Assyrians, the Persians, the Greeks, the Romans—all before the star shone over the stable at Bethlehem.

—C. W. CERAM,
Gods, Graves and Scholars (1951)

How did myths "rule" in ancient Egypt?

Why was Egypt the "gift of the Nile"?

What do we know about Egyptian myth and how do we know it?

Who was the first family of Egyptian myth?

How does "creation by masturbation" work?

Who was Re?

Which god became Egypt's lord of the dead?

Who was Egypt's most significant goddess?

What did Christians think of Isis?

What was the "weighing of the heart"?

Who's Who of Egyptian Myths

Why are there so many animals—real and imaginary—in Egypt's myths?

What did the pyramids have to do with the gods?

What's so great about the "Great Pyramid"?

What is an Egyptian pyramid doing on the U.S. dollar bill?

Was the ruler of Egypt always a pharaoh?

Did a pharaoh inspire Moses to worship one god?

Does Egyptian myth matter?

MYTHICAL MILESTONES

Egypt

All dates are BCE, Before the Common Era. Egyptian history covers thousands of years, and while the order of kings is reasonably well established, many precise dates are more problematic and are often approximated. Many of the dates in this chronology are drawn from *The Oxford History of Ancient Egypt*.

5000 to 4001 The Egyptian calendar is devised, regulated by sun and moon; 360 days; divided into twelve 30-day months.

4000 Sails are used.

3300 First walled towns are built.

3200 Earliest hieroglyphic script appears.

Early Dynastic Period c. 3100–2686

3100 King Narmer/Menes(?) unites Upper and Lower Egypt.

Memphis is founded as the capital of unified Egypt.

Beginning of systematic astronomical observations in Egypt.

3050 Introduction of the 365-day calendar.

Old Kingdom 2686–2160

2667–2648 Third Dynasty ruler Djoser rules with counselor (vizier) Imhotep, who makes the first known efforts to find medical as well as religious methods for treating diseases.

2650 Beginning of period of pyramid building; the first monumental building in stone is the Step Pyramid of Djoser at Saqqara, initiated by Imhotep.

2575 Great Pyramid of Khufu (Cheops), largest of the Egyptian pyramids, is built at Giza.

2550 The Great Sphinx at Giza is carved under the reign of Khafra (Chephren).

2500 to 2001 Division of the day into twenty-four units.

Cult of Isis and Osiris develops.

First use of mummification.

2375–2300 In the pyramid of King Unas, the first known use of "Pyramid Texts"; these are funerary texts inscribed on walls of pyramids; they are the oldest known religious writings in the world.

First Intermediate Period 2160–2055

2150 Series of floods brings famine and discontent; collapse of the Old Kingdom.

Middle Kingdom 2055–1650

c. 2055 Egypt is reunited under Middle Kingdom pharaohs.

1991 Book of the Dead is collected; it is known to Egyptians as "The Chapters of Coming Forth by Day."

c. 1965 Nubia (modern Sudan) is conquered by Egypt.

c. 1800 Horse is introduced to Egypt.

1700–1500 Biblical patriarch Joseph in Egypt (?).

Second Intermediate Period 1650–1550

c. 1660 Invasion of Semitic Hyksos from Palestine, Syria, and farther north. They are excellent archers, wear sandals, and use horse-drawn chariots to conquer the Nile Delta; eventually they rule much of Egypt.

New Kingdom 1550–1069

1567 Expulsion of the Hyksos by Ahmose.

1550 Rise of the New Kingdom; the capital founded at Thebes, which becomes center of the Egyptian Empire. The New Kingdom dynasties usher in a period of stability and rule for nearly five hundred years, expanding Egypt's power into Asia.

1473 Queen Hatshepsut rules as regent for her infant stepson, who will become Thutmose III.

1479 Thutmose III takes the throne and the title of pharaoh. Thutmose III attempts to obscure all references to his aunt Hatshepsut by constructing walls around her obelisks at Karnak.

1470 Massive volcanic eruption on the isle of Thera is viewed as responsible for destruction of an advanced Minoan civilization based on Crete.

1352–1336 Pharaoh Amenhotep IV, also known as Akhenaten, introduces sun worship as a form of monotheism; his religious reforms, called the "Amarna Revolution," plunge the country into turmoil.

1336–1327 Brief reign of famed boy-king Tutankhamun, whose tomb survived virtually intact until discovered in 1922.

1295–1200 Speculative date of Jewish Exodus from Egypt.

1286 Hittites almost defeat the Egyptians at the Battle of Kadesh in modern Syria. Following this battle, Ramses II marries a Hittite princess, cementing a peace treaty between the two powers.

1279–1213 Ramses II rules; widely believed to be the pharaoh during the biblical Exodus.

1245 Ramses II moves Egyptian capital to new city, Pi-Ramesses.

1153 Death of Ramses III, Egypt's last great pharaoh.

1070 End of Twentieth Dynasty.

Third Intermediate Period 1069–664

1005–967 Reign of King David in Israel; Jerusalem established as capital.

967–931 Reign of King Solomon in Jerusalem.

945 Egyptian civil wars; a Libyan dynasty is installed, and the first non-Egyptian line rules Egypt for the next two hundred years.

814 Foundation of Carthage, Phoenician colony in North Africa.

753 Traditional date of the founding of Rome.

747 Rule of Egypt by Nubians.

671 Assyrian king Esarhaddon attacks Egypt, captures Memphis, sacks Thebes, and leaves vassal rulers in charge.

c. 670 Introduction of iron working.

Late Period 664–332

664 Egypt regains independence from Assyria.

525 Persian army led by Cambyses occupies Egypt, which becomes part of the Persian Empire.

490 The Battle of Marathon marks the beginning of the Persian Wars between Greece and Persia.

457 The Golden Age of Athens under Pericles.

450 Greek historian Herodotus visits Egypt and describes customs and history, sometimes quite fancifully, in *The Histories*.

Ptolemaic Period 332–30

332 Alexander the Great conquers Egypt; founds the city of Alexandria.

323 Death of Alexander the Great.

305 Ptolemy I Soter, one of Alexander's Greek generals, becomes king of Egypt; adapts pharaonic titles and Egyptian worship.

290 In Alexandria, Euclid sets out principles of geometry in *Elements*.

250–100 In Alexandria, Hebrew religious texts are translated into Greek, the version of the Old Testament known as the Septuagint.

c. 200 Alexandria is the scientific capital of the world, famed for its museum, library, and university.

146 Rome conquers and destroys Carthage.

49 Roman civil war. Julius Caesar in Egypt with Cleopatra.

46 Caesar returns to Rome with Cleopatra as his mistress and is made dictator of Rome.

44 Cleopatra murders Ptolemy XIV, coruler of Egypt.

Julius Caesar assassinated in the Roman Senate.

41 Marc Antony meets Cleopatra and follows her to Egypt.

31 Battle of Actium; Octavian defeats Marc Antony.

30 Deaths of Marc Antony and Cleopatra; annexation of Egypt by Rome.

4 Death of King Herod; widely accepted date of birth of Jesus.

For the next five centuries, Egypt remained a province of the Roman Empire. But the rise of Christianity, and later the ascendancy of Islam in the Arab world, marked the final end of the old religions of Egypt. According to Christian lore, St. Mark, a Christian missionary, founded the Egyptian (Coptic) church in Alexandria around 40 CE. The city, which already had a large community of Jews, soon also developed a thriving Christian community. During the early years of the Christian Church, the bishops of Alexandria exercised enormous influence in defining Christian beliefs and practices.

Following the conversion of the Roman Empire to Christianity under Emperor Constantine in 313 CE, the Roman emperor Theodosius ordered the closing of all pagan temples throughout the empire in

383 CE. Later imperial decrees by Theodosius and Emperor Valentinian in 435 CE called for the complete destruction of these temples, many of them replaced with Christian churches or shrines. (This was also the fate of the Olympian temples in Greece, site of the Olympic Games for more than 1,200 years.) Vestiges of the old Egyptian religion were permitted to continue in Egypt, even though Christianity was now the official religion in Egypt.

As the Roman Empire went into its decline, Arab armies claimed Egypt and introduced Islam. In 642, Arab Muslims conquered Egypt. The Arabs moved the capital from Alexandria to what is now Cairo. Modern Egypt is primarily Sunni Muslim (94 percent); Coptic Christians and other groups represent a small minority.

Ancient Egypt. Say the words and conjure the images. For movie lovers, it may be the buff and bald Yul Brynner in a chariot chasing down a white-bearded Charlton Heston as Moses in *The Ten Commandments*. For devotees of pseudoscience, it might be the premise of the best-selling book *Chariots of the Gods*, which argues that alien astronauts landed their spaceships in the desert and built the pyramids long, long ago. A younger generation of music fans might be forgiven if the best vision of Egypt they can muster up comes from Michael Jackson's 1992 video "Remember the Time," featuring comedian Eddie Murphy as a pharaoh, with supermodel Iman imperiously enthroned beside him as his queen.

Let's face it. As victims of the myth-making mass media, we have been served up more than a fair share of badly distorted images of ancient Egypt, a culture that stood longer than any other in history. That's unfortunate, because in reality, the Egyptians created a society that prized both morality and beauty—physical and artistic—and expressed those ideals through one of the most unique and rich systems of mythology in the ancient world. A very old set of gods and goddesses formed the soul of one of the most grandiose and unsurpassed civilizations in history. The Egyptian stories of animal-headed deities, sun gods sailing through eternity, and a pair of divine lovers named Isis and Osiris dominated that civilization, inspired its greatest accomplishments, and made an indelible mark around the world for centuries to come.

Springing up on a thin strip of fertile land along the Nile River, hedged in by unforgiving deserts, the great Egyptian culture of priests, pyramids, and papyrus was a remarkable one that dawned more than five thousand years ago and lasted until Rome emerged and Jesus was born. Over the course of more than three thousand years, Egypt's people built a world of epic grandeur that was unparalleled in ancient times for its longevity, prosperity, and magnificent architectural and artistic achievements, all of which profoundly influenced its neighbors—including the celebrated Greeks.

But, as Egypt's vast collection of art and antiquities attest, the pulsing heartbeat of this great civilization was its mythology and religion. From the dawn of Egypt's history, it was a world in which the power of the gods was felt daily, at almost every level of society. In the temples of the sun god at Karnak, where priests tended to the gods and their flocks of sacred animals. In the lives of everyday Egyptians who mummified their family members and pets in the hope of helping them attain eternal life. In great cities like Memphis, where supplicants came each day to the corrals in which sacred bulls were used to divine the future. This was the true ancient Egypt, an extraordinary land of monuments, magic, and—most of all—myths.

How did myths "rule" in ancient Egypt?

We toss around the concepts of "god" and "country" quite easily, without giving much thought to how they got started. Both ideas have been fairly significant throughout human history. For centuries, people have believed that to serve god or country—or both—was a noble calling. But few of us may realize that Egypt—land of the pharaohs, sphinxes, and mummies—essentially invented both concepts.

Going back to a time before history, when Egypt was established as the first true nation along the banks of the Nile, it was a complete **theocracy**—a place where religion and government were inseparably linked in the minds of rulers, priests, and people. Not only were Egypt's royalty the leaders of the nation, they were actually thought to be gods. The pharaohs' status as gods incarnate was what motivated tens of thousands of workers to lift and arrange millions of blocks of stone that weighed more than two and a half tons apiece. These laborers were not beaten under the lash of oppressive overseers. They worked willingly in the belief that the king must have a proper resting place from which he could ascend to the heavens, joining the other gods in his eternal life. Making sure that the pyramids and other tombs were properly constructed and well provisioned with the "grave goods" required for a comfortable life in the afterworld was no small concern. Only then could the resurrected king help ensure that the Egyptian world and its timeless order would continue, uninterrupted by drought, flood, or foreign invaders.

The nearly obsessive interest in ritual and order in ancient Egypt was not limited to the affairs of the king. From birth to death, and covering nearly everything in between, rank-and-file Egyptians lived under a highly structured set of customs and beliefs that were designed to keep them and their blessed land in the good graces of the vast pantheon of gods they worshipped. Proper care for these gods—and their earthly manifestation, the pharaoh—ensured the cosmic order, a concept that the Egyptians called *maat* and that was personified in the goddess named **Maat**, beloved daughter of the sun god Re. It was *maat* that made the sun rise each day and brought the annual flooding of the Nile River, which guaranteed Egypt's plentiful food supply and continued existence. The universal harmony of *maat*—a holy and ethical concept that meant truth, justice, and righteousness, as well as order—was achieved through a religious system in which the gods protected Egypt and held the forces of chaos, destruction, or simple, everyday misfortune at bay, both through proper individual behavior and obeying the ritual laws of the land.

Overseeing those ritual laws was a priestly class—one of the world's first government bureaucracies—whose expertise was in knowing how to please the gods. Whether it was sacrificing an animal to bring the rains that assured a good harvest; collecting taxes for the temple complexes; conscripting workers for three months of each year to build the great stone mausoleums that glorified the king and eased his ascension to the afterlife; or simply shaving one's eyebrows to mourn the death of a beloved cat—the priests saw to the rites that dictated Egyptian life, year in and year out. The rules they articulated and enforced helped Egypt achieve and maintain a remarkable degree of social organization and stability without resorting to draconian punishments, a vast slave economy, grotesque human sacrifices, or a rigid military state. Instead, as author Richard H. Wilkinson writes in *The Complete Gods and Goddesses of Ancient Egypt*, this was a "spiritual world . . . which remains unique in the history of human religion. The character of that spiritual world was both mysterious and manifest, at once accessible and hidden, for although Egyptian religion was often shrouded in layers of myth and ritual, it . . . shaped, sustained and directed Egyptian culture in almost every imaginable way. The deities of Egypt were present in the lives of pharaohs and citizens alike, creating a more completely theocratic society than any other of the ancient world."

And so, to understand ancient Egypt, you must understand its myths. And to know those myths, you must first understand the two great forces that shaped this ancient civilization's history and destiny: the river and the desert, a perfect duality of life and death.

Mythic Voices

Hail to thee O Nile! Who manifests thyself over this land and comes to give life to Egypt! Mysterious is thy issuing forth from the darkness, on this day whereon it is celebrated! Watering the orchards created by Re, to cause all the cattle to live, you give the earth to drink, inexhaustible one. . . .

Lord of the fish, during the inundation, no bird alights on the crops. You create the grain, you bring forth the barley, assuring perpetuity to the temples. If you cease your toil and work, then all that exists is in anguish. If the gods suffer in the heavens, then the faces of men waste away.

—Hymn to the Nile *(c. 2100 BCE)*

Why was Egypt the "gift of the Nile"?

The Greek historian Herodotus, who might also be called the world's first great travel writer, coined the phrase the "gift of the Nile" to describe Egypt. It was a society that utterly fascinated this Greek tourist when he visited Egypt back around 450 BCE. When Herodotus traveled through Egypt, Greece was flourishing in its Golden Age. But Egypt was already three thousand years old, a great trading and military power in the ancient Near East. Having developed the world's first national government, the Egyptians had also created the 365-day calendar, pioneered geometry and astronomy, developed one of the first forms of writing, and invented papyrus—the paperlike writing material that was essential to the birth of the book.

A long, narrow country through which the Nile River flows north into the Mediterranean Sea, Egypt is bordered mostly by vast deserts on its other three sides. The Egyptian word for these hot, sandy wastelands is *Deshret*, meaning "Red Land," and the source of the word "desert." Although the surrounding mountainous areas in the deserts were the source of the gold, gems, and hard stone that provided the raw materials of Egypt's grand buildings and brilliant artistry, these deserts—to the ancient Egyptians—were hellish places that could only bring danger and death.

The lines between these two worlds of life and death were not viewed as metaphoric or symbolic, but were physically tangible realities to the Egyptians. It is literally possible to stand with one foot in the dry desert and the other in the moist soil watered by the river—fertility and life on one side, sterility and death on the other. That clear demarcation between life and death carried over into Egyptian myths and beliefs. Many prehistoric burial sites have been found in the desert, and the obsessive Egyptian preoccupation with death may well derive from the fact that the hot, dry sand created a natural form of mummification that the Egyptians later perfected thorough their elaborate funerary arts. From early times, it seems, Egyptians related the desert with one of their chief gods, **Seth**, who represented the force of chaos and the dangers of the desert. He would enter into a cosmic life-and-death struggle with his brother, the fertility god **Osiris**, which formed one of the core myths of ancient Egypt.

Cutting through the harsh landscape of the desert flowed the Nile, the world's longest river, with its beginnings in the mountains near the equator in central Africa. Gathering the rainfall and snowmelt of the Ethiopian highlands and all of northeastern Africa, and wandering for more than 4,100 miles, the Nile was Egypt's life force. Starting at the end of June, when the rainy season began in central Africa, the Nile flooded its banks each year, leaving a strip of fertile, dark silt that averaged about 6 miles wide on each side of the river. The annual rising of its waters set the Egyptian calendar of sowing and reaping with its three seasons of four months each: inundation, growth, harvest. The flooding of the Nile from the end of June till late October brought down the rich silt, in which crops were planted and grew from late October to late February, to be harvested from late February till the end of June. The ancient Egyptians called their country *Kemet*, meaning "Black Land"*—after this rich, dark, life-giving soil.

Barley, which was baked into bread and brewed into beer, and Emmer wheat—an Asian grain well suited to feed cattle—were the staples, along with lentils, beans, onions, garlic, and other crops that grew in abundance in this moist, fertile soil. At times, there were bad years when drought limited the rains, or flooding rains destroyed the crops. But usually, Egypt's farmers could anticipate and rely upon a surplus that allowed for trading. Trading led to commerce, commerce led to a merchant class, which eventually allowed for the development of the ranks of artisans and craftsmen who didn't need to depend on farming to live. All of this came from the Nile. As historian Daniel Boorstin puts it in *The Discoverers*, "The Nile made possible the crops, the commerce, and the architecture of Egypt. Highway of commerce, the Nile was also a freightway for materials of colossal temples and pyramids. A granite obelisk of three thousand tons could be quarried at Aswan and floated two hundred miles down the river to Thebes. . . . The rhythm of the Nile was the rhythm of Egyptian life."

Since the welfare and existence of the whole country depended on this one central phenomenon—the annual flooding of the Nile—the river became the centerpiece of Egypt's religious ideas. The flooding, or inundation, was personified in the form of different deities. The annual rising of the Nile—which was part of the *maat*—could be fixed to the regular appearance of the "dog star" Sirius, which gave the whole affair a sense of celestial as well as earthly order.

Egypt's history begins with prehistoric villages that grew up along the banks of the Nile more than five thousand years ago. Before that time,

*In recent years, some scholars have interpreted that phrase as evidence that the ancient Egyptians were actually black Africans who then inspired the Greeks and other Western civilizations—a hot topic in the academic world, also called "Afrocentrism." Largely dismissed by most "Egyptologists," Afrocentrism has still made inroads into the American educational system, where it has flourished as a controversial means of endowing primarily African-American schoolchildren with a sense of pride in an African past that was ignored by traditional historians. Unfortunately, for the most part, this approach has replaced one set of simplistic, flawed, and romanticized ideas with another.

Stone Age Egypt was probably settled by people who came from Libya to the west, Palestine and Syria to the east, and Nubia to the south. Adding to this "multicultural" melting pot were traders from what is now Iraq (ancient Mesopotamia) who may have also settled in the area, attracted by the fertility of the land beside the Nile. Grave sites from these early periods show that the dead were carefully buried, often in a fetal position suggesting notions of an expected rebirth, in burial pits that contained possessions needed for an afterlife—a clue to how ancient religious beliefs were formed very early in human history.

Over time, the small farming and herding communities became part of two kingdoms: one controlled the villages that lay in the Nile Delta, where the river spreads out before emptying into the Mediterranean, and which came to be known as Lower Egypt; the other controlled the villages south of the Delta area and was called Upper Egypt. Most of the people in ancient Egypt lived in the Nile River Valley and there may have been between 1 million and 4 million people living in Egypt at various times.

Dark-skinned and dark-haired, the Egyptians spoke a language that was related to the Semitic languages spoken in the modern Middle East—including Arabic and Hebrew—and recent linguistic discoveries place ancient Egyptian among a family of languages called Afro-Asiatic, spoken in northern Africa. By around 3100 BCE, their language was also written in hieroglyphics, a complex system in which more than seven hundred picture symbols stood for certain objects, ideas, or sounds. Recent discoveries show bone and ivory with a form of hieroglyphic script dating as far back as 3400 BCE, and it is supposed that the Egyptian hieroglyphic system may have been invented for administrative and ritual purposes. Some scholars believe that Egypt's writing system developed with Sumerian influence, as certain character types appear in both written languages; others argue that the differences between these two ancient writing systems are greater than their similarities. What is certain is that from the earliest days of Egyptian civilization, hieroglyphics were inscribed on monuments, temples, and tombs, and were set down on official texts, many of which were preserved over the centuries in Egypt's hot, dry climate, providing generations of scholars and archaeologists with a rich array of sources for studying Egypt's past.

Egypt's long history has fascinated many other foreign people, including the Greeks going back to the time of Socrates, Plato, and Herodotus. But serious "Egyptology" began two hundred years ago with the cracking of the secret of the Rosetta Stone, which caused many previous theories and assumptions about Egypt to fall by the wayside. Recent research into Egypt's prehistory has begun to transform a long-accepted version of the civilization's earliest days. There is now evidence that a succession of southern (or Upper Egypt) kings, including one known as Scorpion, grew more powerful in the fourth millennium BCE, and references to well-known gods of Egypt have been found this far back in Egyptian history. The key event in the beginning of that history took place about 3100 BCE, when a king of Upper Egypt, traditionally called Menes the Uniter, but now often identified as Narmer, conquered Lower Egypt. One of the key pieces of evidence for this event is the Narmer Palette, a double-sided carved slate that depicts a king subduing a captive, along with other symbols that suggest a united kingdom.

Merging the two Egypts into one, Narmer and his successors, who may have included Menes (some historians think that they are the same person), began the process of forming the world's first national government. Around 3000 BCE, Memphis was founded as a capital near the site of present-day Cairo. What is also clear from the Narmer Palette and other very old artifacts is that the close connection between gods and kings was well established by the time the country was unified. From its earliest beginnings, Egypt was a theocracy, and its very ancient gods were intricately connected to the Egyptian government throughout its long history.

While doubt has been cast on the existence of Menes, a king named Aha is now counted as the first of Egypt's many kings in a succession of thirty-one dynasties—or families of kings—that ruled the country right down to the time of Alexander the Great in 332 BCE. During the early period of Egyptian history, the Egyptians developed irrigation systems, invented ox-drawn plows, and created the world's first bureaucracy. Based at Memphis and anchored by religious beliefs, the Egyptian national government—in which god and country were not separate entities but completely interlocked—managed the enormous public works, including the construction of the pyramids, and employed an army of scribes to record it all.

While some scholars have questioned whether the Egyptian rulers were actually considered divine from earliest times, it is clear that these kings ruled the first nation-state as the political, military, and religious leaders. It is also clear from the earliest known tombs of these kings that the king was seen as the mediator between his people and the powers of the afterworld, and that the state religion gave legitimacy to the political order. Other documents and artifacts from this very early time show that another significant human invention was securely in place, too—taxes!

The priesthood existed to serve both the deities and the king, who was considered the chief priest of Egypt. The temple complexes run by Egypt's priests were in many ways equivalent to the medieval cathedral towns of Europe. They were not visited on a once-a-week basis or occasional holiday, but were the economic and social center of Egyptian life. As in feudal Europe, most of Egypt's land was in the hands of king and priests. The temples collected and distributed the bounty of Egypt and supported entire populations of civil servants, scribes, craftsmen, and artisans. They collected taxes on behalf of the king—they were the instruments of state power. In one census taken in the time of Ramses III, the two great temples in Thebes employed ninety thousand workmen, owned five hundred thousand head of cattle, four hundred orchards, and eighty ships.

After the earliest dynasties, Egyptian history has been traditionally divided into three major periods, known as the **Old**, **Middle**, and **New Kingdoms**, interrupted sporadically by stretches of social upheaval or foreign rule known as "intermediate periods."* In spite of these interruptions, occasional periods of foreign control and occasional breakdowns in order, Egyptian life maintained its fundamental sense of order and stability with remarkable longevity.

The **Old Kingdom**, or the **Pyramid Age**, began in 2686 BCE and continued for some five hundred years until 2160 BCE. As the name obviously suggests, the period is famed for the construction of the first massive pyramids. During the Old Kingdom, the king's absolute power was solidified, based on the belief in his divinity, his role as chief priest,

*There is a great variety among Egyptian dating systems, and many of the dates presented here are approximate or speculative, but are based on the widely accepted chronology found in *The Oxford History of Ancient Egypt*, edited by Ian Shaw.

and his control of the priesthood, and the promise that only the king would spend eternity with the gods, where he would continue to maintain the cosmic order that blessed Egypt with such plenty. To maintain the status quo, the king wielded unquestioned power. One stunning example of both the stability and total control can be seen in one of the Old Kingdom rulers, Pepy II, who took the throne at age six and ruled for ninety-four years.

The Old Kingdom went into decline and was followed by an unsettled century, called the First Intermediate Period, in which power shifted away from Memphis to Herakleopolis. This time of unrest and disorder was later believed to be a time when the gods withdrew their blessings from Egypt. A new generation of Upper Kingdom rulers restored national order during the **Middle Kingdom** (2055–1650 BCE). This was a four-hundred-year period of peace and prosperity, during which the kings of the Twelfth Dynasty conquered neighboring Nubia (modern Sudan) and began to expand Egypt's trade with Palestine and Syria in southwestern Asia and the advanced Minoan civilization, based on Crete. Often described as a "Renaissance" period in Egypt, the Middle Kingdom saw Egyptian art, architecture, and religion reach new heights. With this exposure to other surrounding cultures, historian Gae Callender explains, "The Middle Kingdom was an age of tremendous invention, great vision, and colossal projects, yet there was also careful and elegant attention to detail in the creation of the smallest items of everyday use and decoration. This more human scale is present in the pervading sense that individual humans had become more significant in cosmic terms. . . ."

After this golden era, atrophy set in and another succession of weak rulers brought an end to the Middle Kingdom around 1650 BCE. While Egypt was in this weakened state, warriors from Asia spread throughout the Nile Delta. Eventually these immigrants, who used horse-drawn chariots and carried improved bows and other more advanced weapons unknown to the Egyptians, seized control of much of Egypt's territory. These invaders, called "Asiatics" by the Egyptians, are better known by their Greek name, the Hyksos kings, and they ruled much of the Delta area of Egypt in a Second Intermediate Period. But rather than attempting to replace Egyptian religion with their own gods and worship, as invaders often do, the Hyksos seem to have adapted

Egyptian forms. Apparently the Egyptians also learned from the Hyksos invaders, adapting their arts of war, and eventually drove the Hyksos out of Egypt.*

A new succession of kings emerged, originally based in the Upper Egypt city of Thebes, and began using the title "pharaoh." These kings developed a permanent standing army that used horse-drawn chariots and other advanced military techniques introduced during the Hyksos period, ushering in the five-hundred-year period of the **New Kingdom**. Beginning in 1550 BCE with Ahmose, the Eigthteenth Dynasty pharaoh credited with expelling the Hyksos from Egypt, this era saw ancient Egypt become the world's greatest power, and it includes some of the most familiar names in Egyptian history—Thutmose III, Queen Hatshepsut, Akhenaten and his wife Nefertiti, Tutankhamun, and a series of pharaohs named Ramses, of biblical fame.

During this era, Egypt also began an aggressive military expansion, and Thutmose I took armies as far as the Euphrates River. His daughter, Queen Hatshepsut, became one of the first known ruling queens in world history, but presented herself publicly and was depicted in art as a bearded man. Egypt reached the height of its power during the 1400s BCE under Thutmose III. Dubbed the "Napoleon of Ancient Egypt," Thutmose III aggressively set out to expand Egypt's boundaries, led military expeditions into Asia, and reestablished Egyptian control over neighboring African kingdoms, making Egypt the strongest and wealthiest nation in the Middle East.

What do we know about Egyptian myth and how do we know it?

History is sometimes mystery. We often "don't know much about" the truth of events taking place in our own lifetimes. So how can we possibly understand or know about a place that existed in a time before books, newspapers, and photographs? In the case of Egypt, fortunately, we have a society that spent a great deal of energy on the idea of posterity. The Egyptians were proud of what they had achieved, and some kings in par-

*The Roman-era Jewish historian Josephus credited the Hyksos with the foundation of the city that later became Jerusalem.

ticular spared little expense in making sure the world knew about what they had done. And much of it was, as the expression goes, "set in stone."

Remarkably well-preserved scrolls, thousands of years old, show Egypt as a highly literate society. We have Egyptian accounts of people doing their taxes, manuals of polite conduct that are 4,500 years old, and letters in which fathers admonish their sons to work hard at scribe school so they won't have to make a living as carpenters, fishermen, or worse, laundry men—a job in which the occupational hazards included washing the garments of menstruating women while dodging Nile crocodiles. Achieving the status of a scribe was a high honor for an upwardly mobile young Egyptian commoner with social aspirations. Ancient Egypt, in other words, was a literate culture that prized learning.

Which makes it all the more surprising that there is no ancient Egyptian Bible, Koran, *Odyssey*, or *Gilgamesh* epic, in which poets would have organized and gathered an "authorized" version of Egyptian mythology. Much of what we know about Egypt's myths, beliefs, and history has been carefully reconstructed from an elaborate array of funerary literature and art uncovered and translated during the past two hundred years. Few ancient civilizations documented their beliefs in as rich a detail and in so many locations as the Egyptians did. Obviously it helps that they had more than three thousand years to create that mother lode of art and architecture. Despite several centuries of grave robberies and plundering by invaders, the world has been left with a vast treasury that includes art, artifacts, and writings found in thousands of tombs, temples, and burial sites located throughout Egypt.

As anyone who has wandered through an old cemetery knows, you can learn a lot from burial plots. Sometimes a simple headstone can provide a world of information about whole families and what they believed and how they died. In Egypt, walking around cemeteries has provided a veritable library of information about thousands of years of Egyptian life and beliefs. For instance, in the carved limestone tombs near the Old Kingdom pyramid of King Unas (c. 2375–2345 BCE), archaeologists found a "house for the afterlife," complete with men's and women's quarters, a master bedroom, and bathrooms with latrines. But perhaps more significant was the discovery of columns of hieroglyphics called the **Pyramid Texts,** considered the world's oldest known religious writings, carved more than four thousand years ago in the tomb of King Unas.

If you grew up on a diet of Walt Disney witches, the word "spell" probably invokes notions of hocus-pocus and "eye of newt." But the Pyramid Texts' collection of "spells" and incantations (the exact Egyptian phrase for them was "words to be spoken") was far less exotic. Actually, in modern parlance, the Pyramid Texts were more like "how-to" manuals—travel guides to the afterlife. Evoking the names of the enormous pantheon of Egyptian gods, the "spells" they contained provided the dead king with the "scripts" that were necessary for his safe passage, survival, and well-being in the land of the dead. They sometimes warned of dangers and included the correct dialogues with gatekeepers and ferrymen he would encounter along the way, providing the deceased with a "cheat sheet" of answers to questions that would vouch for his legitimacy as a king and heir of the gods. Typical of the Texts is this "Utterance," in which the king is ferried across the sky to join the sun god:

The reed-floats of the sky are set down for me,
That I may cross on them to the horizon, to Harakhti.
The Nurse-canal is opened,
The Winding Waterway is flooded,
The Field of Rushes are filled with water,
And I am ferried over
To yonder eastern side of the sky,
To the place where the gods fashioned me,
Wherein I was born, new and young.

More than two hundred of these "spells" were found in the tomb of King Unas, but more than eight hundred others have been discovered since in other tombs dating from this early period. The vast pantheon of Egyptian gods is hinted at by the fact that more than two hundred different gods are mentioned in the various Pyramid Texts. Although once reserved for kings, Pyramid Texts began to appear in the tombs of nonroyalty by the end of the Old Kingdom's Sixth Dynasty, suggesting a fundamental change in Egyptian society that might explain the disorder that sent the Old Kingdom into decline.

Over time, the Egyptian obsession with preparing properly for the afterlife produced ornate coffins painted with hymns and requests to the gods in another collection of spells, known as **Coffin Texts**. This is the modern name for a collection of more than eleven hundred spells and recitations, some of them similar to versions from the Pyramid Texts, which were painted on wooden coffins. Some of these texts included maps showing the safest route for the soul to take as the dead person negotiated the treacherous path through the underworld.

The last—and perhaps best known—form of burial literature is another collection, which was misnamed **The Book of the Dead** when it was discovered and translated in the nineteenth century CE. Used for more than a thousand years, The Book of the Dead, known to the Egyptians as "The Book of Coming Forth by Day," was a New Kingdom innovation, consisting of almost two hundred spells or formulas designed to assist the spirits of the dead achieve and maintain a full and happy afterlife. These spells had such titles as "For Going Out Into the Day and Living After Death," "For Passing by the Dangerous Coil of Apep" (Apep was a terrible serpent in Egyptian mythology), or advice with the ring of a "Hint from Heloise"—"For Removing Anger From the Heart of the God." Another provided the incantation for preventing a man's decapitation in the realm of the dead:

"I am a Great One, the son of a Great One. I am a flame, the son of a flame to whom was given his head after it had been cut off. The head of Osiris shall not be taken from him, and my head shall not be taken from me. I am knit together, just and young, for I indeed am Osiris, the Lord of Eternity."

At one time, these spells and rituals had been for the exclusive use of the pharaohs. But The Book of the Dead became everyman's chance at eternity, and copies of it eventually came to be buried with any Egyptian who could afford one. (See below, *What was the "weighing of the heart"?*)

The many centuries of burials, tombs, temples, palaces, monuments, and statuary left behind by the Egyptians—all with elaborately carved accounts of kings, extolling their achievements—constitute a rough form of the first recorded history. This awesome collection of antiquities documents an Egypt which had a highly developed, unique mythology more than five thousand years ago. This vast record shows that the Egyptians believed from the earliest times that the gods had a profound impact on the shaping of their world and civilization. But the focal point of this complex religion evolved into a near obsession with life after death.

Long before Christianity and its hope of resurrection was born, Egyptian religion was the first to conceive of life after death. At the heart of this religion—and at the center of Egyptian government and society itself—were Egypt's extraordinary gods, a pantheon of breathtaking imagination and totality that found expression in every aspect of the Creation—animal, human, plant, and stone. The beginnings of Egyptian mythology and the elaborate stories of these gods go far back in time, before history, to the time when the first Egyptians imagined Creation.

MYTHIC VOICES

All manifestations came into being after I developed . . . no sky existed no earth existed . . . I created on my own every being . . . my fist became my spouse . . . I copulated with my hand . . . I sneezed out Shu . . . I spat out Tefnut . . . Next Shu and Tefnut produced Geb and Nut . . . Geb and Nut then gave birth to Osiris . . . Seth, Isis and Nephthys . . . ultimately they produced the population of this land.

Extracts from the Papyrus Bremner-Rhind

How does "creation by masturbation" work?

In the Book of Genesis, the Hebrew Bible offers two versions of the Creation. The first is the Seven Day account, in which God speaks and creates the universe. The second tells the story of Adam and Eve and is set in the Garden of Eden. These two biblical accounts differ substantially in details, facts, and style. They were probably composed centuries apart and only merged later on by the early Jewish editors who first compiled the writings that would become the Books of Moses, or the first five books of the Bible. But these two separate and distinct stories have been viewed as one by Jews and Christians for centuries. Raised on a Sunday-school or Hollywood version of biblical events, and not having read the

Bible for themselves, many people are not even aware of the fact that two Creations exist.

Ancient Egypt goes Genesis several times better. There are at least four significantly different Egyptian versions of Creation, some with overlapping details and characters. Each of these Creation stories was connected to a prominent Egyptian city, and each emerged at different times in Egypt's long history. Just as the two Creation stories in Genesis reflect different writers working at different times, the various Egyptian Creation accounts developed over the immense prehistoric time frame that has to be reckoned with whenever talking about Egypt. Early in its history, Egypt had been divided into forty-two separate administrative districts called *nomes*, and each had its own deity. Every town or village also had a temple, often devoted to a localized god, so the number of Egyptian deities grew to the thousands over time.

Keeping this in mind helps to explain why the Egyptian Creation myths defy a simple, "logical" narrative. These are stories that go back to the most distant moments of early human civilization, and then evolve and change over the course of centuries. While some details differ in these various Egyptian Creation stories, there are similarities and recurring characters. All share a central belief that the sun—or more precisely, a sun god—was at the center of the creation, which emerged from a primeval watery chaos called *Nun*, an endless, formless deep that existed at the beginning of time and was the source of the Nile.

The primeval ocean of chaos that existed before the first gods came into being, these waters contained all of the potential for life, awaiting only the emergence of a creator. This watery creation provides an intriguing parallel to the opening lines of the Bible—"In the beginning when God created the heavens and the earth, the earth was a formless void and darkness covered the face of the deep, while a wind from God swept over the face of the waters."

Probably the oldest version of Egyptian Creation myth came from Memphis, the ancient political capital of Egypt. Memphis is the city's Greek name. The original Egyptian name is translated as "White Walls" and referred to the enclosure around the sacred city. Here, the belief held that the world was created by a very old creator god called **Ptah**, temples to whom were built all over Egypt. Most scholars believe that the Greeks translated the Egyptian word *Hewet-ka-Ptah*, which literally means "Temple of the Spirit of Ptah," as *Aeguptos*, and it was eventually transformed into the word we now use as *Egypt*.

A patron of craftsmen, Ptah was able to create the world simply by thought and word alone—"through his heart and through his tongue," as ancient priestly writings put it. Simply by speaking a string of names, Ptah produced all of Egypt, the other gods, including the sun god **Atum** (see below), the cities and temples. In other words, this Creation story was similar to the much later biblical Creation in Genesis 1, in which the Hebrew God speaks and creates the universe. ("God said, 'Let there be light,' and there was light.") For centuries, scholars have speculated and argued about whether the parallels between the Egyptian and Hebrew Creation accounts are just coincidence or whether the Egyptian Creation stories could have influenced the ancient Hebrews. It is an unanswered, and possibly unanswerable, question.

Manifested by the sacred Apis bull in Memphis, the most important of all sacred animals in Egypt, Ptah was seen as a creator deity, and Egyptian kings were crowned in his temple. But Ptah never rose to become Egypt's supreme god and, at a later time, Ptah was merged with other gods to become a god of the dead. The Greeks later equated Ptah with their blacksmith god Hephaestus, known by the Romans as Vulcan. In another minor myth, Ptah was given credit for the miraculous defeat of an Assyrian army when he instructed an army of rats to gnaw through the attackers' bowstrings and the leather on their shields, forcing them to retreat. One of the most recognizable representations of Ptah is a small gilded statue found in the tomb of Tutankhamun.

A second major Egyptian Creation story came from Hermopolis, farther south in central Egypt, a prosperous city built in honor of the god known as **Thoth**. This god of wisdom and transmitter of knowledge is credited with the invention of writing. The ancient Egyptian name of this city was Khemnu, and Hermopolis was its later Greek name, because the Greeks associated Thoth with their god Hermes. But in ancient Egyptian, Khemnu means "Eight Town," and the myth that developed here held that four couples of frog-headed gods and snake-headed goddesses—technically referred to as the Ogdoad, or Group of Eight—were created by Thoth as the different aspects of the universe. **Nun** and his consort **Naunet** personified the original formless waters; **Heh** and **Hauhet** symbolized either Infinity or the force of the Nile floodwaters; **Kek** and **Kauket** embodied darkness; **Amun** and **Amaunet** were the incarnation of hidden power and were also associated with the wind and air.

While the specific details of how this Creation actually takes place are obscure, Thoth was credited as having commanded the Creation, and somehow the eight gods he produced were then responsible for the creation of the sun. In a variant of this myth, a lotus blossom arose from the waters, and from this flower, the young sun god emerged, bringing light and life into the cosmos. After this Creation, six of the gods receded from view, and only Amun and Amaunet joined the other gods of Egypt and continued to play an active role in Egyptian life.

A third Creation story focuses on the making of humans—an aspect of the Creation that is less significant in other Egyptian Creation accounts. This story features the god **Khnum,** an ancient ram-headed creator god who originated in Elephantine, an island in the Nile just above the first cataract at Aswan. In a highly folkloric tale, Khnum made humans by molding people on a potter's wheel, providing the first real link between gods and humans in Egyptian myth. The depiction of Khnum seated at the potter's wheel became a popular motif in Egyptian art. Khnum was especially significant because he controlled the Nile's floodwaters. The inundation of the fields, which produced the grain that allowed Egypt to prosper, was one of the most important aspects of Egyptian life, and Khnum was considered a great fertility god.

The most significant Creation story in ancient Egypt, however, was the one associated with Heliopolis, as Herodotus called it, for it was the City of the Sun (*helio* is Greek for "sun"). One of the most important locations in ancient Egypt, its ruins are near modern Cairo. Sometime around 3000 BCE, near the time of the unification of Upper and Lower Egypt, a Creation account emerged in Heliopolis that became the dominant myth in Egyptian religion and history. This account is prominent in the Pyramid Texts, the collection of hieroglyphic writings found in some of the earliest tombs.

According to the Heliopolis myth, there was the great infinite ocean, described as a primeval being called Nu or Nun. At the beginning of time, the god Atum, "lord of the Heliopolis," father and ruler of all the gods, emerged from these primeval waters. As the sun god, Atum simply came into being and stood on a raised mound—a symbolic representation of the land that rises out of the receding Nile floodwaters. In other words, the essence of Egypt—the sun and water—were merged into this one god. The mound became known as the *benben*, a pyramid-shaped elevation on which the sun god stood. In a temple in Heliopolis, there was a rock, possibly a meteorite, which was venerated as the *benben* stone and was believed to be the solidified semen of Atum. The *benben* stone, the primeval mound from which creation emerged, is considered the inspiration for both the pyramids and the obelisk.

A god of totality and complete power, Atum immediately began to create other gods. (In later times, Atum was linked with the other major Egyptian sun god, **Re** or **Ra**, as Re-Atum. See below.) This is where the story gets tricky, because there are a couple of variations. Clearly, his first act is to masturbate, and by doing so, Atum gives spontaneous birth to his children, the twins **Shu** and **Tefnut.** But in a later passage, Atum is said to "swallow his seed" and then "sneeze" and "spit up" these twins. Shu is the god of air, and Tefnut the lion-headed goddess of moisture. With their creation, there now exists the sun, water, and the atmosphere. The Creation goes on from there, until there is a collection of the most significant gods in Egypt—the nine deities known as the Great Ennead.

MYTHIC VOICES

The glorious god came,
Amun himself, the lord of the two lands,
in the guise of her husband.
They found her resting in the beautiful palace.
She awoke when she breathed the perfume of the god,
And she laughed at the sight of his majesty.
Inflamed with desire, he hastened toward her.
He had lost his heart to the queen.
When he came near to her,
she saw his form as a very god.
She rejoiced in the splendor of his beauty.
His love went inside all her limbs.
The god's sweet perfume
suffused through the palace,
the perfume of Punt, the land of incense.

The greatness of this god
did to the woman what he pleased.
She kissed him
and delighted him with her body.

—*Egyptian Hymn on the*
Birth of Hatshepsut (1490–1468 BCE)

Who was the first family of Egyptian myth?

Most people have to cope with annoying family members and brotherly spats. Egypt's Great Ennead—the first family of the gods—took those fraternal quarrels to cosmic heights and created the core myths of ancient Egypt. All of the most significant deities in the Egyptian world grew out of the Heliopolis Creation story, which continued as the twin brother and sister, Shu and Tefnut, became the first divine couple. They next produced another pair of twins, **Geb** and his sister, **Nut**, the grandchildren of the sun god Atum. Geb was the male earth god and his sister-consort, the female Nut, represented the sky and heavens.

There are two Egyptian versions of how earth and sky were separated. In one, they were locked together in an embrace at birth, and Atum, their grandfather, told Shu to separate the twins. In a second account of their separation, Geb and Nut married, but Atum, the sun god, was angry, since he was not informed of the match and had not approved it. He ordered their father, Shu, to push Nut away from Geb into the sky. Standing on Geb, Shu forced Nut upward to form the great arch of the sky, with her hands and feet resting on the four points of the compass. Nut is usually depicted in Egyptian art in this position—hands and feet straddling the earth, with her back arched to form the heavens. As the sky goddess, Nut was traditionally shown as covered with starlike speckles—and stars were later explained in Egyptian religion as the spirits of the dead who had gone to join the gods in the heavens. Nut's laughter became the thunder, and her tears were the rain.

Apart from Nut's role as mother of other gods, she played a central role in the most essential aspect of Egyptian religious belief—the daily passage of the sun. Every day, the sun god made his journey from dawn to dusk in a boat across the underside of Nut's arched body. At the end of the day, Nut swallowed the sun god and his boat (a symbolic daily death) which then traveled the inside length of her body—equivalent to traveling through the Egyptian underworld, known as the **Duat**. Each morning, she gave birth, and the sun god emerged from her womb. According to this myth, the redness of the sky at dawn was explained as the bloody afterbirth that accompanied the sun god's birth each day.

This myth was the great beating heart, focal point of all Egyptian belief. The sun's daily birth and death symbolized the eternal cycle of life and death. For Egyptians, life and death and the role of the sun as life-giver were all tied together in the regular cycle of the flooding of the Nile, which brought the fertility to the soil and the harvest that sustained Egypt. It carried over into the Egyptians' core belief that humans could also live, die, and be reborn. In that fundamental idea of death and resurrection lay the basic foundation for all of Egypt society and worship.

In the continuation of the Creation story, when Nut became the sky and heavens, her brother-husband, Geb, was forced to lie down and become the earth. As god of the earth, Geb was thought to be the cause of earthquakes, which were attributed either to his laughter or his wailing for his sister-bride. Geb was especially significant, because as earth god, he was responsible for the fertility of the lands, and he is often depicted with his phallus stretching up toward his sister. Some accounts say that the obelisk was designed as a symbol of Geb's phallus pointing heavenward to impregnate Nut. In other versions of this myth, Geb became the first king of Egypt, establishing the divine connection between the king and the gods.

Before Geb and Nut were separated, they produced children, and the Egyptian Creation continued as the four most important children of Geb and Nut were born: Osiris, Isis, Seth, and Nephthys. **Osiris**, their eldest child, became one of the most significant gods in the Egyptian pantheon. He was widely worshipped in popular religion as a god of fertility, death, and resurrection. Originally a god of vegetation, Osiris was credited with bringing plants and seasons to the earth, teaching humans to farm, and creating civilization. He abolished cannibalism, taught men to use tools, and showed them how to make wine and bread. He also ruled on earth and became the first pharaoh, instituting both religion and the legal system. Most significant, he later became the judge of the dead, a crucial role in a society so concerned with the afterlife. (See Isis-Osiris below.)

Geb and Nut's second child was **Isis**, who was both the twin sister and wife of Osiris, and another of the most significant figures in Egyptian myth. In some versions of the myth, her story begins in the womb, where she first makes love to Osiris, her brother and husband. Credited with creating the Nile River with the tears she wept at the death of Osiris, she taught the Egyptians how to grind flour, spin, and weave and was a healer goddess who could cure illnesses. Isis was also credited with introducing marriage.

One of the most widely worshipped figures in all mythology, Isis became the focal point of a religious cult that survived for thousands of years and was passed on to other civilizations, including Greece and Rome. She was known as the Great Mother, devoted wife and a powerful source of magic, and Isis worship continued to flourish down to Christian times. When the Christianized Roman Emperor Theodosius I officially banned Isis worship in 378 CE, her temples were destroyed, often replaced by Christian churches.

The third child of Geb and Nut was the evil **Seth** (also **Set**), the brother and enemy of Osiris. A storm god who may have originated as a desert deity, he was sometimes regarded as the incarnation of evil, and the force of disturbances and discord in the world. An ill-tempered god, Seth personified rage, anger, and violence.

But Seth also played a positive role in the sun-god ritual. As a powerful deity, he was charged with protecting and defending the sun god during his nightly journey through the underworld. During the night, the sun god's boat, or barque, was attacked by **Apep** (or **Apophis**), the serpent of chaos, sometimes also depicted as a crocodile. In an interesting parallel to the biblical serpent, Apep is called the "great Rebel" and "evil One." Possibly based upon the deadly African python and then merged with the crocodile—two of the Nile's most fearsome and deadly creatures—Apep may be one of the oldest versions of the dragon. Seth is often shown as the one who spears Apep, the lord of darkness, when he attacks the barque each night in its travels through the Duat.

But Seth's most important role lies in the story of his hatred for his favorite brother. The profound jealousy resulting in blood feuds between brothers—sibling rivalry played out on a cosmic scale—is a common theme in mythical and biblical stories. Just as Cain was jealous of Abel and killed him, and Isaac cheated his brother Esau out of his inheritance in Genesis, Seth resented his brother's success and great stature. There is a suggestion that this mythical rivalry may have been a symbolic account of the political rivalry between two regions of Egypt. As a desert god of the "Red Land," Seth was viewed as the force of destruction and chaos that threatened vegetation, and their conflict, played out in the saga of Isis and Osiris (see below), is a central piece of one of the most significant myths in world history.

The fourth child of Geb and Nut was **Nephthys**, who clearly plays second fiddle to her older sister Isis, the superstar of Egyptian myth. First married to her brother Seth, Nephthys deserted Seth for her other brother, Osiris. Seemingly barren with Seth, she conceives a child who becomes the jackal-headed god **Anubis**, another key deity in Egyptian burial rituals. Nephthys also becomes significant as a funerary goddess who protects the dead and is often shown on coffins and jars that held the vital organs of the deceased. In Egyptian funeral customs, two women would impersonate Isis and Nephthys to lament over the mummy of the deceased on the funeral boat that carried the deceased to the western side of the Nile, where it would be buried.

These nine deities—Atum, his children Shu and Tefnut, their children Geb and Nut, and their children Osiris, Isis, Seth, and Nephthys—were responsible for bringing all other life into being. They are traditionally known by the Greek word for nine, *ennea*, as the Great Ennead.

MYTHIC VOICES

Hail to you Re, perfect each day,
Who rises at dawn without failing . . .
In a brief day you race a course
Hundreds, thousands, millions of miles.

—Litany of Re

Who was Re?

The boggling variety of these Egyptian Creation stories, along with the various sun gods and creator gods, point up one of the challenges of Egyptian mythology. Lacking a biblical-style story structure, Egypt's mythology does not follow a single narrative stream but emerged gradually as local customs and beliefs were blended over centuries and finally integrated into a unified Egyptian religion, often reflecting the changing balance of power in Egypt. Most historians believe that as different regions or cities rose to prominence, their patron deities emerged as the most significant gods.

Still, a dominant force underlies all Egyptian myth. As in many other ancient societies, for the Egyptians it was the sun and its life-giving powers. The daily passage of the sun across the sky gave rise to many different metaphors and images. In the morning, the sun was born from the sky goddess Nut. At midday, it was a boat floating on the blue sea of heaven. It was even envisioned as a scarab beetle pushing a ball of dung across the sand. That would seem a profane image for an all-high god, but the Egyptians saw metaphors of life and death everywhere, especially in the animal world. The scarab—or dung—beetle lays its eggs in a ball of dung that it rolls to its burrow. Within the dung ball, the eggs incubate in the warmth of the sun. Even in the life cycle of one of the lowliest insects, Egyptians found the eternal image of life. This was the reason that scarabs became such a significant motif in Egyptian art.

Over time, the sun god Re (also called Ra) evolved into the most important member of the Egyptian pantheon, and for much of Egypt's history, he was the supreme deity. Originating in Heliopolis (City of the Sun), Re emerged as the chief sun god, and his name originally may have meant "creator." Re was considered both the ruler of the world and the first divine pharaoh. Although Re's existence probably goes back much further in time, his name was first recorded during the Second Dynasty (2890–2686 BCE), and by the time of the Fourth Dynasty, Egypt's kings were using the words "Son of Re" as one of their honorific titles. From the time of the Fourth Dynasty's (2613–2494 BCE) King Khafra (also known as Chephren), the pyramids and other sacred buildings were linked with the name of Re. By the Fifth Dynasty (2494–2345 BCE), Re had essentially become the state god, and six of the seven

Fifth Dynasty kings built temples exclusively dedicated to him. These sun temples, built near large pyramid complexes, established Re as the "ultimate giver of life and moving force," according to Egyptologist Jaromir Malek. But these temples were also a statement by the pharaohs of their direct connection to Re, in this world and in the afterlife.

For centuries, the worship of Re had been based in Heliopolis, but gradually he was worshipped throughout Egypt. As a sun god, Re traveled in his boat through the sky and was reborn each day. In one story, man was said to be formed from Re's tears. (The words for "tears" and "man" were very similar in Egyptian, just as in ancient Hebrew the words "earth" and "adam" found in the Genesis Creation story are also related.) Gradually, Re was fused with other Egyptian solar gods, and one way the Egyptians explained this was to identify Re with the sun at different times of the day. For instance, he was called Re-Horus as the morning sun and Re-Atum as the evening sun. In the Creation myth of Heliopolis that produced the Great Ennead, the god Atum was merged with Re into a single deity called Re-Atum. In this manifestation, he emerged as the creator god who fathered the first divine pair.

During the Middle and New Kingdoms, when Egypt reached the pinnacle of its power and wealth, Re and Amun—a powerful god worshipped in the cities of Thebes and Hermopolis—were also joined together to become Amun-Re and were viewed as an even more powerful national god. Amun-Re, king of the gods, creator of the universe, and father of the pharaohs, also became the lord of the battlefield. At the crucial Battle of Kadesh in 1286 BCE, when Ramses II defeated a Hittite army, legend has it that Amun-Re supposedly comforted the pharaoh when the battle was going against the Egyptians and promised, "Your father is with you! My powerful hand will slay a hundred thousand men." Faced with defeat, Ramses II was saved by the seemingly miraculous arrival of reinforcements. After the battle, Ramses II apparently decided to make love, not war. He took the daughter of the Hittite leader as one of his seven wives, cementing a peace between the two ancient rivals.

If that notion of gods intervening in battle strikes modern readers as preposterous, primitive superstition, remember: there are many examples throughout history of similar divinely inspired victories. Various "war gods" have been credited with triumph in battles, especially against overwhelming odds. From the Greeks at Troy, to Joshua at Jericho, David and Goliath, and other biblical battles, the notion continued in the Christian era with Emperor Constantine, who converted to Christianity after a religious vision led him to victory in 312 CE, and Joan of Arc, whose religious visions enabled her to lead French armies. The notion of gods interceding in battle is an old and revered tradition, and more than a few twentieth-century American generals have credited God with victories in America's wars. It's one reason football coaches still make their players pray in the locker room.

MYTHIC VOICES

Hail to you Osiris
Lord of Eternity, king of gods,
Of many names, of holy forms,
Of secret rites in temples.

—The Great Hymn to Osiris

Which god became Egypt's lord of the dead?

After Re, no god was considered more important or greater in Egypt than Osiris, and no story was more important than the myth of his life, death, and rebirth. The son of Nut and Geb, Osiris had succeeded his father as the ruler of Egypt. With his sister Isis as his wife, this divine pair first civilized Egypt, and then Osiris decided to do the same for the rest of the world, leaving Isis in his place as ruler. After several years, he returned and found everything in order, as Isis had ruled wisely in his absence.

But his brother Seth was jealous of Osiris's power and success, and plotted to kill him. In some versions of the myth, Seth's jealousy was compounded when Osiris slept with Nephthys—their sister who was also Seth's wife. In anger, Seth cursed their child, who became the jackal-headed god Anubis.

But Seth was not finished with Osiris. He invited his brother to a banquet attended by seventy-two of Seth's accomplices. At the banquet, there was a beautifully carved wooden chest. In a Cinderella's-slipper scenario, Seth offered the coffinlike box to whoever could fit inside it. All the guests tried to fit in but were unable, until finally, the trusting Osiris climbed into the box and fit perfectly. Quickly Seth had the lid nailed shut by his helpers, and sealed it with molten lead. Then they dropped the box into the river, and it was carried out to sea, coming to rest under a tamarisk tree in Byblos, the Phoenician port city (in modern Lebanon). As time went by, the tree grew around the box, eventually enclosing it—with Osiris's dead body sealed inside. When the Phoenician king later had the tree cut down, it emitted a delicious fragrance and was soon famed throughout the world.

Mourning the loss of her beloved husband-brother, Isis was inconsolable, and her endless stream of tears was said to cause the flooding of the Nile. She began to search for Osiris, accompanied by the jackal-headed Anubis, the son of Osiris and his other sister, Nephthys. Hearing about the tree and its wonderful fragrance, Isis realized its significance. She retrieved the box and hid it in the swampy Nile Delta. When Isis finally opened the lid, she turned into a bird—either a sparrow or a hawk, depending on the version—and the flapping of her wings forced the breath back into Osiris's lifeless body. Her beloved husband was alive briefly, just long enough for them to make love before Osiris died once again. Isis became pregnant and the child she conceived was **Horus**, the falcon-headed sky god. The dead Osiris was returned to his tomb, which Isis guarded.

After killing his brother, Seth had become pharaoh of Egypt. Learning that Osiris was lying in a tomb, Seth was unsatisfied. He discovered the box containing Osiris in the tomb, and, in a rage, cut Osiris's remains into fourteen pieces, scattering the parts all over Egypt. In the myth, however, the distraught Isis searched for all the pieces with the assistance of her mother, Nut, the sky goddess, and the jackal-headed Anubis.

Although Isis was able to gather up almost all the pieces, she could not find Osiris's phallus, which had been swallowed by three kinds of fish. (Eating these varieties of fish was considered taboo by some Egyptians.) In one version of this myth, Isis buried these thirteen parts of Osiris where she found them, and each of these became the site of a major Osiris temple. Temples to Osiris throughout Egypt staked a claim to being the burial site of these remnants of Osiris. They attracted devoted worshippers, just as certain Christian churches that claim to possess "relics," such as a piece of the "true cross" or remains of saints, become pilgrimage destinations.

In the more significant version of the myth, Isis once again resurrected Osiris's body. Alive, but unable to reproduce because his phallus had been lost, Osiris went to the other gods to discredit Seth. Now infertile, Osiris was made lord of the dead, given to rule over the land that existed beyond the western desert horizon. To prepare Osiris for his journey to the land of the dead, Isis invented embalming and mummification, which was carried out by her loyal assistant, the jackal-headed Anubis. The preservation of the body of the dead person was thought to be essential for survival after death. As lord of the dead, Osiris was the god who gave permission to enter the underground kingdom. This was the beginning of the elaborate rituals that formed the essence of Egyptian religion—the burial rites that ensured immortality. In many ways, while Osiris did not supplant Re in power, he became Egypt's most popular deity, with a cult following that lasted over two thousand years.

Who was Egypt's most significant goddess?

The family feud did not end there. This epic story continued with the conflict between Horus, the son of Isis and Osiris, and his uncle Seth. With Osiris in the underworld, evil Seth remained the king. But when Horus reached manhood, he vowed to avenge his father and challenged his uncle for the throne. In one version of the myth, Isis disguised herself and convinced Seth that Horus deserved to be the king. But other versions detail a lengthy series of battles, in which Horus castrated and killed Seth—but not before Seth tore out one of Horus's eyes. Judged the victor by the gods, Horus was given the throne of Egypt and Seth ascended into the heavens, to be the god of storms. Having overthrown Seth, Horus became the king and guide of dead souls, and, ultimately, the protector of pharaohs, who took as one of their several titles the name "the Living Horus." When the pharaoh died, he was thought to become Osiris, the god of the underworld.

Based on the seemingly timeless, crucial, and annual pattern of the flooding of the Nile, this was an elemental myth in Egyptian history. Osiris represented growth and life, and Seth represented death. The forces of vegetation and creation—symbolized by Osiris, Isis, and Horus—triumphed over the evil forces of the desert, symbolized by Seth. But, more significant, with the help of Isis, Osiris had cheated death. The Egyptians believed that if Osiris could triumph over death, so could human beings.

Over the centuries, the beguiling Isis became the most significant goddess in the Egyptian pantheon—mother of god, healer, the powerful goddess with deep knowledge of magical arts and sexual power. In one legend, Isis tricked the aging Re into confiding his secret names to her. Using magic to create a snake that bit Re, Isis healed the god only after he revealed all of these names to her. With this knowledge, Isis acquired unmatched skills in magic and healing. In Egyptian, her name is related to the word for "throne," and she is often depicted in Egyptian art as a throne for the king. In Greece, Isis would become identified with Demeter—the Mother Earth, or Grain Mother of Greek myth (see chapter 4)—and she became even more popular in Imperial Rome. Temples devoted to the worship of Isis were built in every corner of the Roman Empire, including one discovered beneath the streets of modern London. The image of Isis suckling the infant Horus, one of the most familiar themes in Egyptian art, was later adopted by early Christians to represent the Virgin Mary. The traditional blue dress of the Virgin Mary, the title Stella Maris (Star of the Sea), the reference to Mother of God, and the symbol of the crescent moon associated with Mary were all borrowed from the Roman cult of Isis.

While very ancient Egyptian papyruses and other works of art serve as sources for the Isis-Osiris myth, it was best known to the Greeks and Romans as it was recorded in a volume called *Concerning Isis and Osiris* by Plutarch (c. 40–120 CE), a Greek biographer and essayist best known for his work *Parallel Lives of Illustrious Greeks and Romans*. Born in Greece, Plutarch studied philosophy in Athens and later lectured on this subject in Rome. After travels through Greece, Italy, and Egypt, he returned to Greece as a priest of Apollo at Delphi, and it is believed that he wrote his great works there. Drawing on earlier sources, Plutarch retold the Isis story, giving Greek names to the Egyptian gods. In his version, Horus became the Greek Apollo and Seth became the Greek

Typhon, a serpentlike monster who appears in Greek myth (and the source of the word "typhoon").

What did Christians think of Isis?

The story of Isis and Osiris—which shares some features with the Mesopotamian myth of Inanna and Dumuzi (see chapter 3) and may even be historically related to it—eventually reached far beyond Egypt. First adopted by the Greeks and later the Romans, it evolved into a significant story of a resurrected deity who promised salvation, and Isis and Osiris became the central figures in the "mystery religions" that flourished in the first century before the rise of Christianity. In *A History of God,* Karen Armstrong describes how these "Oriental cults" found a ready audience in the international empire that Rome had become by the first century. "The old gods seem petty and inadequate," Armstrong writes in a description that almost seems befitting modern times and fascination with so-called New Age religions of our times. "They were looking for new spiritual solutions, [and] deities like Isis . . . were worshipped alongside the traditional gods of Rome."

To the Roman world, Isis was alluring, holding out the promise of magical secrets and even immortality. Married to one god and mother of another one, she contained all of the female creative power associated with great goddesses.

It was against that backdrop of fading interest in the old gods of Rome and growing fascination with attractive new gods, such as Isis and Osiris, that Christianity also began to take root in Rome. To many religious historians, that searching mood in ancient Rome, combined with myths of dying and rising gods, may have opened the way for Christianity. Early in the twentieth century, scholar Jane Harrison wrote: "Of all Egyptians, perhaps of all ancient deities, no god has lived so long or had so wide and deep an influence as Osiris. He stands as the prototype of the great class of resurrection gods who die that they may live again."

The story raises another troubling question. The marriage of Isis and Osiris, like most other divine relationships in Egyptian mythology, was clearly incestuous. That was not unique to Egypt, as many myths feature such family couplings. There is a practical explanation for this, which is that if you are a god and there is nobody else around, sleeping with your sister is the only option. Did that mean incest was condoned in Egypt? In Egyptian history, it was clear that the ruling families condoned intermarriage and incestuous marriages. Again, it was a practical matter, a means to keep power within the family. That raises the question: Did the pharaohs do it because the gods did? Or was it the other way around? That is, were myths of incestuous gods devised to justify incest? There is considerable scholarly disagreement over whether incest was widespread among average Egyptians. While other myths, including those of the Greeks, commonly feature incestuous doings, taboos against incestuous marriages developed in most societies. Under biblical law, most forms of incest were forbidden and were capital offenses, and by 295 CE they were forbidden in Rome as well—which is intriguing, since some of Rome's emperors were notorious for their incestuous couplings.

MYTHIC VOICES

The elaborate burial rituals of the Egyptians required that the deceased be properly prepared for the challenges of the journey to the afterlife. To ease the way and guarantee immortality, the Egyptians developed a rich tradition of instructions, incantations, and spells. As time went by, any Egyptian who could afford one was able to purchase one of these collections that have come to be called The Book of the Dead. Like any set of ritual prayers in Judaism or Christianity, they were designed to ensure that the correct words were spoken at the entrance to eternity. Among these the central was "Negative Confession," which the deceased used to testify that he had lived a life free of forty-two specific sins, one for each of the gods sitting in judgment.

Hail to you, great god, Lord of Justice! I have come to you, my lord, that you may bring me so that I may see your beauty, for I know you and I know your name, and I know the names of the forty-two gods of those who are with you in this Hall of Justice, who live on those who cherish evil and who gulp down their blood on that day of reckoning of characters. . . . Behold I have come to you. I have brought

you truth, I have repelled falsehood against men, I have not impoverished my associates, I have done no wrong in the Place of Truth, I have not learnt that which is not, I have done no evil, I have not daily made labour in excess of what was due to be done for me, my name has not reached the office of those who control slaves, I have not deprived the orphan of his property, I have not done what the gods detest, . . . I have not caused pain, I have not made hungry, I have made no man weep, I have not killed, I have not commanded to kill, I have not made suffering for anyone. . . .

I am pure, pure, pure, pure! My purity is the purity of the great phoenix. . . .

—*Spell 125, "The Negative Confession,"*
The Egyptian Book of the Dead

O my heart which I had from my mother! O my heart which I had from my mother! O my heart of my different ages! Do not stand up as a witness against me, do not be opposed to me in the tribunal, do not be hostile to me in the presence of the Keeper of the Balance, for you are my *ka* which was in my body, the protector who made my members hale. Go forth to the happy place whereto we speed; do not make my name stink to the Entourage who make men. Do not tell lies about me in the presence of the god; it is indeed well that you should hear.

—*Spell 30B*, The Egyptian Book of the Dead

What was the "weighing of the heart"?

From the earliest of days, Egyptian life revolved around the cycles of birth, death, and rebirth. What was true for the earth itself—with life coming from the Nile floods that allowed the crops to flourish—could be true for humans. Very early on in the Egyptian world, this fundamental duality between life and death and the hope for renewal became an essential part of Egyptian mythology. Initially, it was expressed in the life of the kings and royalty. Descended from the gods, they were destined to be reunited with the gods. But at some point in Egypt's long history, the death and resurrection business went retail.

After death, Egyptians hoped to become one with Osiris, the god of resurrection and the underworld. The elaborate rituals of mummification and burial were all expressions of this desire. And the centerpiece of the elaborate rituals that guided the journey of the souls of the dead to the afterlife was the belief that the dead person would come to be judged by the gods in a ceremony known as the "weighing of the heart."

As the deceased traveled to an encounter with the gods, there were thought to be many trials, mirroring the trials that Re went through each night as he passed through the dangerous underworld before reemerging the next day. Ultimately, however, the deceased was brought into the great hall of judgment, before Osiris, accompanied by Isis and Nephthys and foty-two other gods, each one representing the *nomes*—administrative provinces into which Egypt was divided. Standing before the gods on this judgment day, the dead person would attest to having lived a just life. Then his heart was weighed on the scales of justice against the feather of the goddess Maat, a daughter of Re, who was the personification of the Egyptian idea of *maat*, the philosophy, religious notion, concept of harmony, and code of behavior that served as the basis for the stability of Egyptian society. It was the cosmic order that came through justice and right living.

This was the Egyptian equivalent of coming to the Pearly Gates and standing before St. Peter. Or, in more traditional Christian terms, the Judgment Day at which God would judge "the quick and the dead." If the deceased's heart was heavy with sin and evil deeds, redemption and eternal life were lost. For those who failed the weighing of the heart, the fearsome **Ammut**, "devourer of the dead," waited, eager to ravenously eat the heart of the deceased. An eternity in a sort of limbo followed. If the heart was in balance with the feather of truth, the soul of the deceased was saved and could join Osiris and the other gods.

By the New Kingdom (1550–1069 BCE), the possibility of achieving immortality was opened up to at least the upper and middle classes who could afford elaborate burial—the hope of the afterlife in eternity had gone retail and was available in The Book of the Dead. Produced on papyrus scrolls by scribes, these elaborate books were purchased by the families of the deceased and then entombed with the mummified body. Depending on the wealth of the dead person, they might contain as many as two hundred pages. This manual for immortality ensured that the deceased would know the proper words to say when confronted by the feather of truth.

WHO'S WHO OF EGYPTIAN MYTHS

In addition to the gods already singled out, there was an enormous Egyptian pantheon consisting of hundreds of major and minor deities. Some were ancient local gods, patrons of cities, towns, and villages up and down the Nile. Others were newly minted gods that reflected the changes over the long course of Egyptian history. This list includes some of the other most significant Egyptian deities and the role they played in Egyptian society.

Ammut The goddess known as the "eater (devourer) of the dead" stands by the scales when the hearts of the dead are weighed at the entrance to the underworld. If the dead person has led a wicked life and is not fit to survive into the next world, Ammut eats the heart. There are, however, no accounts of anyone failing that test. Terrifying to behold, Ammut incorporates three of the most feared animals of ancient Egypt, with the head of a crocodile, the body of a lion, and the hindlegs of a hippopotamus.

Anubis The god of embalming and cemeteries, he is the jackal-headed son of Osiris and his sister Nephthys. The connection between the jackal and death probably came from people seeing the desert canines who scavenged in the shallow graves of early cemeteries. Adopted by Isis, Anubis becomes her devoted servant and plays a role in the Isis-Osiris story, wrapping the dead Osiris in bandages and making him the first mummy. After Osiris becomes lord of the underworld, Anubis joins him and presides over the crucial ceremony of weighing the heart. Those who pass this crucial test are then taken by Anubis to be judged in person by Osiris.

Anubis became the patron of embalmers, and priests who supervised the preparation of the mummy wore a jackal-headed Anubis mask.

Bast (or **Bastet**) The daughter of the sun god Re, she is the popular, catlike goddess of love, sexuality, and childbirth. At her cult city, Bubastis in the Nile Delta, thousands of cats, which were prized by Egyptians and thought to be lucky, were mummified in her honor.

Bes A popular household god, he is an ugly but friendly dwarf god who frightens away evil spirits, and his name may have meant "to protect." He was likely a god who developed later in Egyptian history and has similarities with about ten other gods. But as Bes, he is one of the most widely worshipped gods, whose image was often found in households on headrests and beds as well as on mirrors and cosmetic items. Because of his benevolent nature, Bes is often depicted with **Taweret** ("the great female one"), a goddess of childbirth, because he also looks on while women give birth, and is considered a good-luck figure. In spite of her benevolent role, Taweret, who protected women during childbirth, has many fearsome animal attributes and is portrayed with the head of a hippopotamus, a lion's limbs, the tail of a crocodile, a swollen human belly, and breasts—her forbidding appearance is thought to keep away evil spirits.

Bes was so popular and long-lasting that Roman soldiers apparently carried his likeness on amulets when they went into battle.

Hapy God of the Nile floods, he lives in a cave near the cataract and it is his job to keep the land along the river fertile. Although a male god, he is often depicted with long hair, large breasts, and a protruding stomach, all symbols of fertility. His annual feast days were especially important, and one ancient text describes the sacrifice of more than one thousand goats to him.

Hathor A powerful, complex goddess, she is one of the most significant goddesses, the protectress of lovers and women, especially in childbirth. Often shown in human form, Hathor is also depicted as a cow-headed goddess. At times, she was closely connected with Re and said to be both his wife and daughter. Hathor suckled the young **Horus** and came to his aid when **Seth** put out his eyes. In other traditions, she marries Horus, and her milk becomes the food of the gods. In the underworld, Hathor also greets the souls of the dead and offers them food and drink.

As men aspired to "become" **Osiris** in the afterlife, women typically wished to be associated with Hathor.

Imhotep Unlike most of the deities in this chapter, Imhotep is no myth. In fact, he was probably more interesting than myth. He was a real man, whose existence is proved by archaeology and written accounts. Only later did Egyptian-era spin doctors take over and transform him into a legend that had nothing to do with sacred stories. (Or cursed mummies. In the most recent *Mummy* films, the evil Mummy is said to be the reincarnated Imhotep who was once buried alive for trifling with the secrets of the dead.) The historical Imhotep was a multitalented priest and chief advisor, or vizier, to King Djoser. He was also the architect of the Step Pyramid of Djoser at Saqqara, the first colossal stone building in history. Although Imhotep's own tomb has not been located, a limestone bust of him was found in Djoser's funerary complex. A high priest in Heliopolis, Imhotep was eventually deified and regarded as the son of Ptah, the ancient creator god. He was also viewed as the patron god of physicians, and even today, modern medicine honors him as the first physician known by name. (A statue in his honor stands in the Hall of Immortals in the International College of Surgeons in Chicago.)

Mehet-Weret (or **Mehturt**) An ancient cow-goddess and sky goddess, whose name means "great flood," she is also identified with the celestial river or canal on which Re sails his boat. In early traditions, Mehet-Weret is seen as the mother of Re, usually depicted as a cow, or half cow, half human, with a sun disk between her horns. She is also later linked with Hathor, another central cow-goddess.

Neith One of the most ancient deities whose existence is known in the prehistoric and early dynastic periods, she is a mother goddess worshipped in the Nile Delta. According to Egyptian tradition, Neith invented childbirth and brought gods, animals, and humans into existence. Because she is so ancient, she acquired other attributes over time and was also considered a warrior goddess. In one story, she spat into the water and her spit turned into Apep, the dragonlike serpent of the underworld who tries to devour Re each night as he passes through the Duat.

Sekhmet The lion-headed wife of the early creator god **Ptah**, she is a war goddess whose name means "powerful," and she could breathe fire against her enemies. Many Egyptian kings adopted her as a patroness of battle. In another of her roles, Sekhmet is the goddess who delivers punishments to the other gods.

Serqet An ancient scorpion goddess, she is the companion of **Isis**, and is one of the four funerary deities (the others were Isis, Nephthys, and Neith) who protect and guard the coffins and canopic jars that contain the embalmed, mummified organs of the deceased. There is a famous gilded wooden statue of Serqet, wearing the scorpion with a raised tail on her head, discovered in the tomb of Tutankhamun in which she guards one side of the king's shrine. Serqet was also invoked through spells that were meant to protect and heal poisonous bites.

Thoth (or **Djehuty**) Originally a moon god, he is best known as the divine scribe who records the weighing of souls when they arrive in the hall of justice to determine their fate after death. Thoth is usually depicted with the head of an ibis, because the curved beak of that bird was thought to resemble the crescent moon. (He is also sometimes depicted as a baboon.) In The Book of the Dead, Spell 30B invokes Thoth in his role in the weighing ceremony:

"Thus says Thoth, judge of truth, to the Great Ennead which is in the presence of Osiris: Hear the word of very truth. I have judged the heart of the deceased, and his soul stands as witness for him. His deeds are righteous in the great balance, and no sin has been found in him."

As the inventor of writing and patron of scribes, Thoth records the Ennead's "divine words" and documents the passing of kings. Worshipped as a patron of learning and the master of inventions, Thoth is also credited with the creation of writing.

Wepwawet Another funerary god with a dog's body and the head of a jackal, his name means "the opener of ways." A very old deity, represented on the Narmer Palette, he guides the dead person's soul through the underworld and assists in the weighing of the heart.

Why are there so many animals—real and imaginary—in Egypt's myths?

Maybe you've heard the riddle of the sphinx? It was once popular fifth-grade humor.

"Which lion doesn't roar?"

Answer: "The Sphinx—it's made of stone."

In ancient Egypt, animals played a prominent role in myth and religion, apparently from prehistoric times, judging from early art and burial practices. Images and references to hawks, falcons, lions, serpents, crocodiles, and bulls fill the pantheon of Egypt and are vividly illustrated in Egyptian art. While that idea was not unique to Egypt, animal worship may have been more significant in Egypt than almost any other ancient civilization. Springing from the belief that animals were manifestations of the gods—vehicles through which the gods could be worshipped—the Egyptians often buried animals in ritual graves, mummified them, provided them with food on their journeys to the afterlife, and used them in worship ceremonies at temples.

The Apis bull of Memphis, for instance, was considered a manifestation of the creator god Ptah and was used to make divinations. Worshippers could ask "yes" or "no" question of the oracle bull, which provided an answer to the petitioner by moving into one sacred corral or another. Other major religious centers, such as Heliopolis and Elephantine, had stables of sacred bulls and rams, respectively, while flocks of ibises and falcons, and thousands of cats—considered manifestations of Thoth, Horus, and Bastet—were maintained throughout Egypt, vast menageries that were used to make sacrifices by pilgrims seeking a favor from the gods.

Not only did the Egyptians represent their gods in animal forms, they also used a combined animal-human form, of which the Sphinx is the most famous example. A Greek word that is derived from Egyptian words meaning "living statue," a sphinx in ancient times was believed to be a mythical beast with the body of a lion or lioness and the head of a ram, hawk, or reigning king or queen. Sphinxes, which were reported by ancient Greek travelers to have been located all across Egypt, were thought to embody the power of the ruler to defend Egypt and served as visible symbols of the strength and power of the pharaoh.

Located near the pyramids at Giza, the Great Sphinx is one of the most instantly recognizable pieces of art in the world as well as the largest statue in the ancient world. Measuring 240 feet (73 meters) long and about 66 feet (20 meters) high, and carved from an outcropping of limestone, the Great Sphinx at Giza was sculpted with the body of a lion and the head of Khafra, son of Khufu the Great. The Sphinx's head, which served as the guardian of the royal cemeteries outside Memphis, seems to have been positioned so that the setting sun would stream through the temple on the days of the two equinoxes, capturing the moment when day and night were in perfect harmony.

Mythic Voices

But no crime was too great for Cheops: when he was short of money, he sent his daughter to a brothel with instructions to charge a certain sum—they didn't say how much. This she actually did, adding to it a further transaction of her own; for with the intention of leaving something to be remembered by after her death, she asked each of her customers to give her a block of stone, and of these stones (the story goes) was built the middle pyramid of the three which stand in front of the great pyramid.

—Herodotus, The Histories (Book Two)

What did the pyramids have to do with the gods?

No doubt it was repeating stories like this—about a king forcing his daughter to become a prostitute to pay for a pyramid—that made people wonder whether Herodotus was a reliable source. What the first Greek historian might have been passing on as "history" sounds suspiciously like the kind of story disgruntled commoners might tell if they don't like the king. Since Cheops—or more accurately Khufu—lived some two thousand years before Herodotus was in Egypt, the story of the daughter in the brothel has the ring of legend, not history. In fact, although the writings of Herodotus on Egypt profoundly influenced the study of Egypt for centuries, recent scholarship suggests that Herodotus never even traveled to some of the places he claims to have visited. For his inaccuracies and tall tales, Herodotus is sometimes called the "father of lies."

Still, it is understandable that Egyptians might have been a little peeved with Khufu. Most likely, the scale of his pyramid, the Great Pyramid at Giza, must have placed extraordinary demands on the Egyptian farmers and working class who paid the taxes and provided most of the labor force that built the pyramids, which could account for stories like the one that Herodotus told.

Even so, the very existence of these pyramids speaks eloquently to the power of Egyptian religion and an incredibly well ordered society that could have produced such marvels in a time with precious little technology. As Egyptologist Jaromir Malek notes, "For a modern mind, especially one that no longer knows profound religious experience and deep faith, it is not easy to understand the reasons for such huge and seemingly wasteful projects as the building of pyramids. This lack of understanding is reflected in the large number of esoteric theories about their purpose and origin." Those theories, which began in the nineteenth century, inspired the word "pyramidiots," for people who proposed extravagantly fanciful ideas about both the function and construction of the pyramids.

The pyramids we typically associate with Egypt today had evolved from earlier burial sites called "mastaba tombs," simple, rectangular, flat-topped structures built from mud bricks. Observing their profound religious beliefs, the earliest Egyptian kings were buried in these tombs to begin their journey to eternity. Initially, these tombs simply served as a safe place for the remains of the mummified king until he was resurrected to join the other gods.

But others apparently went along for the ride. Recent discoveries suggest that household servants and government officials in Egypt's earliest dynasties were sometimes sacrificed to spend eternity with their kings. In 2004, archaeologists announced finding the remains of human sacrifice in some early Egyptian tombs that predate the pyramids. The practice, while it had been suspected, had never been substantiated until a team from New York University, Yale, and the University of Pennsylvania found a series of graves near the tomb of King Aha, believed to be the first king in the First Dynasty. The graves, as reported in the *New York Times* in March 2004, yielded the remains of court officials, servants, and artisans, all apparently sacrificed to serve the king's needs in the afterlife. Nearby graves held the bones of seven young lions, symbols of kingly power, and one grave also held the bones of donkeys, presumably to help transport the king into the afterlife. "We may think of the ritual slaughter of a large number of retainers as barbaric," one researcher told the *New York Times*. But the ancient Egyptians "may have come to look upon the sacrifices as passports to eternal life, a guarantee of immortality. . . ."

The mastaba tomb became more elaborate with the first Egyptian pyramid, the Step Pyramid of Third Dynasty King Djoser (also Zoser,

2667–2648 BCE), which rose like a gigantic stairway, allowing the king to climb to the heavens and join the sun god.

The magnificence of the pyramids took on extraordinary new dimensions, both in size and decoration, with the Fourth Dynasty pyramids at Giza.* Called the Great Pyramid, the pyramid of Khufu contains an estimated 2.5 million stone blocks that average 2.5 tons each, with a base covering about 13 acres. Originally it was 481 feet (147 meters) tall, but some of its upper stones have fallen away, and today it stands about 450 feet (138 meters) high. A dismantled cedar boat, discovered near the southern face of the pyramid, has been restored, and a second boat has also been uncovered nearby. Undoubtedly these boats were intended for the deceased king to make his journey across the sky to join the gods. The king's body and all the trappings within the burial chamber are long gone, victims of grave robbers.

The ruins of the Great Pyramid are one of thirty-five major pyramids still standing in Egypt, each built to protect the body of an Egyptian king. The pyramids of Giza (Al Jizah) stand on the west bank of the Nile River outside Cairo, where there are ten pyramids, including three of the largest and best preserved. These extraordinary monuments to the power of one man have also been the source of wonder, curiosity, and speculation for centuries.

Besides the colossal dimensions they achieved in the Pyramid Age, the pyramid also became more elaborate in its designs and religious functions. The simple burial chamber of early tombs and periods grew to include attached temples where offerings to the dead king were made, multiple chambers, and granite doors and false passageways intended to deter grave robbers (unsuccessfully, for the most part!). The simplest explanation for the Giza pyramids is that the pharaohs had become obsessed with maintaining their status for eternity, an expression of their divinity. But in almost all aspects of its design and construction, the pyramid was symbolically tied to Egyptian mythology. The four smooth, straight ascending sides of the pyramid were meant to imitate the slant of the sun's rays, a physical representation of the centrality of the sun—and the sun god—in Egyptian religion. The building itself rep-

*At 450 feet (138 meters) high, the Great Pyramid is taller than the Arc de Triomphe in Paris and New York's Statue of Liberty.

resented, or re-created, the primeval mound that had emerged out of the watery chaos at the beginning of time—the *benben* stone on which the first god stood and brought to life all the other gods in the Egyptian Creation story.

More recent theories about the pyramids and their geographical alignment are related to Egyptian astronomy. The Great Pyramid of Khufu was called "Khufu's Horizon" in ancient Egyptian times, meaning that it was the place where the earth met the sky. Since the word for "horizon" was also closely associated with the word for "inundation," Egyptologists now believe that the pyramid went beyond being a physical memorial to the dead god-king and represented the totality of the belief in regeneration. The concepts of sun, horizon, inundation, the primeval mound, and the king's resurrection were all tied together in these monumental buildings and the complexes of temples and burial grounds surrounding them.

In a modern context, a parallel of sorts might exist in America's increasingly controversial presidential libraries. Why do some American citizens willingly contribute millions of dollars to finance the construction of large, expensive, but little-used—at least by the general public—buildings to house presidential papers? Critics dismiss these expensive monuments to former presidents and their papers as extravagant and wasteful. But admirers and the society wish to honor a former leader, even, in some cases, a disgraced one. Although presidents are not buried in their libraries, these new tributes fill a limited social purpose but are an expression of the society's wealth, social legends, and desire for posterity—they may be as close as Americans might get to creating the pyramids of Egypt.

What's so great about the "Great Pyramid"?

Practically since the time of Herodotus (484–425 BCE), there has been considerable disagreement over how the pyramids were built. Based on decades of research, it is now believed that the Egyptians, although lacking machinery or iron tools, cut large limestone blocks with copper chisels and saws. The extremely difficult work of quarrying was done in searing heat by slaves, usually prisoners of war. While most of the stone came from nearby quarries, other blocks were floated down the Nile from distant quarries, during the period of inundation. Not only was the Nile higher at this time, which would make it easier to get the massive stones closer to the pyramid complex sites, but the period of flooding was the time when most Egyptian farmers were unable to work their land and provided a large, available labor force. Unlike the slaves who quarried stones, the laborers on the pyramids were paid, conscripted by the pharaoh to spend three months of the year in service to the state.

The most likely method of construction involved a series of ramps. Without using wheels or pulleys, gangs of men dragged the blocks on sleds or rollers to the pyramid site and pushed the first layer of stones into place. Then they built long ramps of earth and brick, and dragged the stones up the ramps to form the next layer. As each layer was finished, the ramps were lengthened. Finally, they covered the pyramid with an outer layer of white casing stones, laid so precisely that from a distance the pyramid appeared to have been cut out of a single white stone. Most of the casing stones are gone now, but a few are still in place at the bottom of the Great Pyramid.

No one knows for sure how long it took to build the Great Pyramid. Herodotus claimed that the work went on in four-month shifts, with one hundred thousand laborers in each shift. Among Egyptologists who have studied the remains of what were practically small towns that housed and fed the workers, the modern consensus is that a workforce of between twenty thousand and thirty thousand, including the "service people," who baked bread and fixed tools for the builders, completed the Great Pyramid in less than twenty-three years. Most of the labor was provided by farmers during the inundations. But there are still unanswered questions about the workers. As historian Charles Freeman notes, "What incentives were needed to keep so many men toiling for so long can only be guessed at."

What is an Egyptian pyramid doing on the U.S. dollar bill?

There is another peculiarly American mystery that pertains to the pyramids, and there is one in your wallet or pocket. The dollar bill, with its strange combination of pyramids, eyes, and Latin text, has inspired considerable speculation and myth—in the sense of something commonly believed but untrue. Many people think that the symbol represents the powerful influence of the semisecret society called the Freemasons. According to this theory, the symbols in question—the pyramid topped by an all-seeing eye—were put there by the "Masonic president," Franklin Delano Roosevelt, to show that the country had been taken over by Masons.

In fact, these symbols are actually the two sides of the Great Seal of the United States, which dates from the late 1700s. Benjamin Franklin, also a Freemason, is often credited with their use, but even that may be a myth. The "All-Seeing Eye of the Deity" is mentioned in Freemasonry, but the concept behind this image dates back to the Egyptians. The unfinished pyramid symbolized the unfinished work of nation-building. Contrary to much popular myth, the pyramid is not a particularly Masonic symbol. The eye in the pyramid was a common symbol of an omniscient deity that can be seen in Italian Renaissance painting, long before the birth of Masonry, which was not formed until the early 1700s.

The Great Seal of the United States, symbol of the nation's sovereignty, was adopted on June 20, 1782, and the reverse side of the seal is what appears on the back of the dollar bill. A pyramid of thirteen courses of stone represents the Union, and is watched over by the "Eye of Providence" enclosed in a triangle. The upper motto, *Annuit coeptis*, means "He [God] has favored our undertakings." The lower motto, *Novus ordo seclorum*, means "the new order of the ages" that began in 1776, the date on the base of the pyramid. Anti-Mason groups and conspiracy theorists have mistranslated this as "New World Order," attempting to fit the seal into the belief that Masons constitute a vast international conspiracy to create such an "order." When the first President Bush used that phrase during his presidency to describe the changing political map of Europe following the fall of Communism in

Europe, it was quickly seized as further evidence of the "Masonic plot."*

One of the oldest and largest fraternal organizations in the world, Freemasonry was formed in London in 1717 by a group of intellectuals who took over a medieval craft guild and fostered what they called "enlightened uplift." They were dedicated to the ideals of charity, equality, morality, and service to God, whom Masons describe as the "Great Architect of the Universe." The order spread quickly through Enlightenment Europe and included men as diverse as Voltaire, King Frederick II of Prussia, and the Austrian composer Wolfgang Amadeus Mozart. As it developed, Freemasonry was viewed as anticlerical and was later thought to be antireligious by conservative Congregationalists in the United States. An anti-Mason movement took hold in the nineteenth century, and the Antimasonic Party became the first significant third party in American politics. But the fact is that Masonry was a voluntary fraternal order—a kind of eighteenth-century spiritual Rotary Club—and not a sinister cult intent on world domination, as it has often been portrayed.

Was the ruler of Egypt always a pharaoh?

The earliest carvings and written references to kings show that the Egyptians long considered the king as the earthly manifestation of the sky god Horus and the son of Re, the sun god. Yet, while all the kings of Egypt are typically thought of as "pharaohs," the Egyptians did not call the ruler that until around 1550 BCE. The administrative complex around the court at Memphis was known as Per Ao ("the great house"). The word "pharaoh" was attached at first to the royal palace, and only later to the king himself.

In theory, the pharaoh owned all the land and ruled the people and

*This old chestnut of a conspiracy theory got fresh legs with the release of the movie *National Treasure* (2004), an otherwise amusing action-adventure story that combined Masonic conspiracies with a treasure map hidden on the back of the Declaration of Independence.

also served as the high priest of Egypt. But in reality, his power was sometimes limited by strong groups, including the priests and nobility, or local provisional ruler of the *nomes*, called *nomarchs*. Although remarkable for the relatively few coups or assassinations in its long history—perhaps a tribute to the power of the Egyptian religion as a stabilizing force—Egyptian politics could sometimes be Machiavellian. There are cases of royal wives getting rid of their divine husbands, and there is even unproved suspicion that the young King Tut was murdered. The intrigues of the Egyptian court are best seen in the story of Pharaoh Amenemhet I (1985–1956 BCE), who was one of the few pharaohs definitely known to have been assassinated. He is famed for a set of instructions supposedly written posthumously, but most likely the work of a scribe, in which he advises his son to be on guard for intrigues:

Excelling in thy greatness . . . Live apart
In stern seclusion, for the people heed
The man who makes them tremble; mingle not
Alone among them; have no bosom friend,
Nor intimate, nor favorite in thy train—
They serve no goodly purpose.

'Ere to sleep
Thou liest down, prepare to guard thy life—
A man is friendless at the hour of trial . . .
I to the needy gave, the orphan nourished,
Esteemed alike the lowly and the great;
But he who ate my bread made insurrection.

From *Egyptian Myth and Legend*, Gresham Publishing, 1907; cited in Jon E. Lewis, ed., *The Mammoth Book of Eyewitness Ancient Egypt*, New York: Carroll & Graf, 2004.

Mythic Voices

Splendid you rise in heaven's lightland,
O living Aten, creator of life!

When you set in western lightland,
Earth is in darkness as if in death.

How many are your deeds,
Though hidden from sight,
O sole God beside whom there is none!
You made the earth as you wished, you alone.

—The Great Hymn to Aten *(c. 1350 BCE)*

Did a pharaoh inspire Moses to worship one god?

Even as Egypt became the world's greatest power, it fell into disarray over religious politics, an intriguing moment in history that might provide a valuable lesson about the volatile combination of belief and government. During his reign, the pharaoh Amenhotep IV (1352–1336 BCE) made a remarkable and radical—if somewhat mysterious and unexplained—decision. Amenhotep severed all links with the traditional religious capital of Egypt in Memphis and its god Amun-Re, chose **Aten** as the only god of Egypt, and set out to build an entirely new city devoted to this god. Located about two hundred miles north of Thebes, the city is known today by the name "Amarna," and this period is called the "Amarna Revolution." It affected Egypt in its time as profoundly as the Protestant Reformation affected Europe.

Aten had previously been a little-known god worshipped in Thebes. Unlike Re and other gods, Aten, whose name meant "disc of the Sun," had no human characteristics. Aten was depicted only as a sun from which rays emanated, ending in hands that held the *ankh*, Egyptian symbol of life. Amenhotep was so devoted to the worship of Aten that he changed his name to Akhenaten. Akhenaten's wife Nefertiti was his supporter in this transformation, taking on the role of priestess and assisting

Akhenaten in the new religious ceremonies. Supposedly one of the most beautiful women in Egyptian history, Nefertiti is the subject of several sculptured portraits that have survived from ancient times. She and Akhenaten began a full-scale attempt to wipe out references to all other gods. Throughout Egypt, statues to Amun-Re were smashed, and the god's name was literally chiseled out of monuments. State temples were torn down, and the traditional religious festivals and public holidays were no longer celebrated. The reasons for this radical reformation—the equivalent of a modern American president trying to wipe out any reference to Christianity in America and banning Christmas, Easter, and other religious holidays—are uncertain. There may have been political reasons behind Akhenaten's purge of the other gods.

Within a short time, the vast state mechanism of religion had been reduced to worship of a single god led by one man, the pharaoh. Only he and Nefertiti could communicate with this god. As popes and other religious leaders have well understood over the centuries, the professed ability to communicate exclusively with the gods is a great way to consolidate power.

After Akhenaten's death, the Egyptians stopped worshipping Aten. The new pharoah, Tutankhamun, began the restoration of the old gods, and traditional worship was completely restored under Horemheb, a general in Tut's service, who managed to secure the throne for himself after the death of Tut's successor, and then leveled Amarna.

But for years, many scholars have argued that the worship of this one divinity lingered among the people of Israel, who, according to biblical accounts, had lived in Egypt for hundreds of years. And that creates another interesting collision of myth and faith. The concept of one god became an important part of the religion that was developed by the Israelite leader Moses. The history of the cult of Aten has led to the suggestion that the Jewish and Christian belief in one God may have been derived from Egyptian worship. Among the proponents of this idea was Sigmund Freud, who laid out his theory in his final book, *Moses and Monotheism*. Or perhaps it was the other way around. As Bruce Feiler writes in his bestseller *Walking the Bible*, "Might the Israelites have learned to worship one god following the lead of some maverick pharaoh? Or might the Egyptians have learned the same thing by taking an idea from the patriarchs?"

In the traditional Jewish and Christian view, such questions are heresy. But they point to the reason why mythology matters. Cultures collide. Myths are absorbed in the aftermath of that collision. The ideas of one civilization are borrowed and remolded by another. There is no question that the Egyptians profoundly influenced the Greeks in their beliefs and practices. Is it reasonable to ask if they had done the same to the ancient Hebrews? Aten's monotheistic revolution raises a beguiling set of questions. Where do the Hebrews, the twelve tribes of Israel, fit into Egyptian history? And did these Egyptian ideas influence the man who brought the Israelites out of Egypt and delivered God's biblical law on a set of tablets received on Mount Sinai?

This is where myth and history collide—and it is one of the fundamental reasons to understand mythology. Where is one faith or religion—or mythology—born? Whose divinely revealed truth is *the one and only truth*?

Other intriguing questions surface, the foremost of which involve Moses. In spite of his exalted status in Judaism, Christianity, and Islam—Moses is referred to fifty times in the Koran, which credits him with negotiating God down to Islam's five prayers a day—Moses is a mystery man. There is no evidence of his existence in any historical documents outside the Bible or Koran. Extensive Egyptian records contain no reference to a Moses—an Egyptian name; it is Moshe in Hebrew—raised in the house of a pharaoh, as the biblical account and the Hollywood version of *The Ten Commandments* have it. There is also no reference in Egypt's ancient monuments of bureaucratic records to "the children of Israel" working as slaves and then escaping en masse. There is a single reference to a battle with the Hebrews in a victory column—or stela—erected by Pharaoh Merneptah.

This lack of historical records has led many scholars over centuries to doubt the existence of Moses. That is, of course, a radical idea to many believers, since the story of Moses leading the captive Hebrews out of Egypt, miraculously crossing the "Red Sea"—a mistranslation of the Hebrew words for "Sea of Reeds"—and entering the wilderness, where they spent forty years before entering the Promised Land, is the essence of Judaism. It also provides important symbolic connections to the life of Jesus.

The biblical account of the sojourn of the Hebrews in Egypt goes back to the story of Joseph, one of twelve sons of the patriarch Jacob (son of Isaac, grandson of Abraham). The favorite son, Joseph was famed for his "coat of many colors," but was envied by his brothers, who sold him into slavery and told their father that Joseph was killed while tending sheep. Taken to Egypt, Joseph eventually rose to become a counselor to the Egyptian throne because of his uncanny ability to interpret dreams. One biblical account tells the story of how the wife of Potiphar, Joseph's Egyptian master, accused Joseph of attempting to rape her after he had actually rejected the woman's advances. This story, told in Genesis, echoes an old Egyptian folktale called "The Tale of Two Brothers," which contains all of the details that were presumably "sampled," in modern terms, by the authors of Genesis.

The Joseph story continues as, years later, his brothers come to Egypt in the midst of a drought in their land and are brought before Joseph, now a highly placed adviser to the pharaoh. The brothers do not realize who Joseph is, but he recognizes them, and in an act of forgiveness, Joseph is reconciled with the brothers who had sold him. Joseph's father, Jacob—or Israel, as he is called—and all his descendants make the trip to Egypt, where they are welcomed by Joseph.

After hundreds of years in Egypt, in the biblical version, the Hebrews are eventually viewed as a threat by a new pharaoh—unidentified in the Bible—and they are enslaved, put to work building cities and fortifications. Eventually the pharaoh is so worried about these Israelites that he orders the killing of their firstborn. A Jewish woman places her child in a basket of reeds to save his life. Found floating in the Nile by the daughter of the pharaoh, the child—Moses—is raised as a prince in the royal house. Moses later sees an Egyptian beating a Hebrew worker and he kills the Egyptian. Frightened, Moses leaves Egypt, has a divine encounter with God in the form of a burning bush, and returns to Egypt to set his people free. After the ten plagues are visited upon the Egyptians, the pharaoh—usually identified as Ramses II, but there is considerable disagreement over that—consents to let Moses leave with his people, who cross into the Sinai Desert and receive the Ten Commandments;

then, after more tribulations, they eventually enter Canaan, the Promised Land. Moses, however, does not go with them. He dies before entering the Promised Land, his final resting place a complete mystery.*

Does Egyptian myth matter?

What difference do all these stories of thousands of gods with animal heads really make? Was Egypt simply one more great civilization that fell into history's dustbin? After the Ramessid Period, Egypt began a long decline, starting with the Twentieth Dynasty (1186–1069 BCE), as struggles for royal power among priests and nobles divided the country. Egypt lost its territories abroad, and its weakness attracted foreign invaders. The decline accelerated rapidly after about 1070 BCE, and during the next seven hundred years, more than ten dynasties ruled Egypt, but many of them were formed by foreign rulers, including Nubians, Assyrians, and the Persians, whose king Cambyses conquered Egypt in 525 BCE. According to Egyptian accounts, the Persian king respected Egyptian religion and assumed the forms of traditional Egyptian kingship.

After declining for centuries, the glories of the pharaohs finally ended in 332 BCE, when Alexander the Great conquered Egypt and added it to his empire. When Alexander died in 323 BCE, his generals divided his empire, and one of them, Ptolemy, gained control of Egypt. In about 305 BCE, he founded a dynasty known as the Ptolemies, which spread Greek culture in Egypt, with Alexandria becoming Egypt's capital and central city. Famed for its magnificent library and museum, Alexandria emerged as one of the greatest cultural centers of the ancient world. The dynasty of the Ptolemies (305–30 BCE) claimed the title of pharaoh and treated the Egyptian gods respectfully, but the ancient connection between the ruler of Egypt and the gods had finally ended.

In 30 BCE, Egypt's ability to produce vast surpluses of grain made it

*A complete discussion of the history and chronology of the Israelites in Egypt, the Exodus, and the Ten Commandments can be found in my earlier book *Don't Know Much About the Bible*.

a great prize in the intrigues that created the Roman Empire. The period included one of the most extraordinary chapters of history, the brief reign of Cleopatra—the last of the Ptolemies—and her involvement with two of Rome's most powerful men, Julius Caesar and Mark Antony. As Caesar's lover, Cleopatra went to Rome and was there when he was assassinated in 44 BCE. She returned to Egypt, had her brother killed, and placed her son—fathered by Caesar, she claimed—on the Egyptian throne. She then became involved with Mark Antony, coruler of Rome. Antony and Cleopatra hoped that their combined armies could win control of Rome against Octavian, Julius Caesar's nephew and heir and another coruler of Rome. In the sea Battle of Actium in 31 BCE, the navy of Antony and Cleopatra lost to Octavian's fleet. The famed lovers later separately committed suicide, and Octavian, who would be renamed Augustus and complete the transformation of Rome from republic to empire, made Egypt a province of Rome, which ruled it for the next four centuries. Rome's control of Egypt gradually weakened after 395 CE, when the Roman Empire split into Eastern and Western parts. By 642 CE, Muslims from Arabia had conquered Egypt.

Having faded from its glory and majesty, the three-thousand-year empire saw its lights dim. Did its history and beliefs matter? Did the great civilization make a difference? Unquestionably, the answer is "Yes."

Aside from its obvious artistic, cultural, and technical achievements, Egypt had great impact on its neighbors and later conquerors, including Greece and Rome, which both assimilated aspects of Egyptian religion, art, and architecture.

There is also considerable evidence that Egyptian writings may have influenced the Bible, aside from the stories of Joseph and Moses. A series of Egyptian moral precepts called the *Wisdom of Amenenope* (c. 1400 BCE), one of the most famous instructional texts in ancient Egypt, has very close parallels to the biblical Book of Proverbs.

Perhaps most significant for world history is the overlooked role of Egypt in the history of Christianity, which took root in Egypt at a very early date. By the end of the second century, Christianity was already well established in the Nile Valley, and soon came to replace the old religion of the gods.

In *The Complete Gods and Goddesses of Ancient Egypt*, Richard H. Wilkinson concludes: "The spread of the religion was aided by the fact that many aspects of Christianity were readily understandable to the Egyptians in terms of their own ancient myths and beliefs. . . . [The] fact that the Egyptians, since ancient times, had viewed their king as an incarnation of a god meant that the Christian concept of Jesus as the incarnate son of God was far more readily embraced in Egypt than elsewhere in the Roman world. . . . [Even] major Christian motifs may have Egyptian origins. The sacred mother and child of Christianity are certainly foreshadowed in the countless images of Isis—whom the Egyptians called the 'mother of god'—and her infant son Horus, as is even the symbol of the cross which is first attested in Egypt as the 'Egyptian' or *tau* cross—a form of the *ankh* sign."

But is there something else? Does Egypt's extraordinary history speak to any deeper spiritual or cosmic significance? Setting aside theories of ancient astronauts, cursed mummies, the psychic power of pyramids, and dozens of other "New Age" obsessions with things Egyptian, does the "gift of the Nile" matter? For those with a Jewish or Christian background—as well as those people whose exposure to Egypt was limited to the annual airing of *The Ten Commandments* with Charlton Heston as Moses—there has always been a cultural hangover of animosity toward ancient Egypt. Through this Judeo-Christian framework, Egypt existed only as the home of the ruthless pharaohs, a place of servitude and inhumanity. It was an image that carried through to the American Civil Rights era, when the Deep South was symbolically associated with Egypt and American blacks saw themselves as the Hebrews trying to escape the pharaoh's cruelty.

Lost in this somewhat narrow view of the Egyptians as the "bad guys" is a different view of Egypt—a society where the values of truth, justice, charity, and other virtues played a critical role in shaping a civilization that produced extraordinary beauty and a spiritual view of the universe, which, at its best, believed that a just life was justly rewarded.

CHAPTER THREE

BY THE RIVERS OF BABYLON

The Myths of Mesopotamia

It is an old story
But one that can still be told
About a man who loved
And lost a friend to death.
And learned he lacked the power
To bring him back to life.

—*Gilgamesh: A Verse Narrative*
(translated by Herbert Mason)

When on high the heaven had not been named
Firm ground below had not been called by name . . .
. . . When sweet and bitter
mingled together, no reed was plaited, no rushes
muddied the water.
The gods were nameless, natureless, futureless.

—from *Enuma Elish*, The Babylonian Creation

By the rivers of Babylon,
there we sat down,
yea, we wept,
when we remembered Zion.

—Psalm 137:1

"Like other people in the ancient world, the Babylonians attributed their cultural achievements to the gods, who had revealed their own lifestyle to their mythical ancestors. Thus Babylon itself was supposed to be an image of heaven, with each of its temples a replica of a celestial palace."

—KAREN ARMSTRONG, *A History of God* (1993)

What role did myths play in ancient Mesopotamia?

Where did Mesopotamia's gods live?

What's so special about the "cradle of civilization"?

How did a swamp inspire Mesopotamia's myths?

How do we know what the Mesopotamians believed?

When Sumer disappeared, where did its myths go?

What is the *Enuma Elish*?

Was Marduk just another macho man oppressing gentle goddesses?

Who was Hammurabi?

Who's Who of Mesopotamian Myths

How did an angry goddess make the seasons?

Was Inanna's city the first "Sin City"?

Who was mythology's first superhero?

Was the *Gilgamesh* a work of "faction"?

Who came first, Gilgamesh or Noah?

Was the Tower of Babel in Babylon?

Was the Bible's Abraham a man—or another Mesopotamian myth?

Who were El and Baal?

What's a Canaanite demoness doing at a rock concert?

What are three Persian magicians doing in Bethlehem on Christmas?

MYTHIC MILESTONES

Mesopotamia

(All dates are Before the Common Era—BCE)

c. 9000 Early cultivation of wheat and barley; domestication of dogs and sheep in the foothills of the Zagros Mountains.

c. 7000 One of the world's oldest known permanent settlements at Jarmo in northern Iraq; crude mud houses; goats, sheep, and pigs herded; wheat grown from seed.

6000 Farmers from northern areas migrate south to settle in the region between Babylon and Persian Gulf.

c. 5500 World's first irrigation systems used.

Fine pottery is invented.

Trading begins from Persian Gulf to Mediterranean.

c. 5000 First religious shrines in Eridu—called the "first city."

c. 4500 First use of sail.

4000–3500 **Sumerians** settle on the banks of the Euphrates.

First use of the plow.

3500 Emergence of the first city-states.

3400 Clay counting tokens and first written symbols in use.

3200 Evidence of wheeled vehicles in Sumer, along with sailboats, potter's wheels, and kilns.

3100 Development of cuneiform script to record land sales and contracts.

3000–2500 Sumerians grow barley, bake bread, make beer.

Metal coins are used to replace barley as means of exchange.

c. 2700 Reign of Gilgamesh, legendary king of Uruk.

c. 2500 Array of grave goods placed in royal graves at Ur.

2334 Powerful Semitic-speaking **Akkadian** dynasty founded by Sargon I, uniting city-states of southern Mesopotamia; the world's first "empire."

c. 2100 Construction of the ziggurat at Ur.

Hebrew patriarch Abraham leaves Ur (date is speculative).

1800 Ammorites from Syrian desert conquer Sumer-Akkadia.

1792–1750 **Old Babylonian Period.** Hammurabi ascends the throne of Babylon and brings most of Mesopotamia under his control.

Babylon made the Mesopotamian capital.

Hammurabi institutes one of the first law codes in history.

1595 Babylon sacked and occupied by invaders from Iranian plateau known as **Kassites**.

1363 **Assyrian Empire** founded by Ashur-uballit.

1300 Alphabetic script developed in Mesopotamia is a refinement of the simplified cuneiform alphabet.

1295–1200 The Jewish Exodus from Egypt (date is speculative).

1240–1190 Israelite conquest of Canaan (date is speculative).

1200 The *Gilgamesh* epic is composed, the first known written legend.

1193 The destruction of Troy (date is speculative).

1146 Nebuchadrezzar I begins a twenty-three-year reign as king of Babylon.

1116 Tiglath-pileser I begins a thirty-eight-year reign that will bring the Middle Assyrian Empire to its highest point.

1005–967 Reign of King David in Israel; Jerusalem established as capital.

967–931 Reign of King Solomon in Jerusalem.

c. 850 Homer composes *The Iliad* and *The Odyssey*.

722 Conquest of Northern Kingdom of Israel by Assyria—the so-called Ten Tribes, some thirty thousand Israelites, are deported to Central Asia by Sargon II; they will disappear from history and be known as the "Lost Tribes of Israel."

693–689 Assyrian king Sennacherib destroys Babylon.

663 Assyrians attack Egypt, sack Thebes, and leave vassal rulers in charge.

612 Fall of Assyrian capital of Nineveh to the **Chaldeans** (neo-Babylonians).

605 Persian religious leader Zoroaster (Zarathustra) founds a faith that will dominate Persian thought for centuries.

604 King Nebuchadrezzar II revives Babylon and builds the Hanging Gardens, one of the Seven Wonders of the Ancient World, and the ziggurat that inspired the Tower of Babel as a temple to the Babylonian god Marduk.

597 Nebuchadrezzar II conquers Jerusalem.

Judah's king deported to Babylon.

587/6 Fall of Jerusalem and destruction of the Great Temple.

Jewish exile in Babylon begins. During this time, many of the books of Hebrew scripture are first written down.

539 **Persian Empire:** King Cyrus captures Babylon and incorporates the city into the Persian Empire.

538 Cyrus allows the exiled Jews to return to Jerusalem.

522–486 Darius I of Persia is defeated by the Greeks at Marathon in 490.

336–323 The reign of Alexander the Great. In 330, the Persian Empire falls to Alexander, beginning the Hellenistic Era, in which Greek civilization and language spread throughout the Near Eastern world. Alexander dies in Babylon in 323.

The next time you walk into a bar on a Friday night, order up a cold brew, and ask someone what his or her "sign" is, pause a moment and thank the ancient Mesopotamians.

At the dawn of history, these people invented the seven-day week, beer, and astrology. (How they overlooked cocktail nuts is a mystery yet to be solved.) If you nurse your drink for an hour and then scribble down someone's name and number before driving home, consider that the ancient Mesopotamians also deserve credit for the sixty-minute hour, the world's first writing system, and the wheel. The list goes on. Ancient Mesopotamia was an extraordinary place that pioneered pottery, poetry, sailboats, and schools. The Mesopotamians came up with the 360-degree circle, a poem considered the first piece of written literature, formulas to predict eclipses of the sun and moon, and the mathematical concepts of fractions, squares, and square roots that still torment high school kids.

But there is something else we should not leave off this impressive list of legacies. In this so-called "dead civilization," the Mesopotamians created a richly imaginative mythic tradition crowded with warring gods, dragon-slayers, the first superhero, and an enticingly libidinous love goddess. These Mesopotamian myths not only played a central role in the daily lives and history of the people in the "cradle of civilization," but their stories and legends also placed an indelible stamp on the literature and history of the Bible.

The six-day Creation in Genesis, for instance, is widely thought to have been influenced by Mesopotamia's Creation epic, first translated a little more than a century ago and rattling religious teacups ever since with the suggestion that parts of the Bible were—gasp!—cribbed from another source. The genealogies of Adam and Eve's descendants suspiciously resemble the lists of early Mesopotamian kings, unearthed in a royal library in the ruins of Nineveh, a fabled city once buried under centuries of sand and featured in the story of Jonah and the whale. (Actually, it was a "large fish." But that's another story.) Mesopotamia's towering, pyramid-like temples, called ziggurats, made a lasting impression as the inspiration for the Tower of Babel. Perhaps most intriguing of all are their flood stories. Composed more than four thousand years ago and told in *Gilgamesh*—an epic poem written centuries before the Bible was set down—these tales may have influenced the Hebrew storytellers who produced their own flood account featuring a godly man named Noah. All of these ancient Mesopotamian legends would have been familiar to the Hebrews, whose patriarch Abraham came from Mesopotamia, and who often came under the thumb of a collection of aggressive Mesopotamian kings counted among the Bible's "bad guys."

Located mostly in what is modern-day Iraq, Mesopotamia was a desirable patch of real estate that became home to some of the world's first human settlements about ten thousand years ago. Watered by the Tigris and Euphrates Rivers—the word "Mesopotamia" is Greek for "between the rivers"—this otherwise arid, flat plain blossomed as people learned how to control these somewhat erratic rivers with irrigation dikes and canals. Like beads on a necklace strung along the two rivers, small farming settlements grew into the world's first cities, flourishing as their surplus food production allowed for expanding trade opportunities. As they developed in wealth and size, these farming and herding communities eventually became "city-states," with merchants, skilled craftsmen, prostitutes, priests, and tax collectors, and armies of scribes who recorded everything from negotiations over the price of figs to real estate deals, law codes, epic poetry, and the military records and amorous adventures of conquering kings.

Unfortunately, the prosperity of these city-states also attracted attention. Unprotected by the vast stretches of desert that kept Egypt safe from most outsiders, the flat plains of Mesopotamia were like an open chessboard, across which armies moved freely. Mesopotamia became a land of repeated invasions and conquests, and history in the Tigris-Euphrates valley tended to be stormy and full of violent upheavals. Unlike the constant Nile, the Tigris and Euphrates were also less reliable sources of water for crops, and changed their course over many centuries, sometimes turning thriving cities into ghost towns. The unpredictable and uneven flow of the two great rivers combined with local political conflicts to shape a mythology and religion that was as much about strife as it was about universal order.

Over thousands of years, Mesopotamia was occupied and ruled by a succession of small kingdoms—some fairly belligerent—that grew to include some of the world's first empires: the Sumerians, Babylonians, Hittites, Assyrians, Chaldeans, and Persians. As these empires rose and fell, power shifted and civilizations grew. Each time power changed hands, the myths of this very old land changed, too. Each new empire borrowed traditions from the one before, and Mesopotamia's myths evolved and were rewritten and reshaped to reflect new political realities. But always, there was one constant. From the earliest times, the worship of Mesopotamia's many gods—who ruled sun, wind, and water, the weather, earth's fertility, and every aspect of the natural world—played a crucial role in dictating life and society in the world "between the rivers."

What role did myths play in ancient Mesopotamia?

Think of Mesopotamia as the Rodney Dangerfield of the ancient Near East—it has never gotten the respect or star billing accorded to Egypt and Greece. Maybe it was because the people there were considered the villains of the Bible, having sacked Jerusalem, carted thousands of Jews off to captivity or oblivion, and introduced so many of the Bible's pernicious "false gods." Or maybe it was because their ancient attractions were not viewed with the same awe as those that sent worshipful tourists flocking to Egypt and Greece. (One historian describes the achievements in Mesopotamia as "less spectacular art and crumbling mudbrick ruins.") Keats didn't write an "Ode on a Mesopotamian Urn." And New Age trendsetters have not adopted the Mesopotamian ziggurat as a totem of mysterious psychic powers. Also, during much of the twentieth century, what was once Mesopotamia has largely been shut off to the Western world, due to culture, history, and politics. In case you hadn't noticed, Iraq hasn't been topping anybody's list of ten best tourist destinations for most of the past fifty years.*

*The nation of Iraq was formed in the aftermath of World War I, when the British, who then ruled it, called the area by its ancient name, Mesopotamia. But Iraq as it exists today has little to do with the ancient civilizations that rose and fell there. A British-installed, independent kingdom was established in 1923, and the British dominated the oil fields and politics of Iraq for the next quarter-century. In a military coup, King Faisal was killed in 1951. Successive military regimes were

Whatever the reasons, Mesopotamia took a back seat to Egypt and Greece, an oversight worth correcting, because the oversimplified—or overlooked—past of this ancient land, which has become so significant in modern times, is a fascinating piece in the jigsaw puzzle of ancient civilization. Religion, history, and myth all mingled together there, and the story of Mesopotamia's city-states and the empires that grew from them offers another vivid example of that fascinating crossroads where legend and ancient life intersect.

Like Egypt, the successive empires of Mesopotamia were theocracies—societies in which government and religion were inseparably fused. The gods of Mesopotamia didn't just make the rain fall or crops grow. These gods chose the earthly kings—or, at least, that's what the kings and temple priests told their subjects. The people existed to serve the gods—through their earthly representatives, the kings and priests. In each city-state, the local god became the symbol of the city's strength and source of its prestige, wealth, and power. To put it simply—"My god is bigger than your god."

As Mesopotamia's cities eventually expanded to become small empires, the power of their gods increased as well, and the most powerful empire, obviously, had the most powerful god. **Marduk**, the central deity of Babylon, took charge when Babylon became the region's preeminent city-state. Local myths were revised so that his status as a Zeus-like king of the gods was celebrated and made sacred in Mesopotamia's central Creation story. Just as Re became Egypt's state god, or Yahweh later became Israel's national god, Marduk, once an agricultural deity, emerged to lord over the pantheon of Mesopotamian deities, superseding the earlier chief god of the Sumerians, An, and taking control of the weather, the moon, rain, justice, wisdom, and war. (See below, "*Who's Who of Mesopotamian Myths.*")

The other key concept from Mesopotamian myth was the *me* (pronounced "may"), a somewhat abstract collection of divine laws, rules, and regulations that governed the universe from its creation and kept it operating. Unlike the Egyptian concept of *maat*, which was order, truth, and justice, the Mesopotamian *me* was a far more complex list of insti-

increasingly dominated by the Baath Party until Saddam Hussein ultimately seized power in 1979.

tutions, people, rituals, and other elements of a culture that included more than one hundred separate characteristic items forming the basis of Sumerian society. In some respects, it was comparable to the intricate laws of ancient Judaism that went far beyond the basic Ten Commandments, and defined the role of priests and the manner of worship.

But the *me* was, in many ways, even more complex, covering nearly every dimension of Sumerian society. Among the varied aspects of the *me* were a catalog of official institutions, like kingship and the priesthood; certain ritualistic practices, including holy purification; desirable qualities of human character and moral laws; and even lists of occupations that included scribes and blacksmiths. Highly conceptual, the list of what constituted the *me* also included such acts as lamentation, rejoicing, sexual intercourse, and prostitution. Various parts of the *me* could also exist in physical objects, such as the throne—in which kingship resided—or drums that contained rhythm. Like building blocks of an orderly society, all of these basic ideas, institutions, and practices had to be maintained intact to ensure the cosmic order. Possession of the *me* meant to hold supreme power, and **Enki**, the chief god of Sumer, was the keeper of the *me*.

Where did Mesopotamia's gods live?

If the poet Robert Frost was right that "good fences make good neighbors," walls may be even better. To fortify their cities against invasions, the Mesopotamians built high-gated walls around their cities, with temples, palaces, and royal houses enclosed within another set of walls in their centers. Around them were "suburbs," encompassing the fields and orchards. Every city also had a riverfront harbor area, which was the center of commerce.

But the temples provided the focal point of Mesopotamian life and society. Housed within the prominent ziggurat towers that loomed high above the relatively flat plains of Mesopotamia, the temples were more than just symbolic or ritual buildings, or tombs for dead kings. Built for the cult worship of a particular god who was responsible for both the city and the people, the temples were thought to be the actual home of the gods, where they lived with their families and servants. The gods of Mesopotamia's cities might be fearsome and powerful, but they were homebodies, completely tied to their cult city and its temple, and they needed the daily attention of priests and priestesses. Each day, the rituals of feeding, clothing, and washing the god were carried out within the sanctuary. As anthropologist Gwendolyn Leick put it, "Heaven was no further than the temple roof. By providing the gods with lodgings and sustenance, the city partook of the essence of divinity."

Employing large numbers of workers, these city temples drove daily life. They were run by a priestly hierarchy and controlled enormous wealth collected through taxes and offerings, held large tracts of fields and orchards, and even functioned as "banks," making loans. Though the daily worship of the gods was a priestly duty, religion played a great part in the lives of ordinary people. Everyday worshippers attached themselves to a particular god or goddess—just as modern Christians might be especially devoted to a favorite saint—and they offered prayers and sacrifices in return for blessings and protection from evil spirits. Even though they were unable to access the inner sanctums of the temples, ordinary people participated in the great religious processions in which statues representing the gods were paraded through the streets.

Many people also heeded exorcists and diviners for prophecies and advice. In Mesopotamia, divination was a highly specialized art. The Mesopotamians believed that the whole universe was filled with coded messages about the future, and these people sought advice from expert diviners, trained for years in the art of reading the signs in animal entrails and organs, such as the liver of a freshly slaughtered lamb. Dream oracles were also popular, and the practice of astrological readings began in Mesopotamia as soothsayers attempted to find portents in the changing heavens—the beginning of carefully recorded astronomical records. As Daniel Boorstin wrote in *The Discoverers*, "If the rising and setting of the sun made so much difference on earth, why not also the movement of the other heavenly bodies? The [Mesopotamian] Babylonians made the whole sky a stage for their mythological imagination. Like the rest of nature, the heavens were a scene of living drama."

Originally intended to demonstrate that the king's decisions and laws had divine approval, these elaborate ancient Mesopotamian systems of reading signs, omens, and oracles were probably as ubiquitous in ancient Ur and Babylon as the "psychic reader" business is on the streets of many big cities today.

But the chief mythic event in this world came during their New Year, when a great public festival was held. This eleven-day religious observance was not just a spiritual event or festive holiday, but a national drama, a form of political theater meant to solidify the king's role as protector and provider. During the pivotal New Year celebration (which fell in April), when the ancient Creation stories were sung at great public gatherings, the king actually reenacted the role of the great fertility god in a ritual marriage to a priestess representing the goddess **Inanna** (aka Ishtar). This marriage ceremony—which would have been publicly consummated—was meant to ensure prosperity, strength, and order.

Using myth and belief, the rulers of Mesopotamia—and the priestly classes allied with them—created and cemented their political and social power. The importance of this leap in human history can't be overstated. It was a development as significant as the invention of the wheel or writing.

MYTHIC VOICES

In the temple of Babylon there is a second shrine lower down in which is a great sitting figure of Bel, all of gold on a golden throne, supported on a base of gold, with a golden table standing beside it. I was told by the Chaldeans that, to make all this, more than twenty-two tons of gold were used. Outside the temple is a golden altar, and there is another one, not of gold but of great size, on which full-grown sheep are sacrificed. . . . On the larger altar, the Chaldeans also offer something like twenty-eight and a half tons of frankincense every year at the festival of Bel. In the time of Cyrus, there was also in this sacred building a solid golden statue of a man some fifteen feet high—I have this on the authority of the Chaldeans, although I never saw it myself.

—HERODOTUS *describes Babylon in* The Histories

Like the "Where's Waldo?" of the ancient world, the Greek historian Herodotus also popped up in Babylon, which was the great capital city of several Mesopotamian empires as well as the Persian empire of King Cyrus (d. 530 BCE). His description of the inner sanctum of a temple, with its rooms devoted to gods, is well supported by archaeological investigations. The god whom Herodotus referred to as "Bel" (which means "Lord") was actually Babylon's central deity, Marduk, and the title of "Bel" was transformed into "Baal" in the myths that would figure significantly in the history of the Bible.

An idol devoted to Bel is also featured in the story "Bel and the Dragon," a brief addition to the biblical Book of Daniel (of lion's den fame). Set in Babylon during the time of Cyrus, the story satirizes the priests who ate the food set out before the idol of Bel. When Daniel tells Cyrus of this deception, the Persian king has the priests killed.*

What's so special about the "cradle of civilization"?

Okay. You're back in elementary school and your teacher pulls down one of those window-shade maps that tend to snap right back up. The teacher is already in trouble, since this seems more like slapstick comedy than school. Then you open your first World Civilizations textbook to a list of "key words" and see "Fertile Crescent," and "Cradle of Civilization," "Hammurabi's Code," "Nebuchadrezzar's Hanging Gardens," and "Seven Wonders of the Ancient World." The phrases are almost clichés, but they contain more than a nugget of truth, capturing the extraordinary accomplishments of the people and empires of Mesopotamia, where much of civilization began.

So what is so special about this part of the world? Why did these people in this rather dry, hot, and unappealing part of the world produce so many of civilization's "firsts"? Why here?

*This brief piece of a larger biblical tale is also a perfect example of how even Christians don't always agree on their "holy" stories. While Catholics traditionally include "Bel and the Dragon" as part of the Book of Daniel, Protestants do not. In their Bibles, such as the King James and New Revised Standard Versions, the brief narrative is placed in the Apocrypha, a collection of writings that are not considered part of the divinely inspired "canon" of the Bible.

As any real estate broker will tell you, it comes down to three things—location, location, location.

Ancient Mesopotamia spanned a geographic area that now includes most of modern Iraq, eastern Syria, and southeastern Turkey. It extended from the marshy lowlands on the Persian Gulf in the south to the highlands of the Taurus Mountains (bordering modern Turkey) in the north, and from the Zagros Mountains (in modern Iran) in the east to the Syrian desert in the west.

The oldest known communities in ancient Mesopotamia were villages established in the Zagros foothills more than nine thousand years ago. These early sites, such as Jarmo in northern Iraq, were among the world's oldest known human settlements, along with the biblical city of Jericho, near the Dead Sea, Tell Hamoukar in modern Syria on the fringes of the Tigris-Euphrates valley, and Catalhoyuk (also called Catal Huyuk, and pronounced *cha-tahl-hu-yook*) in modern Turkey. With ample water supplies in otherwise dry areas, it was here that people first began to cultivate wheat and barley, domesticate animals, build crude mud houses, and keep herds of goats, sheep, and pigs.

Around 6000 BCE, some of these early farmers moved south, to the region between the future site of Babylon and the Persian Gulf. Drawn to the rivers, they settled in what became the heart of ancient Mesopotamia, in the southern end of the flat plain between the Tigris and Euphrates, roughly the area between modern Baghdad and Basra, which became all too familiar to a world that watched the war in Iraq unfold during the spring of 2003. Like the Nile, the rivers of Mesopotamia also flood, and farmers began to dig irrigation canals that would water their otherwise dry lands. This intensive agricultural undertaking demanded cooperation—and with it, the beginnings of a social order.

As the old story from Aesop put it, "Necessity is the mother of invention." The intensive communal agriculture allowed people to successfully farm, provided a constant source of food that encouraged larger settlements, and led to expanding populations. Without the pressure of needing to hunt and gather food, communities set down permanent roots and grew. Over time, their stability allowed them to produce textiles, pottery, and other inventions that marked the beginnings of civilization. The first wheel, for instance, was not used by Fred Flintstone in the Stone Age, as generations of cartoon lovers may believe, but more likely by anonymous Mesopotamians around 6500 BCE.

As their agricultural improvements succeeded, populations flourished, and the division of labor became more complex. A social hierarchy developed, in which a ruling class emerged that was responsible for organizing production and trade. The region was also lacking in many basic natural resources, such as wood, stone, and metal ores. Again, necessity led to the invention of dried mud bricks for construction. This shortfall in raw materials also made trading for other resources more important, and trade routes gradually grew along the course of the rivers. Eventually, control of key river-crossings became a source of economic, military, and political power.

How did a swamp inspire Mesopotamia's myths?

Around 5000 BCE, settlements sprang up in a place called Eridu on the Euphrates River, a southern site near the marshes that mark the transition from land to sea. The people here, who are considered the first city-dwellers, also built some of the first known religious shrines, and ruins of a small temple with an offering table and a niche for statues have been found here. This marshy, or swampy, area, where freshwater mingled with salt water, would inspire the core of the Mesopotamian Creation myths, in which the freshwater and salt water were actually imagined to be deities who created the world. Water, especially freshwater, was the key to existence in this otherwise arid, hot plain. Not surprisingly, then, in some of the world's earliest Creation stories and myths, the earth, the gods, life, and humanity emerged from these primordial Mesopotamian waters.*

Sometime before 3500 BCE, a new group moved into the region and settled on the banks of the Euphrates. Although it is not known where these people originated—most historians surmise that they came

*The literal heirs of these people are Iraq's so-called Marsh Arabs. During the 1990s, Saddam Hussein tried to destroy these people—who had rebelled at the urging of the first President Bush after the Gulf War in 1991—by systematically draining the marshes on which their way of life depended. The UN has described this as the "environmental crime of the century."

from the east—the area they settled became known as **Sumer**, and the civilization they built is called Sumerian. The Sumerians began to build cities that gradually became city-states, including Ur (presumed home of the biblical Abraham), Uruk (the biblical Erech), Kish, and Nippur. The preciousness of water led to "water-wars" between these city-states until the more powerful ones gradually swallowed the smaller ones. Over the next fifteen hundred years, the Sumerians gradually harnessed animals to plows, drained marshlands, and irrigated the desert to extend areas of cultivation. Their increased agricultural efficiency eventually led to the first "leisure class," allowing for the development of commerce, and with it merchants, traders, artisans, and priests, to make sure that the gods approved of everything that was going on. By 3000 BCE, the first walled cities were built in Mesopotamia, always including temple complexes within the city walls.

Although political power in these cities was initially held by free citizens and a governor, as the city-states grew and vied for power, the Sumerians also may have developed one of the world's first systems of monarchy, headed by a priest-king. Sumer's first known king—in Sumerian, the word was *lugal* and meant "big man"—was Etana of Kish (c. 3000 BCE), described in ancient writings as "the man who stabilized all the lands." But in one of these Sumerian cities, long before the Greeks coined the word "democracy," the "first bicameral congress" met in 3000 BCE. Prominent Sumerologist Samuel Noah Kramer points out that in the city of Uruk, one council made up of elders and another of arms-bearing men both met to decide whether or not to go to war with the neighboring city of Kish. This "congress" voted for war and the king approved.

The Sumerians are also credited with inventing a form of bureaucracy around the same time that the Egyptians did. Devised to manage the land, Sumer's bureaucracy consisted of a priesthood that was responsible for surveying and distributing property and collecting taxes. To make sure that everything worked, the Sumerians also invented every bureaucrat's best friend—records. That required the invention of writing, and the Sumerians are also credited with introducing the world's first system of writing around 3200 BCE, word-pictures that developed into wedge-shaped characters known as cuneiform, which comes from the Latin word *cuneus*, meaning "wedge." Cuneiform characters consisted of small indentations made with a wedge-shaped tool called a stylus, impressed in wet clay. The Sumerians used about six hundred characters, which ranged from a single wedge to complicated signs consisting of thirty or more wedges. The clay hardened, and the cuneiform tablets became the first known "official records" in history.

There is even a Sumerian legend to explain this invention. A messenger of a king of the city of Uruk arrived at the court of another king, but was so winded from his journey that he was unable to deliver his message. The clever king wanted to make sure that didn't happen again, so when he needed to send another message, he patted some clay and set down the words of his next messages on a tablet. The king of Uruk had invented writing. How the person at the other end could read this message was not explained in the legend.

There is still disagreement in the scholarly world as to why writing developed in Sumer, and whether it happened in other places, such as Egypt or China, either earlier or at the same time. One leading theory holds that Sumerian writing grew out of accounting, as molded clay "tokens" were used to represent quantities of different trade goods, such as oil, grain, or livestock. Early evidence does suggest that Sumerian cuneiform was used almost exclusively for these accounts for its first five hundred years. But eventually, writing evolved to express the spoken word, and among its earliest uses in Sumer was to record the ingredients of beer. There is no evidence that the ancient symbol for beer was two women in bikinis wrestling in mud.

How do we know what the Mesopotamians believed?

Here's a sobering thought. Long after most of the books we produce today are gone, the writings of these ancient Mesopotamians will still be around. Why? Because their literature, business accounts, and other writings were literally "written in stone," the hardened clay tablets that have been found in the tens of thousands in a variety of sites in what once was Mesopotamia.

Like their Egyptian neighbors, the Mesopotamians created an enormous trove of art, architecture, and, most significant, written records that have survived the ages and thousands of years of conquest, right down to modern times. The widespread chaos and subsequent looting of Baghdad's museums and ancient archaeological sites in the days following Saddam Hussein's ouster in 2003 provided a vivid reminder of the area's extraordinary history. Some of the world's oldest artworks, including a five-thousand-year-old sculpture of a woman's face called the *Sumerian Mona Lisa*, were stripped away by looters, along with thousands of antiquities. Fortunately, many of the country's most valuable pieces had been stored out of harm's way in the run-up to the war, and thousands of other stolen items have since been returned. But there are still many missing relics, and the art world was put on high alert for these looted artworks, many of which may likely end up in a secretive and lucrative black market.

In spite of those losses, and with hopes that Iraq will eventually be reopened to a new era of scholarly archaeology, much is already known about Mesopotamia's past. The surprising secrets of Mesopotamia's history were first opened in the mid-nineteenth century, when the first great cache of cuneiform tablets was unearthed in Nineveh. An ancient city located near the modern city of Mosul in northern Iraq, Nineveh was once the capital of the Assyrian Empire. Discovered in the ruins of the library of a king named Ashurbanipal were more than 24,000 clay tablets that included business documents, personal letters, and some of the world's oldest known literature, including the epic of *Gilgamesh*. This treasure trove gave the world its first look at the myths and history of Sumer. Since then, many more thousands of tablets have been found in the sites of such ancient cities as Nippur, Ur, and Ebla (in modern Syria), giving archaeologists a comprehensive source of written materials from very early times in Mesopotamia, as well as their first hints of the astonishing connection of Sumer's myths to the Bible.

When Sumer disappeared, where did its myths go?

Around 2350 BCE, the peace and relative tranquility of the Sumerian civilization was dealt a blow when people from the west (probably the

Arabian Peninsula) swept in, settled in the northern area of Mesopotamia, and eventually conquered Sumer. These invaders were Semitic—people who spoke a language related to modern Arabic and Hebrew.* Under a king known as Sargon I, Sumer and the northern region of Akkad were united around the year 2340 BCE, and Sargon built the city of Akkad (or Agade), established an enormous court there, and then built a new temple in the city of Nippur, an ancient city located about a hundred miles from modern Baghdad. An outstanding military leader and administrator, Sargon gained control over much of southwestern Asia. He reigned for fifty-six years, and during his rule, Semites replaced the Sumerians as the most powerful inhabitants of Mesopotamia. These Semites and their language came to be called **Akkadian**, after Sargon's capital.

But curiously, while the Sumerians basically disappear from history, their civilization, culture, and most of their myths and religion do not. Their gods, including the Creation goddess named **Tiamat** and a love goddess named Inanna, survived. The entire Sumerian pantheon of nature deities was absorbed by the new arrivals, and the Sumerian religion was largely adopted by the Akkadians, who added the Sumerian gods to the list of deities who protected their own cities, only changing their names to Akkadian ones. After Sargon's death, the Akkadian Empire was torn apart by infighting and rebellion. While a few of the Akkadian city-states maintained independence for a short while, they were soon absorbed into a rising **Babylonian Empire** beginning around 1900 BCE.

Built near the earlier site of Akkad, the city of Babylon—located south of modern Baghdad—emerged as the greatest of Mesopotamia's city-states, becoming an urban center that would have enormous impact, especially in biblical history. The very word "Babylon," which translated as "gate of the gods," meant that people there believed this was the spot where the gods actually came down to earth. The idea was

*The widely misunderstood term "semite" is adapted from the biblical name of Shem, a son of Noah. Although the term is often equated exclusively with Jews—as in "anti-Semitic"—Shem was thought to be the ancestor of all the Semitic peoples, who included, in the ancient world, Babylonians, Canaanites, Hebrews, Phoenicians, and Arabs.

not unique to Babylon. Almost every culture has a sacred spot it considers an "omphalos," a Greek word for navel, meaning the spot at which the gods appear on earth. The first great Babylonian civilizations—called "Old Babylonian"—flourished between 1900 and 1600 BCE under a series of kings, including Hammurabi, who made Babylon his capital.

MYTHIC VOICES

When on high the heaven had not been named,
Firm ground below had not been called by name,
When primordial Apsu, their begetter,
And Mummu-Tiamat, she who bore them all,
Their waters mingled as a single body,
No reed hut had sprung forth, no marshland had appeared,
None of the gods had been brought into being,
And none bore a name, and no destinies determined—

Then Tiamat and Marduk joined issue, wisest of gods.
They strove in single combat, locked in battle.
The lord spread out his net to enfold her,
The Evil Wind, which followed behind, he let loose in her face.
When Tiamat opened her mouth to consume him,
He drove in the Evil Wind while as yet she had not shut her lips
As the terrible winds filled her belly,
Her body was distended and her mouth was wide open.

He released the arrow, it tore her belly,
It cut through her insides, splitting the heart.
Having thus subdued her, he extinguished her life.
He cast down her carcass to stand upon it.
After he had slain Tiamat, the leader,
Her band was shattered, her troupe broken up;
And the gods, her helpers who marched at her side,
Trembling with terror, turned their backs about,

In order to save and preserve their lives.
Tightly encircled, they could not escape.

—*The* Enuma Elish, *the Mesopotamian Creation epic* (from Tablet One)

What is the *Enuma Elish*?

The large number of competing groups and city-states that lived and ruled in ancient Mesopotamia meant that there were competing stories of the gods and Creation—as had been true in Egyptian mythology. But the most complete, best-known—and most significant—Mesopotamian Creation story is an epic Akkadian poem named after its opening words, "Enuma Elish," traditionally translated as "When on high."

Discovered on seven clay tables found in the mid-nineteenth century in the ruined palace of Ashurbanipal in Nineveh, the poem was first translated by George Smith in 1876 as *The Chaldean Genesis*. Smith's suggestion that the authors of the Bible—whom most Victorian-era Europeans believed to be divinely inspired—had borrowed from "pagan" Mesopotamians was not well received. Since Smith's translation of the Nineveh tablets, fragments of even earlier versions of the *Enuma Elish* in the Sumerian language have also been discovered—evidence that this Creation tale is a very ancient account that went through generations of editing and retelling.

Unlike the Greek *Iliad*, the *Enuma Elish* is not an adventure story but a religious poem, somewhat like the opening chapter of Genesis, that describes in richly poetic language the beginnings of the world. But while it undoubtedly had some influence on the author of Genesis, the *Enuma Elish* is very different in its tone and the events it recounts. A story of warring gods battling for supremacy over their Creation, the poem tells of the emergence of one supreme god, Marduk, the Babylonian agricultural god. Combining religion with a political agenda, the poem and the mythology it contained were designed to establish Marduk, the patron god of Babylon, as chief among the gods, and to establish Babylon as the most powerful city-state in Mesopotamia.

The epic opens at the very beginning of time with the creation of the gods themselves. In the beginning, the gods emerged two by two from a formless watery waste—a substance which was itself divine. This sacred raw material existed through all eternity, and as Karen Armstrong writes in *A History of God*, "When the Babylonians tried to imagine this primordial divine stuff, they thought that it must have been similar to the swampy wasteland of Mesopotamia, where floods constantly threatened to wipe out the frail works of men."

At first, there were just two gods—**Apsu**, personified as the primordial freshwater under the earth and in the rivers, and Tiamat, who symbolized the salt water of the sea. Biblical scholars believe that the Hebrew word for the "abyss" in the opening of the Creation account in Genesis is a corruption of the word "Tiamat." The Creation in the *Enuma Elish* also plays out in six stages, mirrored by the six days of the Genesis Creation, again reflecting the influence of the Mesopotamian epic on the Hebrew version. These two gods, Apsu and Tiamat, then joined to produce the other gods, all identified with different aspects of nature: **Lahmu** and **Lahamu**, a pair whose names meant "silt," or water and earth mixed together; another pair identified with the horizons of sky and sea; and then **Anu**, the heavens, and **Ea**, the earth.

But these young gods, as all parents of small children know well, were too noisy. Eager to get some sleep, Apsu decided to destroy all of the children. One of the child gods, Ea, discovered the plot, put Apsu to sleep, killed him, and then took his place as god of the waters. With his spouse **Damkina**, Ea later sired Marduk, a perfect god who is "highest among the gods."

Once Tiamat realized what her children had done, she decided to avenge her dead husband. She took the form of a fearsome dragon and created a small army of monstrous creatures to battle the other gods, who were her own children. Marduk, the sun god, came before an assembly of the gods and promised to fight Tiamat—who was in essence his grandmother—on the condition that he would become their ruler. They agreed, which led to an epic battle between Marduk, who had many weapons and powers, and Tiamat, with her troop of fearsome monsters. Marduk dispatched the monsters and then faced Tiamat in face-to-face combat.

The Lord trampled the lower part of Tiamat,
With his unsparing mace smashed her skull,
Severed the arteries of her blood . . .

So much for taking care of Grandma.

Slicing Tiamat in half, "like a fish for drying" or "opening a shellfish," Marduk uses the two halves of her body to create the sky and the earth. From Tiamat's eyes, he opened the Tigris and Euphrates Rivers, and he transformed her breasts into the mountains from which freshwater springs flowed. Promised that he would rule the gods if he defeated Tiamat, Marduk then organized the rest of the universe, naming the months of the year and creating the stars and moon. He devised laws and then established his home in the city he named Babylon:

When you descend from heaven for assembly,
You will spend the night in it, it is there to receive all
of you.
I will call its name Babylon, which means the houses of
the great gods,
I shall build it with the skill of craftsmen.

As the last act in this creation, almost as an afterthought, Marduk also created man. First he killed Kingu, who was Tiamat's consort, and then mixed his divine blood with dust. According to the myth, as far as Marduk and the other gods are concerned, man's purpose is simple: man will do all the work so the gods can relax. The story is significant because, as Karen Armstrong points out in *A History of God*, "The first man had been created from the substance of a god; he therefore shared their divine nature. . . . The gods and humans shared the same predicament, the only difference being that that gods were more powerful and were immortal."

Before they could rest, however, the gods decided to build a proper shrine for Lord Marduk. For one year, they manufactured bricks and after a second year, they had built a temple—a ziggurat—to honor Marduk as king of the gods.

For much of Mesopotamian history, each year, this Creation epic became the focal point of worship, as the *Enuma Elish* was read as part of the New Year celebration in every city. The myth also confirmed Babylon's special status as a sacred place, home of the gods, and the center of the world.

Was Marduk just another macho man oppressing gentle goddesses?

Besides underscoring Babylon's ascension as the greatest power of the time, the Marduk-Tiamat conflict has taken on another significant spin. Recently, a movement of scholars has advanced the notion that most prehistoric cultures revered female deities—usually a benign but all-powerful mother goddess—as the dominant deity. Seen as a nurturing force of fertility in a world that was mostly dependent upon the return of the crops and the continuity of life through birth, this goddess was thought to be more significant than the male deities, who basically existed as studs, to service the goddess and sire children. But, according to this in-vogue theory, a great change took place when male gods were elevated above the goddess, not just in Mesopotamia but in almost every society. The victory of Marduk over Tiamat is widely considered one stark and particularly violent example of the conquest by a warlike, macho-male god and the demise of goddess worship.

This so-called Goddess movement was partly inspired by the writings of Jane E. Harrison, who had suggested in 1903 that "The Great Mother is prior to the masculine deities." More recently, the field has been led by such scholars as Maria Gimbutas, whose 1974 book, *The Gods and Goddesses of Old Europe*, argued that there had been an ancient, more peaceful time in the world, in which the supreme deity was a Mother Earth, creator and ruler of the universe.

Historian Karen Armstrong concurred with this notion, which she contended was also true in ancient Israel, where the Hebrew god Yahweh, "a jealous god," forced the "chosen people" to get rid of their idolatrous but popular goddesses. Writes Armstrong in *A History of God*, "The prestige of the great goddesses in traditional religion reflects the veneration of the female. The rise of the cities, however, meant that the more masculine qualities of martial, physical strength were exalted over female characteristics. Henceforth women were marginalized and became second-class citizens. . . . The cult of the goddesses would be superseded, and this would be a symptom of a cultural change that was characteristic of the newly civilized world."

"Mars" had pushed "Venus" off the pedestal.

There is an intriguing archaeological mystery lying behind this theory: the many thousands of prehistoric figurines from all over the world—some of them from the Stone Age, and dated from eighteen thousand to twenty-five thousand years old. Their widespread existence hints at more than just a fascination with the female form. Often loosely described as prehistoric "Venuses," these statuettes and figurines come in many shapes and forms, but are often full-bodied and sexually enticing, with exaggerated breasts. Some, but hardly all, are pregnant. As author Nancy Hathaway put it, "We don't know if they were erotic images, religious icons, household objects, or charms meant to promote fertility. We do know that there are thousands of them. . . . with [an] absence of an equivalent number of male figures. . . ."

There is, however, broad disagreement about whether the "Venuses" actually represent another age and a different mind-set in human worship—a matriarchal, nonviolent, vegetarian epoch in which the female deity was dominant. In a *New York Times* article on "Venus" figurines, some archaeologists offer that these small figurines were pendants to be worn by men on the hunt, a Stone Age "picture of the wife" to carry while away from home. There is even a hint—widely dismissed—that they were Stone Age "porn." But there is no real evidence to suggest that they were all "goddesses."

The "Goddess movement" has blossomed during the past thirty years, prompted largely by social changes and the shift in attitudes brought about by feminism, the advent of women's studies on university campuses, the contemporary transformation of traditional sex roles—and a rejection of male-dominated orthodox religions. At about the same time, the environmentally oriented "Gaia movement" emerged, a theory that suggests the earth itself is a living "entity," named after a Greek earth goddess. The new wave of Goddess worship also awakened enormous popularity in the "Wiccan movement," said to be among the fastest-growing religions in America. (It is apparently even acknowledged by the U.S. Department of Defense as a legitimate religion that can be practiced on military bases.) Also called "the Craft" or even "Witchcraft," the practice of Wicca as a religion developed in the United Kingdom in the mid-1900s. Essentially, it is a fertility religion with roots in the ancient myths, which celebrates the natural world and the seasonal cycles that were central to farming societies in Sumerian and Babylonian mythologies, as well as those of the Egyptians, Greeks, Romans, and Celts. Contemporary Wicca is an "equal opportunity" borrower and also draws on Buddhist, Hindu, and American Indian rites.

MYTHIC VOICES

When Marduk sent me to rule over men, to give the protection of right to the land, I did right and righteousness in . . . , and brought about the well-being of the oppressed.

2 If any one bring an accusation against a man, and the accused go to the river and leap into the river, if he sink in the river his accuser shall take possession of his house. But if the river prove that the accused is not guilty, and he escape unhurt, then he who had brought the accusation shall be put to death, while he who leaped into the river shall take possession of the house that had belonged to his accuser.

3 If any one bring an accusation of any crime before the elders, and does not prove what he has charged, he shall, if it be a capital offense charged, be put to death.

25 If fire break out in a house, and someone who comes to put it out cast his eye upon the property of the owner of the house, and take the property of the master of the house, he shall be thrown into that self-same fire.

129 If a man's wife be surprised [in flagrante delicto] with another man, both shall be tied and thrown into the water, but the husband may pardon his wife and the king his slaves.

130 If a man violate the wife [betrothed or child-wife] of another man, who has never known a man, and still lives in her father's house, and sleep with her and be surprised, this man shall be put to death, but the wife is blameless.

131 If a man bring a charge against one's wife, but she is not surprised with another man, she must take an oath and then may return to her house.

—*from* The Code of Hammurabi
Translated by L. W. King (1910)

Who was Hammurabi?

The most famous and significant king of Old Babylon, Hammurabi (c.1792–1750 BCE), was an Ammorite, or "Westerner," whose family invaded Sumer sometime after 2000 BCE. Hammurabi conquered several Sumerian and Akkadian cities and founded the empire based in Babylon, raising it from relatively small town to major power center. But Hammurabi is even more renowned for a code of laws that is considered one of the oldest and most significant in human history. Although Hammurabi's code is often cited as the "first" law code, that designation rightfully belongs to the code of an otherwise obscure Sumerian king called Ur-Nammu, who preceded Hammurabi by about three hundred years, according to the leading Sumerian historian Samuel Noah Kramer. But not much of Ur-Nammu's law code is readable, and it exists only in fragmentary pieces. That is why Hammurabi gets so much credit. We have his complete works.

Carved on stone columns discovered in 1902 in the city of Susa (now Shush in Iran), Hammurabi's code is now on display at the Louvre in Paris. The columns show the sun god Shamash handing the code of laws to Hammurabi, laws derived from even older Sumerian codes, including those of Ur-Nammu. Severe by modern standards, with the death penalty prescribed for even relatively minor offenses, the code addressed commonplace issues such as business and family relations, labor, private property, and personal injuries.

While Hammurabi's code was seemingly ruthless when it came to punishment—literally calling for "an eye for an eye"— it still represented a great step from the lawlessness of pre-civilization era. The laws governed everything from traffic regulations on the Euphrates River to the rights of veterans, but also provided protections for the weakest members of society—including women, children, the poor, and slaves. This shift from arbitrary violence and clan vengeance marked a startling step toward civilized norms of justice.

This is one more example of how myth and history sometimes blend. In Egypt, a pharaoh tried to make one mythical god the only god and failed to change the people's minds. Hammurabi was a man who used the gods to give his laws the weight of worship. It was no different from the biblical Ten Commandments, which were said to have been handed to Moses by his one god on Mount Sinai. In a few more centuries, the Greeks would also create new law codes, but they would be the creation of men, not gods—civilization was slowly being born from barbarism, and sometimes myth was the midwife to that long labor.

WHO'S WHO OF MESOPOTAMIAN MYTHS

Just as the Greek gods and goddesses were later borrowed and renamed by the Romans, the chief gods of the early Sumerian myths were adopted by the later Akkadians and Babylonians. As cities grew, new gods were added to the pantheon, weaving a complex web of sometimes competing deities.

This list includes most of the chief deities of Mesopotamia, with their Sumerian names followed by their Akkadian or Babylonian names. As with many other mythologies, there are often variant stories and differing versions of the Mesopotamian gods, reflecting the different people who moved through this part of the world over the course of thousands of years of conquest, and then adapted and reshaped the local myths to suit their needs and political agendas.

An (**Anu**) The Sumerian sky god, originally presided over the assembly of gods. With his mate, the earth goddess **Ki**, he is the father of other gods, including Enki. In the Sumerian view, the stars are his soldiers and the Milky Way his royal road. Originally the source of rain, An is the father figure who makes seeds sprout, but later evolves into the chief god. An has the power to proclaim the Sumerian kings, who were believed to be chosen by the gods when they met in a sort of democratic forum. Sumerian royalty was then supposed to carry out the duties that An and the other gods had determined.

When Sumer was eclipsed by the Babylonian Empire, An was demoted and transformed into a grandfather figure. However, in a particularly violent version of this power shift, An was deposed by Marduk when Babylon became the predominant city. Marduk first destroyed An by flaying him alive, cutting his head off, and tearing his heart out. He then also dispatched An's son Enlil. This violent mythical demise may have carried over into the actual religious practices under the Assyrian king Ashurbanipal (668–627 BCE). During his reign, human sacrifice was apparently relied on to appease the gods, and there are accounts of human flesh being fed "to the dogs, pigs, vultures, eagles—the birds of heaven and fishes of the deep." Excavations of royal tombs in Ur also revealed many bodies other than the king's, suggesting either a mass suicide or human sacrifice, in which wives, concubines, musicians, and entertainers were killed and entombed with the dead king.

Apsu (Abzu) The Sumerian-Akkadian deity who embodies the primordial freshwater ocean; is one of two original gods whose waters surrounded the earth, which floated like an island. First conceived of as the water itself, Apsu later became a male deity and united with his mate Tiamat to create all the other gods and goddesses. In the *Enuma Elish*, he is supplanted of by one of his offspring, Enki, who killed him.

Dumuzi (Tammuz) The god of herders and seasonal fertility; is not only one of the most significant figures in Mesopotamian mythology but was adopted by many later civilizations, including the Greeks and Romans.

As the dying and rising-from-the-dead husband of the love goddess Inanna in a central Mesopotamian myth (see below, *How did an angry goddess make the seasons?*), Dumuzi gets in big trouble with his wife, who banishes him to the underworld. He later escapes, and the myth of his death and resurrection is one of the earliest parallels to the annual cycle of fertility and harvest. Songs lamenting his death were typical of the fertility celebrations in Mesopotamia, and the veneration and worship of Dumuzi-Tammuz carried over to biblical times.

In the prophetic biblical Book of Ezekiel, among the sins committed by the Israelites is weeping for the dying god Tammuz (Ezekiel 8:14). This was considered an abomination by Ezekiel, and one of the reasons that Israel was destined to fall. Another biblical connection to Dumuzi-Tammuz is found in the Old Testament's Song of Solomon (or Song of Songs). A series of erotic poems celebrating the physical love between a man and a woman, the Song of Solomon closely parallels the sacred marriage texts celebrating the union of Inanna and Dumuzi. The male figure—or bridegroom—in the Song of Solomon appears as a shepherd, which was also the role of Dumuzi.

Dumuzi-Tammuz also had the title "lord," which was later translated by the Greeks into the word "Adonis." As the cult worship of Tammuz moved westward into Greece, the title and name were merged, so the Greek Adonis (see chapter 4) is actually based on Dumuzi-Tammuz.

Enki (**Ea**) The most clever of gods, he becomes the god of freshwater by killing Apsu. In the *Enuma Elish,* he is born from the union of the sky god An and Ki, the earth goddess, and is the god who slays Apsu—who is both a god and the actual underground reservoir of freshwater. Enki eventually subdues Apsu, puts him to sleep, and kills him. Having done that, Enki takes his place as chief god, and with his wife Damkina gives birth to Marduk.

Sometimes depicted as half-man, half-fish, Enki is responsible for creating the world order and is the keeper of the *me*—the divine laws, rules, and regulations that govern the universe. Possession of the *me* meant to hold supreme power, and in one tale Inanna (see below) visits Enki, and after getting him drunk, convinces the high god to give her the *me*, which she then takes to her cult city of Uruk.

The source of all secret magical knowledge, Enki is responsible for giving arts and crafts to mankind. He also invents the plow, fills the rivers with fish, and controls the freshwater. The relationship between earth fertility and his virility is evident in the close connection between the Babylonian words for "water" and "semen."

In another story, Enki broght water to the barren isle of Dilmun, which many biblical scholars associate with the real Persian Gulf island of Bahrain, off the coast of Saudi Arabia. After this, Dilmun was transformed into an idyllic paradise where animals did not harm each other and there was no sickness or old age. Many scholars feel that this Mesopotamian paradise might have provided some inspiration for the biblical Garden of Eden, but there are also significant differences between these earthly paradises.

In a somewhat obscure but intriguing story, Enki went on to father a group of goddesses through a series of incestuous unions with his daughters and granddaughters. But when his wife, the mother goddess Ninhursaga, discovered Enki pursuing their daughters, she became angry, and cursed him so sickness attacked eight parts of his body. He was then cured by having sex with Ninhursaga. This myth is believed to be a warning against incestuous rape and unbridled sexuality.

As the water god, Enki also figures prominently in a pair of Mesopotamian flood narratives. (See below, *Who came first, Gilgamesh or Noah?*)

Enlil (**Ellil**) The son of An and brother of Enki, the god of wind and air, and for a time replaces his father as king of the gods and chief god of the Sumerians. As lord of the wind, he can be either destructive or benevolent. In one story, he watches **Ninlil** (see Ninhursaga), the grain goddess, as she bathes, and, unable to resist, rapes her. For this sexual assault, he is banished from his cult city Nippur by the assembly of gods.

Enlil descends to the underworld, but the pregnant Ninlil follows him there so their son can be born in his presence. The child becomes the moon god Nanna, but before they can leave the underworld, they must have other children who can survive there, which would allow Nanna to leave. This story of Enlil "dying" and then returning to earth is another of the earliest examples of the widespread concept of a dying and reborn god—the concept that so transfixed James Frazer in *The Golden Bough*—which is repeated many times in other myths.

Inanna (**Ishtar**) "Lady of Heaven" is the most complex and in many ways influential Mesopotamian deity. The Sumerian goddess of love, sex appeal, and battle, she is significant not only in Sumer but in other, later mythologies. She is described in one text as the one whom not even 120 lovers could exhaust. She is adapted in later myths of Western Asia and reappears in other cultures as *Astarte* (Canaan), *Cybele* (Anatolia), *Aphrodite* (Greece), and *Venus* (Rome).

Patron goddess of the city of Uruk, she is also identified with the planet Venus, the brightest object in the night sky, and the disappearance and reappearance of the planet were explained by Inanna's descent into the underworld in one of the oldest versions of the universal myth of the journey of souls from the land of the living to the land of the dead. (This central Mesopotamia myth of Inanna and her husband Dumuzi is told in detail below. See *How did an angry goddess make the seasons?*)

Inanna also figures prominently in *Gilgamesh*, and as the patron goddess of prostitutes, she is the most significant figure in the annual rite on New Year's Day in which a priestess representing Inanna couples with the real-life king. This ritual was probably enacted by the living king with a temple prostitute who represented the goddess.

Marduk The son of Enki; later emerges as the chief god of Babylon after defeating Tiamat, the she-dragon, in the epic battle described in the *Enuma Elish*.

Known in the Bible as **Merodach** and later as **Bel** (or **Baal**), Marduk essentially became one of the chief adversaries of the Hebrew God. Several actual kings with names related to Marduk (Evil-Merodach, Merodach-baladan) appear in biblical records.

After the conquest of Sumer by the Akkadians, Marduk became the supreme god of Mesopotamia, and his temple at Babylon contained the great ziggurat associated with the biblical Tower of Babel. (See below, *Was the Tower of Babel in Babylon?*)

Nanna (**Sin**) The Sumerian moon god, also called Sin by the Akkadians. He is the firstborn son of Enlil, who had raped Sin's mother, the grain goddess Ninlil. In some traditions, Nanna is the father of Inanna (Ishtar). He is revered as the god who measures time, and because he shines at night, is considered the enemy of wrongdoers and dark forces. Renowned for his wisdom, Nanna is consulted by the other gods when they need advice.

Ninhursaga The Sumerian goddess of bounty, whose name means "lady of the stony ground." As the Mesopotamian earth goddess, she takes several forms. As Ninlil, she is the wife of Enlil. As **Ninki**, she is the wife of Enki and bore his children on Dilmun, the island of paradise, and as **Nintur**, she is worshipped as a midwife. As a fertility goddess, she has power over birth and nourishes the Sumerian kings with her milk, giving them a measure of divinity.

Ninurta The Sumerian war god and patron of the hunt; another son of Enlil and Ninhursaga. Called "lord of the earth," he is god of the thunderstorms and spring floods, and began as a great bird but was later humanized. In one story, nature rises up against Ninurta, but some parts of the natural world, including some stones, take his side. The stones that side with Ninurta became the precious stones.

Tiamat The Babylonian she-dragon of chaos; represents the saltwater ocean, as opposed to the freshwater of Apsu. In the Creation epic *Enuma Elish*, Apsu and Tiamat mingle and give birth to **Lahmu** and **Lahamu**, whose names mean "silt." From them came all the other gods. Tiamat plays a central role in the *Enuma Elish* and the myth of Marduk, who kills her and turns half of her body into the sky and the other half into the earth. But her legacy goes past Mesopotamia's myths.

The word for the "deep" in Hebrew is *tehom*, believed to be a version of "Tiamat," and the conflict between the creator god and chaos, in the form of the sea, plays a role in later Canaanite religious ideas, which also influenced the Israelites. The "myth" of the Hebrew god defeating the chaos monster of the sea appears in several places in the Bible. Exodus 15:1–18, believed to be one of the oldest pieces of literature in the Bible, is a hymn or Song of Moses, which uses the ancient metaphor of the Divine Warrior and his victory over the sea. Psalm 74:13 is a poem that describes the Hebrew God dividing the sea and breaking the heads of the "dragon in the waters," called

Leviathan, before the Creation begins. The depiction of the primordial chaos as a dragon or sea serpent is one of the most universal metaphors in mythology.

Utu (**Shamash**) The benevolent sun god of justice who gives the law code to Hammurabi. The son of **Nanna** and the brother of Inanna, he is thought to cross the heavens by day and traverse the underworld by night, in the same fashion as Egypt's sun god Re. Utu constituted part of a divine triad of sun, moon, and the planet Venus with his father, the moon god Nanna, and his sister, Inanna.

How did an angry goddess make the seasons?

Men: have you ever gotten in trouble when you didn't notice that the wife or girlfriend was gone all day? Was she a little peeved at being overlooked? Then you know what kind of trouble the shepherd king Dumuzi was in when he enjoyed his wife's absence a bit too much.

Like most ancient cultures dependent on agriculture, the Mesopotamians were preoccupied with fertility—both in their lives and their myths. Just as the death of Osiris was the central myth in Egypt and was tied into Egyptian views of the seasonal crop cycle, the story of fertility goddess Inanna and her husband Dumuzi was the focal point of Mesopotamia's view of the world. While the Osiris myth featured feuding brothers, Seth and Osiris, the family dispute in Mesopotamia starts out between sisters.

Inanna goes to visit her sister Ereshkigal in the underworld, where she is queen of the dead. The ambitious Inanna was in a constant quest for greater power, and she coveted her sister's throne. She leaves behind her beloved Dumuzi, the shepherd god and her "honey man." In a celestial striptease, at each of seven gates leading into the underworld, the beautiful and bejeweled Inanna must remove an article of jewelry or clothing, and finally, at the seventh gate, her remaining garment. Naked at last, Inanna stands before her sister, who stares at her with the "eyes of death," and she dies instantly. For three days, Inanna's corpse hangs, rotting on a hook.

Back in the land of the living, with Inanna in the underworld, sex takes a holiday. People stop coupling.

> *No bull mounted a cow, no donkey impregnated a jenny,**
> *No young man impregnated a girl on the street,*
> *The young man sleeps in his private room,*
> *The girl slept in the company of her friends.*

There are competing versions of how Inanna is released. But once restored to life, she must promise to send someone back in her place—which is another common mythic theme. When Inanna returns to Uruk, her cult city, she finds her husband, Dumuzi, sitting on his throne, looking far from mournful. Enraged that her husband did not weep for her, Inanna gives Dumuzi the same withering gaze of death that she had gotten from her sister, and he is taken to the underworld in her place.

Inanna later realizes that she misses her "honey man," and begins to grieve for him. Her laments for her husband become popular Mesopotamian songs, and are the very songs that Ezekiel hears the women of Israel later singing, in what he considers an abomination. Inanna pleads with her sister, and Dumuzi is finally released for half the year, his place in the underworld taken by his compassionate sister, so he can spend half the year with Inanna.

To the Mesopotamians, the disappearance and reappearance of Dumuzi were connected to the seasonal cycles of fertility and crops—just like the story of Persephone in Greek myth, or Amaterasu in Japanese myths. (The later Akkadian version of this myth features Ishtar and Tammuz with slight variations, but the ultimate outcome is the same, and the story was a popular one, well known throughout the Near East.)

Was Inanna's city the first "Sin City"?

"What do women want?" Sigmund Freud famously asked. In one modern pop-anthem, the answer is that girls just want to have fun. And

*A jenny is female donkey.

Inanna may have been one of the first "girl power" goddesses to live that creed. Accounts of life in the cult cities of this hard-living, hard-loving goddess also give us a very different picture of ancient city-life. It wasn't all-work-no-play back then. Mesopotamians, we know, were party people.

Best known as the goddess of sexual love, the Sumerian Inanna (and her Babylonian counterpart, Ishtar) was also a goddess of war, and she enjoyed battle as though it were a dance. Aggressively sexual, she knew few boundaries, and in poems she says, "Who will plow my vulva? Who will plow my high field? Who will plow my wet ground?"

Inanna was also, not surprisingly, the patroness of prostitutes and ale houses. According to historian Gwendolyn Leick, her city, Uruk, one of the oldest in Mesopotamia, was an ancient "Fun City," and Inanna was a beguiling figure who "stands for the erotic potential of city life, which is set apart from the strict social control of the tribal community or the village." Inanna prowled the streets and taverns in search of sexual adventure, and, according to Leick, sex in the streets was not an unusual thing in ancient Mesopotamia—in Inanna's Uruk and perhaps in Babylon. The idea of sex as "immoral" was not widely held in ancient civilizations, including the Egyptian and, to some extent, Greek worlds. In many cultures, sex was viewed as part of the natural order, and was routinely made a part of the fertility rites that were celebrated openly. Many of the restrictive codes about sexual conduct began with the institution of Mosaic Law in Israel, which is one reason why Babylon gained such a reputation as a sinful place in the view of both the Israelites of the Old Testament—who were also held captive in Babylon—and New Testament Christians, who called Babylon the "great whore," although they meant the hated Rome in the time of the Emperor Nero.

In his recent, modern-English version of *Gilgamesh*, translator Stephen Mitchell hints at this atmosphere in the streets of old Mesopotamia:

Every day is a festival in Uruk,
with people singing and dancing in the streets,
musicians playing their lyres and drums,
the lovely priestesses standing before
the temple of Ishtar, chatting and laughing,
flushed with sexual joy, and ready

to serve men's pleasure, in honor of the goddess,
so that even old men are aroused from their beds.
—GILGAMESH

In every city of ancient Sumer, and other Mesopotamian cities, pairs of temples were dedicated to Inanna and her husband Dumuzi. In the annual marriage ceremony on New Year's Day, the king of each city would impersonate Dumuzi, and a priestess—or possibly a cultic prostitute—would portray Inanna in a ritual intended to ensure fertility and prosperity. In some early accounts, this rite was actually a sacrificial one, and the figures representing Dumuzi and Inanna were killed every eight years. Some archaeological finds suggest that the king was killed along with a large group of family members and retainers. But over time that idea undoubtedly proved unpopular with the kings, who were stand-ins for Dumuzi, and the ritual evolved, with the "death" becoming ceremonial. In later times, the sacrifice was performed symbolically, and the king—or his stand-in—was merely struck.

The city of Uruk and Inanna play central roles in the most enduring work of Mesopotamian literature, an epic poem called *Gilgamesh*.

MYTHIC VOICES

As king, Gilgamesh was a tyrant to his people.
He demanded, from an old birthright,
The privilege of sleeping with their brides
Before the husbands were permitted.

—*from* Gilgamesh: A Verse Narrative
translated by Herbert Mason

Who was mythology's first superhero?

Kids today, as always, grow up in a world saturated with superheroes. Comic books, movies, cartoons, and video games all provide a steady diet of Superman, Spider-Man, Hulk, and a host of heroes—some old, some new. With supernatural powers that allow them to defeat evil and danger, these superheroes are also almost always tempered by some flaw, some bit of humanity that hints at the weakness and faults that lie within every mortal.

The first such character in literary history is probably the hero—or antihero—of *Gilgamesh,* which is widely considered the oldest epic poem in world literature. Its central character, a semidivine king named Gilgamesh, who possesses unusual powers and an oversized ego to go with them, is arguably the world's first superhero. Model for many successive flawed heroes, Gilgamesh is the man who seemingly has it all, but sets off on a series of quests, seeking to become more noble, or enlightened—or immortal—in the process.

A powerful king of Uruk, Gilgamesh claims that he is two-thirds god and one-third man. A perfect physical specimen, a skilled athlete and sex machine, Gilgamesh forces the young men of his city to work building the walls of the city and routinely rapes all the young maidens in Uruk, a tradition that continued into European feudal history as the *droit du seigneur* ("the right of the lord"). Worn out by his demands, the people of Uruk pray for help, and the gods fashion a creature named Enkidu—a mythical prototype for Frankenstein, the Golem, and other mythical monsters—to challenge Gilgamesh. Covered in shaggy hair, Enkidu is more beast than man, eating and drinking with the gazelles and cattle.

A young hunter sees Enkidu in the woods and tells his father about this wild man. His father says they must tell King Gilgamesh about him. Instead of going to fight the man-beast, Gilgamesh enlists the aid of Shamhat, a prostitute from the temple of Ishtar, to do the work of taming Enkidu.

Shamhat is no ordinary "streetwalker." In Stephen Mitchell's description in his modern-English translation of *Gilgamesh,* "She is a priestess of Ishtar, the goddess of love, and as a kind of reverse nun, had dedicated her life to what the Babylonians considered the sacred mystery of sexual union. . . . She has become an incarnation of the goddess, and with her own body reenacts the cosmic marriage. . . . She is a vessel for the force that moves the stars."

Shamhat eagerly and provocatively introduces this savage man to the arts of lovemaking. After seven days (!) of fairly nonstop and wild sex, Enkidu is tamed—all that sex has civilized him. As Stephen Mitchell translates the poem, "He knew things now that an animal can't know."

Told that Gilgamesh is sleeping with all the young maidens before they are married, Enkidu is outraged and sets off to challenge the much-reviled king. The two wrestle, but then realize that they are meant to be friends—some authorities suggest that their friendship, like that of Achilles and Patroclus of the *Iliad* and of the biblical David and Jonathan, may be homosexual. They join forces to fight the giant of the pine forest, a fearful creature named Humbaba. With the help of the gods, Gilgamesh and Enkidu kill the forest monster, decapitate him, and put his head on a raft that floats back to the city.

Back home, the freshly bathed and robed Gilgamesh catches the eye of the love goddess Ishtar (Inanna), who wants this hunky hero for a lover. But Gilgamesh is all too aware of the unfortunate fates that have befallen most of Ishtar's other lovers. He turns her down, politely at first, but later calling her an "old fat whore."

It's not nice to call the love goddess names like that. In a "woman scorned" rage, Ishtar demands that her father, Anu, destroy Gilgamesh. So Anu sends Inanna back down to Uruk with the Bull of Heaven. The bull roars, and the earth opens, swallowing hundreds of Uruk's young men. When the divine bull roars a second time, hundreds more fall into the chasm, including Enkidu. But Enkidu grabs the bull by the horns—literally—and tells Gilgamesh to kill it with his sword. Gilgamesh kills the divine Bull of Heaven, and the two friends ride in triumph through the streets of Uruk.

In a series of dream visions, Enkidu foresees his own death, which comes after he falls ill and suffers for twelve days. Distraught over the loss of his friend, and obsessed by his own fear of death, Gilgamesh sets off in search of the secret of immortality. After more adventures, he meets his distant ancestor, Utnapishtim, who, with his wife, had been the only survivor of a great flood, and is now immortal. He reveals to Gilgamesh the secret of a plant that grows underwater and gives eternal life. Gilgamesh finds the magical plant and retrieves it, but sets it down while he bathes. Drawn by its scent, a serpent devours it and is rejuvenated—a mythic explanation for why the serpent sheds its skin.

Gilgamesh realizes that immortality is not to be his, except in posterity through the achievement of the great city walls he has built.

The poem *Gilgamesh* was unknown until it was discovered in the mid-nineteenth century. It was unearthed in the temple library and palace ruins in Nineveh, the ancient capital of the Assyrian Empire, which had taken control of most of Mesopotamia. Based in the northern valley of the Tigris River, the Assyrians were powerful warriors who came to dominate the Mesopotamian area in the ninth century BCE. Their military innovations included mail armor, armored charioteers, and the earliest use of siege warfare. They soon conquered most of the modern Middle East, dominated Babylon, and even subdued Egypt (in 669 BCE). This was the temple of the Assyrian king Ashurbanipal (668–627 BCE), last great king of the Assyrians, who were ultimately defeated by an alliance of their enemies. Nineveh and another great Assyrian city, Nimrud, were destroyed in 612 BCE.

Approximately three thousand lines long, written on twelve tablets—some of them only found as fragments—the poem may have been composed in southern Mesopotamia before 2000 BCE. Fragments of copies found elsewhere in Syria and Turkey seem to show that this text was popular throughout the ancient Middle East, and was probably used by student scribes who were learning to write, just as typing out the phrase "The quick brown fox jumps over a lazy dog" was once used to test typewriters and teach keyboard technique to typists. (Why? The sentence contains every letter in the alphabet!)

Following the discovery of the tablets, in 1872, George Smith delivered a paper before the London Society of Biblical Archaeology which included a partial translation of the cuneiform texts, along with an analysis of several episodes in the *Gilgamesh* epic, especially its flood narrative. The material rocked the world of biblical scholarship, suggesting that the story of Noah and the flood, as recorded in Genesis, might have been "borrowed" from this earlier—and worse, "pagan"—source.

Was the *Gilgamesh* a work of "faction"?

The Gilgamesh story has it all. Sex, love, monsters, whores, friendship, battles, more sex, more battles, the search for immortality—and, finally, disillusioning truth. Although it is filled with Mesopotamian mythology, the Gilgamesh epic may also be based on some misty history.

The notion that myths are based in real events ("euhemerism") is an old idea. But the story of Gilgamesh may be one of the earliest examples of that possibility. Here are a few reasons why:

- The character of Gilgamesh was apparently based on a real king who ruled the city-state of Uruk around 2600 BCE.
- The Sumerian King Lists, discovered among the tablets in Nineveh, record him as the fifth ruler in Uruk's First Dynasty.
- He was known as the builder of the wall of Uruk; his mother was said to be the goddess Ninsun, and his real father was, according to the King Lists, a high priest.

Then again, harsh facts do intrude on the legend. Although the poem credits Gilgamesh with building Uruk's walls, these walls actually predate his lifetime by at least one thousand years, according to archaeological evidence. That means Gilgamesh may have also been the first politician given credit for something he hadn't actually done. Isn't that novel!

Who came first, Gilgamesh or Noah?

Apart from the significance of *Gilgamesh* as one of humanity's first works of literature, one aspect of this epic has caused great controversy since its translation into English: its inclusion of a flood story that is remarkably similar to the biblical story of Noah.

During his adventures, Gilgamesh goes to visit his ancestor Utnapishtim, who possesses the secret of immortality, given to him by the gods. In *Gilgamesh*, the gods are annoyed by the humans and their growing numbers and all the noise they make, so they decide to send a flood to destroy humanity. The water god Enki is forbidden to reveal this plan to humans, but he realizes that if there are no humans, there will be no sacrifices to the gods and no people to do all the work. Enki cleverly reveals the plan of the coming flood to Utnapishtim, and instructs him to build a boat and fill it with the seeds of all living things. After a storm that lasts six days and seven nights, the boat comes to rest on a mountaintop and, like the biblical Noah, Utnapishtim releases birds to see if there is any dry land. When the birds do not return, Utnapishtim knows the floodwaters have receded. While many points between the

Sumerian legend and Hebrew Bible diverge, the close parallels in details seem more than coincidental. The description of the boat and the storm, the coming to rest on the mountains, and the release of birds are all highly similar narrative features.

Complicating matters is the fact that the flood story of Utnapishtim is not the only deluge account in Mesopotamian myth. There are actually two other stories of a great destructive flood. One is an old Sumerian tale about Ziusudra, who is told that the gods plan to destroy all mankind. The details of this old story are unclear, as a complete version has never been found. But it is very similar in feeling to both *Gilgamesh* and another flood tale, about a man called Atrahasis, which is found in the Babylonian Creation myth *Enuma Elish*.

In this story, Enlil was in charge of the minor gods who were digging the Tigris and Euphrates Rivers. When they complained about the work and revolted, the gods decided to create mankind to do the work instead, and Nintur, the goddess of birth, mixed some clay with blood and created man. But when the population of men grew too large, the noise they made kept Enlil awake at night, and he asked the gods to send a plague to thin out mankind. A wise man, Atrahasis—whose name literally means "exceedingly wise"—got wind of this divine plan and consulted Enki, the Sumerian water god, who was a bit of a trickster and a friend of mankind.

Enki told the people to keep quiet and make offerings to the plague god to avert the disaster. But as time went by, Enlil again wanted to destroy noisy mankind, this time with a drought. Again, disaster was averted by Enki's intervention. When the people's noise disturbed him a third time, Enlil ordered an embargo on land bounty, but Enki saved mankind from starvation by filling the canals and rivers he controlled with fish.

Finally Enlil decided to send a great flood, and this time Enki advised Atrahasis to build a boat and take his family and animals on board. All mankind was destroyed, except Atrahasis and his family, who survived and repopulated the earth.

So, did the biblical authors "sample" these ancient Mesopotamian stories, conveniently borrowing them, perhaps while they were in captivity in Babylon? Or were these just common stories that were "floating around" the ancient Near East? This question has troubled scholars and archaeologists since it was first raised in 1872. Of course, biblical purists completely reject that notion, holding that Noah's story, like the rest of the Bible, is the divinely inspired word of God. But the many parallels are too striking to ignore.

Whether the Hebrew story is borrowed or original, the existence of so many flood stories around the world raises a larger question: was there ever one cataclysmic flood in earth's history that would explain these many myths?

Among the most intriguing new insights into the old flood question have come from the research done by Columbia University geologists William Ryan and Walter C. Pitman 3d, authors of *Noah's Flood: the New Scientific Discoveries About the Event That Changed History* (1998), and the man who discovered the *Titanic*, deep-sea explorer Robert Ballard. The two scientists theorized that sometime around 5600 BCE, there was a major inundation of the Black Sea—then a freshwater lake—by water rushing in from the Mediterranean through the present Bosporus Straits. In 2000, Ballard, using his famous underwater equipment, bolstered their argument with the discovery of remains of wooden houses beneath the Black Sea near the Turkish shore. Their theory, simply stated, is that this cataclysmic event destroyed everything for some sixty thousand square miles, killing tens of thousands of people. This ancient deluge then provided the historical memory for all of the flood narratives that later emerged, including those of Mesopotamia, ancient Israel, and Greece.

It would not necessarily explain flood accounts of many other civilizations. To that, writer Ian Wilson has a six-word answer in his recent book *Before the Flood*: "The end of the Ice Age." The earth has experienced numerous ice ages in its 4-billion-year-plus history. The last of these occurred about sixteen thousand years ago—well within the time scale that modern humans have been around. Arguing that the sudden and cataclysmic rises in sea levels from melting ice would have struck many human populations clustered along seashores, Wilson argues, "It stands to reason that these events must have been responsible for at least some of the Flood stories that are commonplace in the folk memories of so many people around the world."

Beyond the obvious interest in explaining the biblical story, the Ryan-Pitman thesis—enthusiastically endorsed by Ian Wilson—is that a widespread antediluvian civilization once existed in and around the Black Sea. The survivors of the Black Sea inundation then spread out, taking civilization with them. It is an audacious idea, which would essentially require rewriting archaeological—and human history—textbooks. But recent archaeological work done in Turkey, at Catalhoyuk, and more recently in the northeast corner of Syria near Turkey's Taurus Mountains, have provided some support to this revision in long-held ideas of how and where civilization developed.

Was the Tower of Babel in Babylon?

History, myth, and biblical traditions all come crashing together in the largely familiar account of the Tower of Babel, a story which appears in Genesis 11. In this tale, men have come from the east and settled in Shinar—a biblical place understood to be the kingdom of Sumer in ancient Mesopotamia. The men all speak the same language and decide to build a city and a tower to "make a name for ourselves." But when God comes down and sees this activity, in which men are making bricks to build a tower with its "top in the heavens," He is not happy. Men are making their way heavenward. Threatened and annoyed, God decides to confuse their speech so that the tower-builders cannot understand one another. After the construction of the tower is thrown into chaos, God scatters these people all over the face of the earth. The biblical account has traditionally been viewed as an explanatory myth that accounts for the world's many different languages and the spread of different nations.

But in exploring life and myth in the "cradle of civilization," it should be noted that many more languages were spoken over the centuries in Mesopotamia than elsewhere, due to the successive waves of people who conquered or moved through the area, including the Sumerians, Akkadians, Hittites, and later, Persians. The conquerors of Mesopotamia all spoke different languages, so Babylon's "multicultural" history, and Israel's place in that history, need to be taken into account when considering the Bible story and its historical background.

The first great Babylonian civilization had flourished between 1800 and 1600 BCE, under Hammurabi and other kings. The Babylonians later fell to the Kassites, who ruled Babylon from the sixteenth to the twelfth century, in what is called the **Kassite Period**. Speaking a little-known language, they came from the Caucasus region, the mountainous area between the Black and Caspian Seas (an area that today includes Georgia, Armenia, and Azerbaijan, as well as Chechnya, the restive and very troubled region of Russia).

The Assyrian Empire, based in northern Mesopotamia, took control of Babylon during the 700s BCE, but the city resisted Assyrian rule, and King Sennacherib of Assyria destroyed Babylon in 689 BCE. The New Babylonian, or **Chaldean, Empire** began in 626 BCE, when the Babylonian military leader Nabopolassar became king of Babylon. Nabopolassar won control of Babylon from the Assyrians, and under his reign, the Chaldean Empire grew to control much of what is now the Middle East. Nabopolassar and his son Nebuchadrezzar II rebuilt the city on a grand scale. During the reign of Nebuchadrezzar, from 605 to 562 BCE, workers built huge walls almost 85 feet (26 meters) thick around the outside of Babylon, and people entered and left the city through eight bronze gates. The grandest of these was the Ishtar Gate, decorated with figures of mythical dragons, lions, and bulls made of colored, glazed brick. The Ishtar Gate opened onto a broad paved avenue connecting the Temple of Marduk inside the walls and the site of the great New Year festival. Nebuchadrezzar's main palace stood between the Ishtar Gate and the Euphrates River, an area that may have included the famed Hanging Gardens. The ancient Greeks described these gardens, which grew on the roof of a high building, as one of the Seven Wonders of the World. The Temple of Marduk stood in an area to the south and included the famed ziggurat tower.

The function of the ziggurat has inspired several theories, including the idea that these towers may have originated as burial mounds—just as the Egyptian pyramids started with kingly tombs—from which the god-king Marduk could be resurrected. Another idea is that each of these towers served as a symbolic "sacred mountain," typical of many mythologies, as home of the gods. Pointing out that Mesopotamia was flat and had no natural high places—such as the Greek's Mount Olympus—to serve as home of the gods, Daniel Boorstin argues that in many cultures, "where there were no natural mountains, people built artificial mountains. . . . 'Ziggurat' means both the summit of a mountain and man-made stepped tower."

But Boorstin's interpretation is not shared by everyone, and there are disagreements over the rationale for the ziggurats. Historian Gwendolyn Leick argues that there is nothing in Mesopotamian literature to substantiate that the ziggurat was meant to imitate or evoke a natural mountain. Instead, she writes, "In areas prone to flooding this was a practical device, and the towering sanctuaries must have been reassuring sights as high and therefore safe places, not necessarily to keep the people safe, but to protect the gods, upon whose benevolence all life depended."

By all accounts, the grandest of these ziggurats was the temple complex in the city of Babylon that may have been first built around 1900 BCE, then expanded by Nabopolassar and continued by his son Nebuchadrezzar, a project that took forty-three years. Designed to signify the triumph of Babylon over its enemies, Nebuchadrezzar's ziggurat was clearly awesome, involving the production of at least 17 million bricks. Many historians and archaeologists agree that this was the same tower described in Genesis.

After Nebuchadrezzar conquered Jerusalem, he took Judah's king as a captive to Babylon, and, in 586 BCE, destroyed Jerusalem's Great Temple. Thousands of Jews, among the nation's elite, were taken into captivity in Babylon, one of the most significant events in the history of Israel and the development of the Old Testament or Hebrew Bible. During this time, many of the books of Hebrew scripture were first written down.

In the bustling capital city, the captive Israelites would have heard many languages—with hints of ancient Sumerian, Akkadian, Babylonian, Egyptian, and Persian filling the air of the ancient bazaars. Clearly, the story of the Tower of Babel had great significance to the exiled Israelites, because it provided an interesting play on the name of the city of Babylon. In Babylonian, the city's name means "the gate of the gods," but in Hebrew, the word for Babylon is related to a word meaning "to confuse." The author of the biblical Tower of Babel story was essentially using a bilingual pun, a typical Hebrew literary device, to disparage the people who had captured the Israelites and held them captive.

Finally, the Tower of Babel reflects a classic story line, in which the gods become annoyed when people get a little too full of themselves. The theme that the gods—or God—don't want competition from mankind is a common one in myths, and usually it does not end up well for humanity. Mankind overreaching—whether by building high towers, trying to fly, or stealing fire—has been a mythic concept opposed not only by the God of the Hebrew Bible but by gods in other mythologies. However, it just may be part of human nature to strive for the heavens, whether that means building towers in the desert, erecting skyscrapers in the city, or sending rockets into space.

Mythic Voices

Now the Lord said unto Abram, Get thee out of thy country, and from thy kindred and from thy father's house, unto a land that I will shew thee. And I will make of thee a great nation, and I will bless thee, and make thy name great; and thou shalt be a blessing. And I will bless them that bless thee, and curse him that curseth thee: and in thee shall all families of the earth be blessed.

—Genesis 12:1–3 (King James Version)

Was the Bible's Abraham a man—or another Mesopotamian myth?

Mesopotamia was a land in which myths and men mixed. Gilgamesh was both real and a myth. Marduk was a myth, but one with great impact on biblical—and therefore on Western—history. But one of Mesopotamia's most famous men ever is a mysterious figure whose very existence is an open question.

The biblical patriarch Abram—his name was later changed to Abraham, Hebrew for "father of a multitude"—is one of the most revered men in Judaism, Christianity, and Islam. What generations of believers may not know about this man, hailed as the "father of all nations," is that he was from ancient Mesopotamia. According to Genesis, Abraham came from the city of Ur in Sumer, and was born in a direct line descended from Adam. According to Genesis, he then received a divine message to go to Canaan, a land that God promised to Abraham and his descendants. He dutifully obeyed God's every command and was richly rewarded with many children and great herds.

But there is no specific proof outside the Bible or Koran that such a person existed. His name and exploits appear nowhere in Mesopotamia's surviving tablets. While some scholars maintain that there was an actual Abraham, it is generally believed that he was a legendary figure. Supporters of the position that Abraham truly existed say that certain aspects of his life and travels fit within the framework of Mesopotamian history. References to many of the specific customs mentioned in the biblical story, including the idea that a man could have a legitimate heir conceived by his wife's servant—as Abraham did—buttress their position.

Of course, to believers, the "historical" Abraham doesn't matter as much as what he represents—a pioneer of faith. That faith is underscored in a crucial biblical event heavy with mythic overtones—the story of Abraham's willingness to sacrifice his son. By agreeing to sacrifice his beloved son Isaac at God's request, Abraham passes what is viewed as the supreme test of individual faith.

The aborted sacrifice of Isaac and the substitution of an animal in his place is, in the view of most scholars, the symbolic moment in which the ancient Jewish people rejected human sacrifice. In many of the cults and religions of the ancient Near East, including Mesopotamia, human sacrifice was practiced. In most of these cultures, the victims were ritually killed in a way that was meant to appease the gods, often to assure the fertility of the crops. In Mesopotamia, the whole purpose of human existence was to provide the gods with the necessities of life.

But back to the essential question: Did Abraham of Mesopotamia exist? Or was he the invention of Hebrew writers who created a "foundation myth" to justify Israel's presence in the lands it eventually conquered? Doubts linger. And perhaps—barring some remarkable archaeological discovery—it is an unanswerable question. Given the historical background of Mesopotamian life and society, many of the details of Abraham's story certainly have the historical ring of truth. Chances are that a man named Abraham—or an ancient Semitic version of that name—may well have existed and, like Gilgamesh or King Arthur, he was turned into a national legend over the centuries.

The story of Abraham ultimately stands as one more example of how one man's myth is another man's faith.

MYTHIC VOICES

Then the Israelites did what was evil in the sight of the Lord and worshipped the Baals, and they abandoned the Lord, the God of their ancestors, who brought them out of the land of Egypt, they followed other gods from among the gods of the people who were all around them, and bowed down to them and they provoked the Lord to anger. They abandoned the Lord, and worshipped Baal and the Astartes.

—*Book of Judges 2:11–13*

Who were El and Baal?

Any discussion of the myths of Mesopotamia would not be complete without discussing Canaan, the "Promised Land" that Abraham and his descendants had been granted by God. Located as a sort of land bridge connecting the great empires of Mesopotamia and Egypt, Canaan was both a battleground and a bustling bazaar that felt the influence of the great empires around it. Set in this crossroads of the ancient Near East, Canaan gave rise to a body of myths that drew heavily on the earlier Mesopotamian legends, and the Canaanite stories violently clashed with the beliefs and ideas of the people of Israel.

The people of what was Canaan—today comprising Syria, Lebanon, Jordan, Israel, and Palestine—were Semites, but over time, Canaan became a true Middle Eastern melting pot. Canaanites, and groups called Edomites and Moabites, settled the area and were later joined by the Philistines, who may have migrated from the Mediterranean islands of Crete and Cyprus. (The contemporary word "Palestine" is derived from the word "Philistine.") Another group that moved to the region was the Phoenicians, who had been based in such Mediterranean coastal cities as Tyre, Sidon, and Byblos. Extraordinary sailors and dyers of cloth, they also get credit for devising the alphabet adopted by the later Greeks, which influenced Western writing.

With these many groups and influences merging and mingling, Canaan was a land of many gods and cults, but one group of Canaanite gods was most widely worshipped. Its chief deity was known as **El**—which means "god" in ancient Semitic languages. The supreme god El, often depicted as a man wearing bull's horns, was creator of the universe. Benevolent and all-knowing, El was a somewhat remote god. His consort was Asherah, a goddess who is related to Ishtar (Inanna).

At some point in this region's history, the Canaanite El was merged with the one god of the Hebrews, called **Yahweh**, who gave Moses the Ten Commandments. The significance of the divine name El is clear from the Hebrew words that include it, such as Beth-El, which means "house of God." This was the name given to the spot where the biblical Jacob dreamed of a stairway to heaven—often interpreted as a ziggurat. Jacob took the stone on which his head was lying when he dreamed, stood it upright, and anointed it with oil. Piles of standing stones were traditionally constructed in Canaanite fertility cults. In a later scene in Genesis, Jacob wrestled with a mysterious stranger—also a typical mythical theme—who proves to be El himself. After this wrestling match Jacob's name was changed to Isra-el, interpreted as "he strives with God." He became the father of twelve sons, each of whom headed one of the twelve tribes of Israel. Again, the influence of "pagan" mythology on biblical faith is no small matter.

While the Canaanite El would be merged into the Jewish ideal of one god, Canaan's other chief deity did not fare so well. Canaanite mythology was especially concerned with fertility, and to the Canaanites, **Baal** was the most significant abundance god. Baal (which means "lord"), whose name is connected to the Mesopotamian Bel (another name for Marduk), would go down in biblical history as a figure of supreme evil.

In a story that mirrored the victory of Marduk over Tiamat, Baal defeats Yam, another dragonlike sea god. With that victory, Baal—like his Babylonian counterpart Marduk—assumed the role of chief god. When his archenemy, **Mot**, lord of the underworld, invited Baal to come to the underworld, Baal accepted the invitation, hoping to overcome Mot and take control of the underworld, too. But when Mot forced Baal to eat mud, considered the food of the dead, Baal died. With the death of the god of plenty, all of the crops on earth died as well.

While El and the other gods in heaven mourned the dead Baal, Baal's wife **Anat** (the Canaanite counterpart to Inanna-Ishtar), descended to the underworld, and killed Mot with a sickle. Anat then burned him and ground him, treating the god of the dead like harvested wheat. With Mot's death, Baal was revived, life returned to earth, and the crops grew once more. But Mot returned to life and fought with Baal until the latter agreed to return to the underworld for a few months of the year, establishing the mythical reason for the season change.

Canaanite religion centered on worship of Baal, who was also responsible for the rain. According to their beliefs, the rains came when Baal had sex, with his semen falling in the form of life-giving rain. Most likely, Canaanite rituals included priests having sex, apparently coupling with women, men, and even animals, if the Hebrew accounts are to be believed. Many of the Mosaic Laws of the Old Testament included laws against incest, bestiality, transvestitism, temple prostitution, and idol worship, all of which may have been typical elements of Canaanite worship. The famous scene in *The Ten Commandments* in which the Jews melt their jewelry, which is then turned into a golden calf, was a reference to Baal worship—and, by extension, to Marduk, who was also called the "bull of heaven."

Many of the Old Testament accounts of Jewish history continue the theme of two contending beliefs—the one god of Judaism against the many false gods of the Canaanites. The final insult paid to Baal by the Hebrews was a pun that changed his name and connected it to another familiar biblical word, Beelzebub. In an Old Testament story, an evil Jewish king, who is sick, requests help from the god he calls "Baalzebub." This is a wordplay on the Canaanite name meaning "Lord Baal," because in Hebrew, it was mockingly translated as "lord of the dung" or "lord of the flies." By New Testament times, Beelzebub became associated with the name of Satan.

What's a Canaanite demoness doing at a rock concert?

The third intriguing figure from Canaanite myth is a character who entered the American pop-culture pantheon in the 1990s through a series of female-oriented summer concerts called the Lilith Fest. The name Lilith comes from the Canaanite storm demon Lilitu, with perhaps deeper connections to the Mesopotamian goddess Ninlil. She has a most intriguing side story—Lilith was once thought to be Adam's first wife, the predecessor to Eve.

There is no mention of Lilith in Genesis—and the only biblical reference to her at all is a single mention in the Book of Isaiah. But in folklore and in the Talmud—which is a collection of ancient commentaries on the Bible made by Jewish rabbis over many centuries—Lilith has a rich history.

The issue of Adam's first wife arises because there are two Creation stories in Genesis. In the first, which describes the six-day Creation in poetic terms, men and women are created simultaneously, both in God's image. The second version tells the longer, more folkloric story set in the Garden of Eden. In this version, man is formed first and woman only later, out of the man's rib. Troubled by this discrepancy or contradiction in the two biblical accounts, early Jewish commentators suggested that the wife in Genesis 1 was a different woman from Eve. Some of these early biblical scholars proposed that she was Lilith.

In this old folk version told outside the Bible, God made Lilith, like Adam, out of clay, but he used unclean earth. When Adam and Lilith had sex, Lilith balked at being on the bottom all the time. Since she thought that they had been created equally, she wanted to be on top and made the mistake of uttering the unspeakable name of God. For this crime, Lilith was sent away and turned into a demon who haunted men in their sleep, causing the nocturnal emissions, which drained their fertility for ordinary women. She was also thought to cause barrenness and create miscarriages, and frighten babies in their sleep—perhaps a mythic explanation for the modern term "crib death" (or Sudden Infant Death Syndrome).

With the sexually adventurous Lilith gone, the Jewish folklore went on, Adam was lonely, and God created Eve, a more docile woman. There is even the suggestion in some accounts that it was Lilith who came into the Garden of Eden in the guise of the serpent and tempted Eve to taste the fruit of the Tree of Knowledge.

What are three Persian magicians doing in Bethlehem on Christmas?

Perhaps the most beloved of Christian holidays is Christmas, the day celebrating the birth of Jesus, the divine son of God, in Christian belief. It is also a holiday loaded with pagan trappings, hung, like too many ornaments, on the Christmas tree of lore and legend. One vestige of this pagan past is the familiar Nativity tale of "Three Wise Men"—also called "Three Kings"—from the East, who come to honor the newborn Christ child, lying in a manger in Bethlehem, with three famous gifts: gold, frankincense, and myrrh. The Gospel of Matthew describes these visitors as "magi," translated from the Greek as "wise men," who follow a miraculous star to Bethlehem. Though their number and names are never specified in the biblical account, "three wise men" are presumed because there are three gifts. Only in medieval times were they given the names Melchior, Balthasar, and Gaspar.

So who were these "wise men"?

Magi were the hereditary members of a Persian priesthood, known for interpreting omens and dreams, for their astrological skills, and for practicing magic ("magi" is where the word "magician" comes from). This vast knowledge of rituals gained them the reputation as the true priests of **Zoroastrianism**, a religion founded by a Persian prophet named Zoroaster. Little is known of Zoroaster (the more widely used Greek for the Persian name Zarathustra), except that Zoroastrian tradition places him as living around 600 BCE. But many scholars believe that he lived between 1400 and 1000 BCE.

Zoroastrianism holds a belief in one god, **Ahura Mazda**, who created all things. Zoroaster also taught that the earth is a battleground where a great struggle is taking place between the spirits of good and evil. Ahura Mazda calls upon everyone to fight in this struggle, and each person will be judged at death on how well he or she fought. It is not known whether the magi, whose practices predate Zoroaster, influenced the prophet Zoroaster, or if they became his followers. But the magi became part of Zoroastrian belief and were said to keep watch upon a "Mount of the Lord" until a great star appeared that would signal the coming of a savior. In other words, centuries before Jesus was born, a Middle Eastern religion flourished with one god, a battle between good and evil, a judgment day, and resurrection.

The connection between ancient Persia and Christmas doesn't end there. **Mithra** was an ancient sun god of the Aryan tribes who settled in ancient Persia. According to Zoroastrian traditions, Mithra was said to be an ally of the supreme god Ahura Mazda, and under Ahura Mazda's leadership, Mithra and other gods fought against the Zoroastrian god of evil.

The Persians spread the worship of Mithra throughout Asia Minor during the period when they dominated Mesopotamia and the Middle East, from about 539 BCE, under King Cyrus, until they were defeated by the Greeks in two wars fought between 490 and 480 BCE. The Persian Empire later fell to Alexander the Great in 330 BCE. Mithraism survived the Persian Empire's fall and eventually became popular in Rome, especially among Roman soldiers. Shrines to Mithra often showed the god slaying a bull, a rite that symbolized the renewal of creation. This ritual supposedly bestowed immortality on Mithra's worshippers, one reason it appealed to soldiers facing combat.

During Roman times, Mithraism ranked as a principal religion competing with Christianity, until the 300s CE. Among its several similarities to Christianity—including a resurrection, judgment day, a Satan-like figure, and guardian spirits much like angels—was a holy day celebrated in Rome on December 25. In 350 CE, Pope Julius I chose this day as the official date of the celebration of Jesus' birth. (Other connections between Roman pagan traditions and Christmas can be found in chapter 4, *What were the Bacchanalia and the Saturnalia?*)*

*The spread of Mithraism under the Roman Empire extended into Spain, and some authorities suggest that the killing of a bull was part of a Mithraist ritual that gradually evolved into the practice of bullfighting, whose conventions were formalized in Spain in the 1700s, and the famed "running of the bulls" each spring in Pamplona. However, others view the *corrida de toros* as a vestige of much older traditions of bull worship and sacrifice as evidenced in many cultures, including those in Crete, Anatolia, Mesopotamia, ancient India, and ancient Israel. These connections include the sacred Apis Bull of Egypt, the famed bull riders of Knossos discussed in chapter 4, and the hero Gilgamesh, whose conquest of the Bull of Heaven was discussed earlier in this chapter. In either case, men have been taking the bull by the horns, literally, for a very long time.

CHAPTER FOUR

THE GREEK MIRACLE

The Myths of Greece and Rome

Having the fewest wants, I am nearest to the gods.

—SOCRATES

Prayer indeed is good, but while calling on the gods a man should himself lend a hand.

—HIPPOCRATES

Not one of them who took up in his youth with this opinion that there are no gods ever continued until old age faithful to his conviction.

—PLATO

Whatever it is, I fear Greeks even when they bring gifts.

—VIRGIL, *Aeneid*

That is the miracle of the Greek mythology—a humanized world, men freed from the paralyzing fear of an omnipotent Unknown. The terrifying incomprehensibilities which were worshipped elsewhere and the fearsome spirits with which earth, air and sea swarmed, were banned from Greece.

—EDITH HAMILTON, *Mythology*

Where did the Greeks get their myths?

Was Greece ever a theocracy?

Who kept the "family tree" of the gods in ancient Greece?

How do you get Creation from castration?

Who's Who of the Olympians

How did man get fire?

What was in Pandora's "box"?

Why does Zeus send a great flood to destroy man?

Which mythical monster has the worst "bad hair day"?

What kind of hero kills his wife and children?

Which great hero gets "fleeced"?

Which Argonaut was a god of healing?

Was Hippocrates a man or myth?

Was Atlantis ever discussed in Greek myth?

Is Theseus and the Minotaur just another "bull" story?

What was the Delphic Oracle?

Do all little boys want to kill their father and sleep with their mother?

Is Homer just a guy from *The Simpsons*?

How did Homer fit a ten-year war into a poem?

Is the *Iliad* all there is to go on when it comes to the Trojan War?

Was there really a Trojan War?

Which crafty Greek hero can't wait to get home?

Did the Romans take all their myths from the Greeks?

Who were Romulus and Remus?

Was Homer on the Romans' reading list?

What were the Bacchanalia and the Saturnalia?

MYTHICAL MILESTONES

Greece and Rome

Before the Common Era (BCE)

c. 3000 The early Minoan civilization is established on the island of Crete.

c. 2000 Greek-speaking Indo-European peoples begin to migrate into the Aegean Sea area.

Palace of Knossos is built on Crete.

Egyptian-influenced hieroglyphic script used on Crete.

1900 Potter's wheel introduced to Crete.

1750 Linear A, an early form of script, used on Crete.

1600–1400 Height of **Minoan** civilization on Crete.

1628 Volcanic eruption on the island of Thera (modern Santorini).

c. 1600 Rise of **Mycenaean** civilization on Greek mainland.

1450–1400 Fall of Minoan civilization on Crete after invasions and volcanic disasters; Mycenaeans take control of Crete.

1400 Mycenaean civilizations dominate the Greek mainland.

Mycenaeans adapt Linear B script.

c.1280–1184 Trojan War with Mycenaean Greeks.

1150–1100 Collapse of Mycenaean dominance.

Possible Dorian invasions from the North.

1100–800 Beginning of the so-called **Dark Ages** in Greece.

c. 1000 Worship of Zeus grows at Olympia.

c. 900–800 **Archaic Age** begins; growth of the Greek city-states, or poleis—independent cities ruled by a variety of governments.

900–700 Early books of Hebrew Bible composed.

776 First documented Olympic games are held at Olympia.

753 Traditional date for the founding of Rome by Romulus.

750–700 Homer's *Iliad* and *Odyssey* are first written down.

750 Greek colonization of Mediterranean spreads.

First evidence of use of Greek alphabet.

700 Hesiod's *Theogony*, *Works and Days* composed.

621 Draco and the first written law code in Athens.

600 Thales of Miletus; birth of philosophy (Ionian School).

First Greek coins used in Lydia.

594 In Athens, Solon is given extraordinary powers; he reforms government, establishes rules for public recital of Homeric poems.

580 Sappho and the flowering of Greek lyric poetry.

570 Anximander develops systematic cosmology.

525 Pythagoras begins philosophical-religious brotherhood; develops mathematical, scientific, and mystical ideas.

520 Xenophanes, philosopher-poet, develops ideas of human progress, philosophical monotheism, skepticism toward deities.

509 Foundation of Roman Republic.

508 Democratic reforms instituted in Athens.

490 First Persian invasion of Greece; Greeks defeat the Persians at the Battle of Marathon.

480 Second Persian invasion, led by Xerxes. The Persians win at Thermopylae; Athens is sacked; the Persians are defeated in the naval Battle of Salamis; Persian troops withdraw after loss at Plataea in 479.

480–336 The **Classical Age**—the culminating years of Greek achievement.

476 Massive new Temple of Zeus built at Olympia to celebrate Greek freedom, combined with Olympic Games. The temple, including a massive statue of Zeus, is completed c. 420. The temple and statue are among the Seven Wonders of the Ancient World.

460–430 The **Golden Age** of Pericles in Athens.

The three great tragedians flourish: **Aeschylus** (525–456), Sophocles (496–406), and Euripides (485–406).

447 In Athens, work begins on the Parthenon. Completed in 432, the temple, dedicated to the goddess Athena, stands on a hill called the Acropolis, overlooking the city. It is the crowning achievement of the Golden Age under Pericles.

431 The Peloponnesian Wars commence, with Athens and Sparta as main rivals.

430 Great plague strikes Athens; Pericles dies in 429.

404 Athens surrenders; a period of Spartan domination; oligarchy returns to Athens.

399 Suicide of Socrates, accused of corrupting the youth of Athens.

385 Plato returns to Athens to open his academy; writes *The Symposium*.

364 Fighting between rival cities during Olympic Games, a traditional time of truce.

359 Plato's *Republic* completed.

338 Macedonia, led by Philip, takes control of Greece, ending independence of city-states.

336 Philip of Macedonia is assassinated.

Alexander the Great, Philip's son, begins his conquests, extending Greek rule from the Mediterranean to the Himalayas.

335 **Aristotle** (384–322) founds the Lyceum.

332 Zeno founds the Stoic school of philosophy—based on the idea that virtue is the only good.

323 Death of Alexander in Babylon after a drunken feast. His empire is broken up into kingdoms controlled by Greek generals, such as the Ptolemies, who rule Egypt as Greek-speaking pharaohs.

229 First Roman incursion into Greece.

146 Romans defeat Greek rebellion. In the city of Corinth, all men are killed and women and children sold as slaves.

80 The Roman general Sulla pillages Olympia during the civil wars fought in Greece.

31 Battle of Actium off the west coast of Greece: Octavian, the future Emperor Augustus, defeats Mark Antony and Cleopatra, ending Rome's civil wars.

27 Octavian proclaims himself "first citizen," and takes the name Augustus. Begins a new era of *Pax Romana*, and Greek culture spreads throughout Roman Empire.

Common Era (CE)

312 Emperor Constantine's Christian vision before Battle of Milvian Bridge.

313 Constantine's Edict of Milan permits Christianity.

337 Constantine is baptized on his deathbed.

330 Byzantium becomes capital of the Roman world and is renamed Constantinople (modern-day Istanbul, Turkey).

393 Last official Olympic Games.

394 Theodosius II, the Christian Roman emperor, bans all pagan festivals, including the Olympics. The statue of Zeus at Olympia is carted to Constantinople, where it is later destroyed in a fire.

426 The Temple of Zeus is burned on the orders of Theodosius II; Christian fanatics destroy the rest of the sanctuary at Olympia.

If your introduction to Greek history and classic mythology came from watching the opening-night ceremonies of the 2004 Olympics in Athens, you may be understandably befuddled. The panoramic view of these myths and the achievements of one of history's most remarkable civilizations unfolded in a sort of Macy's Thanksgiving Day Parade merged with a "toga party." In a badly costumed pageant of gods and warriors, centuries of myth and some of the most significant moments in Western history rolled by on floats. Above it all floated the figure of Eros, god of love, suspended like an awkward Peter Pan wishing he were somewhere else.

Those opening Olympic moments may be blissfully forgotten. But set against the magnificent remains of Athens' Golden Age, the 2004 Olympic games did provide a stunning reminder of an extraordinary moment in world history—a breathtaking glimpse of the glory that had been Greece. Those marble ruins in downtown Athens and the modern Olympic games are vestiges of that glory. The remnants of the Parthenon pulse with what was once the heart of Athens, a magnificent temple celebrating its patron goddess, Athena. And while they barely resemble their ancient predecessors, the modern Olympics are a bloated version of the competitions once dedicated to Zeus, king of the gods of Mount Olympus. The ancient world's longest-running show, the original Olympics—which, like the modern Super Bowl, attracted thousands of spectators for several days of sports, drinking parties, and whoring— took place every four years for more than twelve hundred years. And you thought *Cats* had a long run!

Yet the pinnacle of Greek civilization that produced those exquisite ruins on that hill in Athens was but a passing moment in human history some 2,500 years ago. It was a relatively brief episode in the march of humanity, but one that changed everything. Ancient Greece had a profound, unique, and lasting impact on the Western world. Like it or not, Western civilization was born in Greece.

As Greek-born writer Nicholas Gage described it in *The Greek Miracle*, "In the fifth century before Christ, an unprecedented idea rose from a small Greek city on the dusty plains of Attica and exploded over the Western Hemisphere like the birth of a new sun. Its light has warmed and illuminated us ever since. . . . The vision—the classical Greek idea—was that society functions best if all citizens are equal and free to shape their lives and share in running their state: in a word, democracy. . . . The concept of individual freedom is now so much a part of our spiritual and intellectual heritage that it is hard to realize exactly how radical an idea it was. No society before the Greeks had thought that equality and freedom of the individual could lead to anything but disaster."

The Greek—or more precisely, Athenian—concepts of government by the people, trial by jury, and the first real notion of human equality (limited, to be sure; women and slaves, for the most part, didn't count) mark the true beginnings of the Western democratic tradition. What we call science and the humanities—including biology, geometry, astronomy, history, physics, philosophy, and theology—were also essentially invented by the Greeks. In the spoken and written arts, these ancient people introduced and perfected epic and lyric poetry, as well as tragic drama. In their art and architecture, the Greeks created an ideal of beauty that has dominated the Western world. And these ideals of beauty were reflected in the mythology they created.

"The world of Greek mythology was not a place of terror for the human spirit. It is true that the gods were disconcertingly incalculable. One could never tell where Zeus's thunderbolt would strike. Nevertheless, the whole divine company . . . were entrancingly beautiful with a human beauty, and nothing humanly beautiful is really terrifying."

These are the words of Edith Hamilton, perhaps the greatest promoter of Greek myth in our lifetime. Largely due to generations of students having had her book *Mythology* on their school reading lists, many people instinctively think of Greek myths when they hear the word "myths." Hamilton's 1942 introduction to these classic stories provided the standard for a long time.

In those two words—"human beauty"—Edith Hamilton may have best summarized what we consider the Greek ideal. But Hamilton's romanticized notion of Greek myth, as well as the traditional vision of Greek culture and history, have undergone serious revision of late. Recent scholarship has shone a light on some other aspects of the classical world. And Edith Hamilton's worshipful tone ignores some nasty realities. The Greek gods of Olympus may have been "entrancingly beautiful," as Hamilton wrote. But their stories were filled with as much cruelty, violence, incest, adultery, sibling rivalry, and venality as any of the earlier myths of Egypt or Mesopotamia—myths which the Greeks clearly borrowed and then revised to suit their needs.

The vision of idealized Greek marble figures, perfectly painted urns, and beautiful gods and goddesses delighting in cups of ambrosia is only part of the picture. It overshadows a more complex view of Greece, in which war and conquest, human sacrifice and slavery all played a part. This dark underbelly of the Greek past shows up in its myths, just as it does in the great Greek literature that emerged from them.

There is, however, another reality reflected in the story of Greece and its myths. More than in any other ancient culture, the mythic tradition in Greece is a grand story in which, in the words of Voltaire, "men passed from barbarism to civilization." The Greek poets and playwrights reshaped and recast the brooding, violent ancient stories of feuding, spiteful gods and flawed heroes into the poetic epics and drama of an emerging social order that profoundly influenced Western civilization. The Greek myths permeated the *Iliad* and the *Odyssey* of Homer, which form the core of the Western literary tradition. Also based largely in the Greek mythic traditions were the great Greek dramas, highlighted by the three playwrights of the Athenian Golden Age —Aeschylus, Sophocles, and Euripides, whose works have influenced writers for more than 2,500 years and are still staged around the world.

Before the "Greek Miracle"—as this extraordinary period of cultural and social ferment is called—sculpture in such places as Egypt and Mesopotamia mostly showed stiff, unapproachable gods and kings, often on a monumental scale. But in the hands of Greek sculptors, the divine became human, for example, in the form of a discus thrower perfectly frozen in action. On pottery and vases, Greek painters depicted not only the gods and heroes, but ordinary women of delicacy and beauty serving food and drinks. (They also made an art of obscenely painted drinking cups, of a sort not usually displayed in modern museums, portraying the popular wine-and-sex parties called symposia. But that's another story.)

Greek architects created a classical sense of scale and beauty still considered the standard for great and important buildings. These poets, playwrights, and sculptors transformed Greek arts, and in doing so, changed the basic view of humanity, elevating the human form to the nearly divine.

At the same time, their philosophers and early scientists, rooted in the same ancient ideas of gods and religion, pushed the envelope of what human reason could discern. In this Greece, humanity was no longer helplessly trapped in a world in which people existed to serve the gods. At this unique moment in human history, the gods were glorified. But the Greeks also realized that, as the philosopher Protagoras put it, "Man is the measure of all things."

That was the glory of Greece.

Where did the Greeks get their myths?

Two of the most famous goddesses in Greek myth make their debuts on the mythical stage as fully formed and perfect adults—one usually naked, and the other in battle armor. Aphrodite—you know, the one on the half-shell—is the goddess of love, and she emerges full-blown from the sea, au naturel but with strategically placed long locks, in one of the most famous artistic depictions of her birth. Athena, the goddess of wisdom and war, is born in full battle regalia, emerging from the head of her father Zeus when another god hits him on the head with an ax.

People seem to have the same idea about Greek myths—that somehow, they were created full-blown, in just the form we know them today, crafted from the genius of some anonymous poet or philosopher. But ancient myths, as their history in Egypt and Mesopotamia prove, aren't that simple. Over the course of centuries, myths are invented, told, and then start to travel. As they make the rounds, they are borrowed, reshaped, and retold—often to fit a very local agenda. Like old wine in new bottles, or reality shows that originate in England and get picked up by American networks, ancient myths sometimes resurfaced with a different name and a changed face. It was no different in Greece, where the origins of the myths—and the religion they spawned—serve as a fascinating reflection of Greek history.

Recent discoveries from the worlds of archaeology and literature make it clear that what evolved into Greek mythology was a mélange—like some fusion cuisine—of some existing local stories with bits and pieces borrowed from other Near Eastern civilizations, including the Egyptians, Mesopotamians, and the Phoenicians—from whom the Greeks also appropriated a writing system, later adapted as the Greek alphabet. Forced by their geography to turn to the sea, the Greeks had early on mastered trade and travel. As they ventured out around the Mediterranean ports of call, they encountered these other ancient neighbors. Eventually they brought back not souvenirs but samples of these foreign cultures and religions to what was then Greece—the hilly, rocky, northern mainland jutting out into the Mediterranean, Aegean, and Ionian Seas; a southern peninsula called the Peloponnesus; the many islands that dotted the surrounding waters; and the west coast of Asia Minor (now Turkey).

These "foreign influences" found their way into a Greece that was already a mythological melting pot. Beginning around 2000 BCE, waves of conquering invaders had swept into Greece and merged the stories of their gods with those that were already established on the Greek mainland. These bloody invasions go back to a time long before the brief Golden Age of Greece that so many students falsely equate with Greek history. The story of Greece actually plays out over a much longer span, and its civilization and mythology can be separated into five distinct periods.

The earliest known civilization to flourish in what came to be thought of as Greece was not actually Greek but a sophisticated and rather extraordinary culture called Minoan. Based on the Mediterranean island of Crete, the Minoan Period may have begun as early as 3000 BCE—around the same time as Mesopotamia and Egypt—and then suddenly and somewhat mysteriously disappeared from history around 1400 BCE. In the early twentieth century, the long-abandoned capital of Crete's early civilization was discovered by English archaeologist Sir Arthur Evans in one of the most dramatic finds in history. At Knossos (or Cnossus), Evans uncovered the remains of a huge, luxurious, and graceful palace, whose walkways were paved with cobblestones. The palace was complete with ceramic bathtubs and fully functioning, flushable indoor plumbing serviced by an elaborate system of drains. Its walls were decorated with brightly painted frescoes depicting handsome, naked young men and women somersaulting over the backs of bulls, an ancient Mediterranean "rodeo" that was clearly tied into the worship of an elaborate bull cult, a vestige of the Minoans' origins in Asia Minor. The palace may have also provided the source of one of the most significant myths in Greece, the story of the Minotaur, a fearsome half-man, half-bull that demanded human sacrifices.

Although the Minoan written language, Linear A, has not yet been fully deciphered, we know it was most likely used—as the early cuneiform was in Mesopotamia—for keeping track of trade and commercial accounts. The Minoans were among the first seagoing traders, and their ships sailed to Egypt to do business in the land of the pharaohs. Most likely the Minoan deities included a sea god whom the Greeks later called Poseidon, and an earth goddess who later became the Greek goddess Rhea.

The Minoan Age flourished until about 1400 BCE, when it more or less disappeared from history, perhaps partially crippled by a devastating volcanic eruption nearby, or conquered by new arrivals, known as the Mycenaeans. These Aryan, or Indo-European, warlords had swept into mainland Greece about five hundred years earlier, presumably coming from the steppes of the Caucasus Mountains (between the Black and Caspian Seas). A warrior race, the Mycenaeans rolled over the existing inhabitants of the Greek mainland—whose own origins are similarly mysterious—and began to fuse their own stories and beliefs with those of the people they conquered, as well as those of the Minoans they encountered on Crete. This era is called the Mycenaean Age, after Mycenae, one of the most significant cities of the period—first excavated by the famed Troy-discoverer Heinrich Schliemann in the late nineteenth century. The Mycenaean Age lasted from 1600 to about 1110 BCE, and is generally considered to be the period in which the small Greek kingdoms and the events described by Homer in the *Iliad* may have taken place. Apparently, the "Mycenaeans" may have called themselves "Achaeans," one of the names used by Homer to describe the men who attacked Ilium—or Troy. Most scholars place the destruction of Troy around 1230 BCE, but there is considerable disagreement on that date—others argue for a later destruction, around 1180 BCE.

More widely accepted is the idea that these war-loving, chariot-driving Mycenaeans were responsible for what the modern business world might call a "hostile takeover." When they came crashing into the Greek mainland, they apparently brought with them a set of their own, very old gods, such as the sky father, Zeus; the Earth Mother, Demeter; and Hestia, the virgin goddess of the hearth. The local farmers they encountered and subdued on mainland Greece probably worshipped an ancient Earth Mother, who became Hera. And the very stormy marriage of Zeus, the conquerors' sky god, and Hera, the fertility goddess of the conquered locals, may actually symbolize the merger of these two ancient mythologies. The concentration of power in such cities as Mycenae, Tiryns, and Thebes, all of which are featured prominently in Greek myths, is another clue that many of the Greek myths and legends may have originated in their familiar form during this Mycenaean Period.

Mycenae and most other settlements on the Greek mainland were destroyed sometime after 1200 BCE, ushering in a Greek Dark Age, the third major period in Greek history, which lasted until about 800 BCE. Historians do not know why Mycenaean Greece fell into chaos. Perhaps climate change led to famine. Suspicion also falls on the invasion of another group, Greek-speakers called Dorians, from northern Greece, who moved south into the region, forcing many Mycenaeans to flee to Asia Minor. One reason the lights went out during this Dark Age was that somehow Greek knowledge of writing (which used a form called Linear B adapted from the Minoans) was lost, and the Greeks only began to write again sometime after 800 BCE.

That is about the time that someone familiar with Phoenician writing invented the Greek alphabet. Phoenician writing only had signs for consonants; some clever but anonymous Greek added indications for vowel sounds. For the first time—experts generally agree—writing could approximate the sound of the human voice (and that system is the basis for the writing you are now reading). With that development, the two great epics, the *Iliad* and the *Odyssey*, were presumably written down for the first time sometime after 800 BCE, along with the works of a Greek poet named Hesiod, who conveniently catalogued the history and exploits of the gods.

The Dark Ages gave way to a fourth historical period, called the Archaic Age (800–490 BCE), with the emergence of the written Greek language, the return of those who had moved away, and the spreading of colonies in the west—in Southern Italy and Sicily. This period marked the beginnings of the polis, or Greek-city states, which would usher in the greatest developments in Greek history. Established as centers of trade and religion, each city-state was surrounded by walls to protect it against invasion. Within the city, there was usually a fortified hill—an acropolis—and at the center of each city was the agora—an open area that served as a market area and city center.

Finally, Greece flowered spectacularly in the Classical Period (490–323 BCE). This Golden Age was centered in Athens and had its earliest flowering with the Athenian lawmaker Solon's democratic reforms in 594 BCE. It continued to grow over the next few centuries, bursting into full bloom in the Greece many of us think of when we think of the ancient world. A key moment came with the defeat of the Greeks' great foreign rivals, the Persians, in a series of wars fought between 490 and 479 BCE. Shared religion, language, and culture played a central role in Greek life, and served the Greeks well when the Persian Empire threatened. The usually fiercely independent city-states joined under Athenian leadership to defeat two separate Persian invasions in one of the most fascinating turning points in Western history.* These wars were the central subject of Herodotus's *Histories*, in which he proudly wrote, "This proved, if there were need of proof, how noble a thing is freedom." Freedom is a good thing, but so is heavy, bronze armor—helmet, shield, and breastplate. Which is what was worn by the hoplites, the citizen-soldiers who were the Greek city-states' version of the National Guard and who fought in tight, well-organized formations—another key to the victory over the lighter-armored Persians.

With the victory over Persia by a united Greece, Athens emerged as Greek's leading city and reached its pinnacle. Over the next century and a half, the great philosophers Socrates, Plato, and Aristotle walked the streets of Athens, the agora, or marketplace, and established their schools—the model for the university—in which the ideas that form the basis of Western philosophy were discussed and debated. This was also the time of voting-rights experiments and the flourishing of the great trio of playwrights—Aeschylus, Sophocles, and Euripides—whose tragedies were performed in front of tens of thousands of Athenians in a dramatic

*In the first Persian invasion of Greece, in 490 BCE, the Greeks defeated the Persians at the Battle of Marathon, inspiration for the race of the same name. The second Persian invasion, led by Xerxes, came in 480 BCE. The Persians won at Thermopylae, and Athens was sacked. But the Persians were then defeated in the naval Battle of Salamis; Persian troops withdrew after their loss at Plataea in 479.

competition that had its roots in a religious festival honoring the agricultural god, Dionysus, who was also credited with the invention of wine.

In all of these periods, myths played a central role in Greek life and society; they were at the core of religious observances and entertainment. Along with language and a common culture, myths provided a bond that no central Greek government ever could. But clearly, sometime around 800 BCE, as the so-called Dark Ages began to give way, something changed. A switch was thrown. And from then on, some Greeks began to abandon the notion that the gods controlled the universe. It was perhaps a singular moment in human history. Before this tipping point, most other ancient civilizations viewed life as the work of the gods, who needed to be served and worshipped, and their divine representatives on earth, kings and pharaohs—who also demanded to be served and worshipped.

Suddenly, in Greece, the fundamental understanding of the universe and man's place in it was transformed through a revolution in thinking. A range of Greek thinkers began to search for natural explanations—a humanistic mind-quest to discover a rational system of creation in which order was not dependent on sacrificing animals to the gods and invoking magical oracles.

Of course, not everyone liked those notions, which challenged the status quo. That was one reason that the philosopher Socrates would eventually be placed on trial and sentenced to death in 399 BCE. His concepts were actually not so much antireligious as they were threatening the Athenian powers-that-be. But there was no turning back the sweeping tide of change. An unstoppable series of ideas had been set in motion, and history would never be the same.

Mythic Voices

The continual buzz of conversation, the orotund sounds of the orators, the shrill shouts from the symposia—this steady drumbeat of opinion, controversy, and conflict could everywhere be heard. The agora (marketplace) was not just a daily display of fish and farm goods; it was an everyday market of ideas, the place citizens used as if it were their daily

newspaper, complete with salacious headlines, breaking news, columns, and editorials.

—THOMAS CAHILL, Sailing the Wine-Dark Sea

Was Greece ever a theocracy?

Unlike Egypt and Mesopotamia, Greece never produced a heavy-handed, monolithic central government ruled by divine kings. Even in Greece's earliest times, there is no evidence that Minoan or, later, Greek kings ruled with the sort of divine sanction that Egyptian and Mesopotamian rulers claimed—and Rome's emperors would later attempt to claim. Over the course of its history, Greek civilization had developed chiefly in small city-states, like Athens and Sparta, consisting of a city or town and the surrounding villages and farmland. Small, fiercely independent, and often quarrelsome, these city-states were strongly patriotic, and full of citizens—free and male, that is—who made that great leap into participating proudly in public affairs, both as a ready army of hoplites, and then as participants in the decision-making. The ancient Greek city-states were never united as a "nation" in the modern sense of the word. More to the point, they were never "nations under gods," or theocracies.

Even so, a shared "public" religion played a central role in Greek life. The cult practices and deities among the different Greek communities had enough in common to be seen as one system. Herodotus later characterized this shared religion as "Greekness," by which he referred to common temples and rituals. Chief among these rituals were various forms of sacrifice. (There is some evidence of human sacrifice in prehistoric Greece and on Crete, but the practice disappeared.) And a typical ritual was the "libation"—the pouring of water, wine, olive oil, milk, or honey—in honor of the gods, heroes, or the dead, usually before a meal.

With temples dedicated to the favorite patron god or goddess in every city-state, the Greeks believed that certain deities actively watched over them and directed daily events. Both priests and priestesses served in the temples to perform rituals. Families tried to please household deities with gifts and ceremonies that included animal sacrifice and offerings of food. Like Athens, which was protected by its namesake, Athena, each city-state honored one or more deities as the patron deity of the community, and held annual festivals in their honor.

Large crowds also gathered in ancient Greece for religious festivals that included feasts, colorful processions, and choral performances, which evolved into the first Greek drama. In Athens, for instance, there was a great civic festival called the Panathenea in late summer, during which there were sacrifices and a large procession by groups representing the different segments of Athenian society.* Every four years there was "greater Panathenea," which included major athletic and musical competitions open to all Greece. Athletic festivals were also popular, and the Olympic games, the most famous of these festivals, involved all of the Greek city-states. Held in honor of Zeus, the first recorded Olympics took place in 776 BCE and continued every four years for more than a thousand years. Even wars were usually—but not always—halted during the Olympic festival.

The Greeks also believed that their deities could help them foresee the future, and people flocked to shrines called oracles to consult seers, both male and female, who played a central role in the lives of Greeks, whether highborn or common. One of these was Dodona, a sacred site where priests interpreted the sounds of the wind blowing through the leaves of a sacred oak tree. The most important shrine in ancient Greece was Delphi, home of the sacred oracle of the god Apollo, site of the omphalos, or sacred "navel stone," believed to be the center of the world. A conical stone thought to be part of the Greek Creation myth (see below, *How do you get Creation from castration?*), the navel stone was the mystical connection to the navel of Mother Earth. The Greeks, like other ancient civilizations, were also devoted to the possibility of magic, and sacred objects, such as amulets and household idols, and spells and other magical rituals were all part of everyday Greek life.

But as classicist Barry Powell points out, "The Greek gods had per-

*Again, not such a "foreign" or primitive idea. Just think of a small-town, Main Street Memorial Day parade or a St. Patrick's Day parade in New York, in which every political party, civic group, association, and union usually marches together. And men burning sacrificial meat on open fires? That's called a "tailgate party" or a July 4th backyard barbecue. Just modern vestiges of an ancient rite.

sonalities like those of humans and struggled with one another for position and power. They did not love humans (although some had favorites) and did not ask to be loved by them. They did not impose codes of behavior."

Clearly that was starkly different from the state religions of Egypt, Mesopotamia, and Israel, where long traditions established the nearly unbreakable connection between worship, personal conduct, and the politics of the state. The Greeks had their public and private rituals, and they were important, but no Greek ruler ever tried to elevate himself to pharaoh or introduce a single deity, as Akhenaten had done in Egypt. Nor were the Greek gods believed to be in complete control of the universe or human destiny. As historian Charles Freeman noted about the Greeks, "They never pretended that their gods were always benevolent or omnipotent in human affairs, and so bad fortune could be rationalized as a natural element of existence."

And the concept of a "natural element" would soon be viewed by many Greeks as far more important—and interesting. Judging from their extraordinary achievements across the many disciplines—literature, art, science, mathematics, philosophy—these ancient Greeks clearly came to prize a way of life that stressed the importance of the individual, encouraged creative thought, and elevated the power of observation. Greek thinkers laid the foundations of science and philosophy by seeking logical explanations for what happened in the natural world. They could see that the eclipse of the moon wasn't the capricious act of a god, but the shadow of the earth, and from that they discerned that earth was round. And although the classical Greeks had Homer and Hesiod and the great works of the tragedians, it is also important to note that they had no Bible and no *Enuma Elish* to dictate their lives and behavior, no strict concepts like the Egyptian *maat* or the Mesopotamian *me* rigidly ordering their existence. To them, the gods were not the source of truth, justice, and laws. Quite to the contrary, writes Barry Powell, "The Greeks invented ethics, a way to tell right from wrong without divine authority, and secular law, which together make up humanism."

MYTHIC VOICES

Rhea, surrendering to Kronos, bore resplendent children. . . . The others great Kronos swallowed, as each of them reached their mother's knees from her holy womb. His purpose was that none but he of the Lordly Celestials should have the royal station among the immortals. For he learned from earth and starry Heaven that it was fated for him to be defeated by his own child, powerful though he was, through the designs of great Zeus. So he kept no blind man's watch, but observed and swallowed his own children. Rhea suffered terrible grief.

—HESIOD, Theogony

A man fashions ill for himself who fashions ill for another, and the ill design is most ill for the designer.

God and men disapprove of that man who lives without working. . . . You should embrace work-tasks in their due order, so that your granaries may be full of substance in its season.

If your spirit in your breast yearns for riches, do as follows, and work, work upon work.

—HESIOD, Works and Days

Who kept the "family tree" of the gods in ancient Greece?

Homer gets most of the glory, but Hesiod did the heavy lifting. When it comes to understanding the origin and genealogies of the gods, and some of the most familiar stories in Greek myth, we have a Greek shepherd to thank. Hesiod's books haven't been optioned by Hollywood like Homer's, but they are among the best sources for many of the ancient tales of the Greek gods.

Much of what is known about the Greek myths is derived from two principal sources: Homer's two epic poems, *Iliad* and *Odyssey*, and two far less famous books of poetry called *Theogony* (from *theo*, the Greek word for "god") and *Works and Days*. These last two were supposedly inspired by the mythical Muses, who appeared to a shepherd named Hesiod, a farmer from a region northwest of Athens called Boeotia. The Muses "breathed a sacred voice" into Hesiod's mouth, and he began to describe the creation of the world, and the succession of heavenly rulers who made up the complex genealogy of the Greek gods.

Though he probably lived around 700 BCE, shortly after or around the same time as Homer, Hesiod is far less famous and accomplished a poet than Homer is. But we do know a bit more about him, because his writings actually include some autobiographical clues. His father had been a merchant sailor, and after living in Cyme, on the coast of Asia Minor (Turkey), had moved back to mainland Greece and started a farm in a time of growing prosperity in Greece. The family's estates were small, and when Hesiod and his brother, Perses, inherited them, the brothers apparently quarreled over their shares. It also seems apparent that Hesiod was a somewhat cranky country gentleman, and no fan of women—as best represented in his telling of the familiar story of Pandora, the first woman (see below, *What was in Pandora's "box"?*). The ills of the world, in Hesiod's words, are all due to a female, created by the gods to torment men—"a calamity for men who live by bread."

By this time in history, the Greeks had borrowed and adapted the Phoenician writing system. As the scholar and translator M. L. West writes, "The existence of writing now made it possible for poems to be recorded and preserved in a more or less fixed form. Hesiod and Homer were among the first to take advantage of this possibility, and that is why . . . they stand at the beginning of Greek literature."

After the Muses appeared to Hesiod on the sacred Mount Helicon and presented him with a staff, he was told to sing of the gods and became a poet, or a man who entertained at private gatherings and feasts, an ancient "wedding singer," reciting the familiar old stories and songs as well as composing them. *Theogony* was the first result of this "divine" inspiration. Relatively brief, compared to Homer's two major epics, the poem contains the names of more than three hundred gods, some of them obscure and insignificant in the Greek pantheon.

Theogony also tells of the birth of the first gods, their stormy family relationships, the story of Prometheus, and ends with the marriage of Zeus and Hera, king and queen of the Greek gods.

Works and Days, Hesiod's even more popular work, was a poem addressed to his brother, Perses, in which he examined human life and set forth his moral values. Also fairly brief, it is nonetheless a rambling compendium of myths, moral philosophy, proverbial wisdom, and practical advice that makes Hesiod sound like an ancient Greek Ben Franklin, offering *Farmer's Almanac*–style advice on cultivating crops, what should be sown, and when the harvest should take place. But *Works and Days* also expresses Hesiod's philosophy that life is difficult and people must work hard in spite of the just rule of Zeus, the king of the gods.

It may be that this advice was aimed directly at his brother, Perses, who got the larger share of the family farm, apparently after bribing some local officials Hesiod called "bribe swallowers." But Perses was not, apparently, sufficiently industrious and Hesiod constantly upbraided him for his laziness. Perhaps more curious, some of his advice went so far as to explain the proper way to relieve oneself:

> Do not urinate standing towards the sun; and after sunset and until sunrise, bear in mind, do not urinate either on the road or off the road walking, nor uncovered: the night belongs to the blessed ones. The godly man of sense does it squatting, or going to the wall of the courtyard enclosure. . . . And never urinate in the waters of rivers that flow to the sea, or at springs—avoid this strictly—nor void your vapours in them; that is not advisable."
>
> —*Works and Days*, M. L. West, translator

Not exactly what we have in mind when we think about the glories of Greece.

But Hesiod's two collections offer a treasure trove nonetheless, both in understanding the early stories of the gods and as a valuable source of insight into common life in the Greek world of the Archaic Age.

MYTHIC VOICES

Great Heaven came bringing on the night, and desirous of love, stretched out in every direction. His son reached out from the ambush with his left hand; and with his right he took the huge sickle with its long row of sharp teeth and quickly cut off his father's genitals, and flung them behind him to fly where they might. They were not released from his hand to no effect, for all the drops of blood that flew off were received by Earth. . . . As for the genitals, just as he first cut them off with his instrument of adamant [a hard stone] and threw them from the land into the surging sea, even so they were carried on the waves for a long time. About them a white foam grew from the immortal flesh, and in it a girl formed.

—HESIOD, Theogony

How do you get Creation from castration?

The Egyptians managed Creation out of masturbation. The Mesopotamians envisioned freshwater and salt water having sex. The Greeks go that one better and base their chief Creation account on a rather painful story of the violent castration of a god.

The most important Greek Creation myth, an elaborate account of the violent birth of the gods, is found in Hesiod's *Theogony*. It is a story filled with crude and bizarre twists, acts of outright brutality, and—as in the other Near Eastern myths—feuding families that span generations. Translator and scholar M. L. West even argues that this "succession myth" was not the "product of Hesiod's savage fancy," but a Greek version of earlier texts, including the Babylonian *Enuma Elish*.

Whatever its mythic origins, this Greek Creation story centered on primordial forces awakening out of Nothingness and bringing alive a succession of gods, giants, monsters, and finally, the seemingly divine figures who all possess suspiciously human failings.

The Creation begins in a state of emptiness called Chaos—literally, "a yawning (or gaping) void"—out of which the five original "elements" simply appear and are then personified as the first gods:

- Gaia (also Ge or Gaea), the primordial earth goddess
- Tartarus, both a god and the bleak, deepest region of the underworld located within the earth
- Eros, the force of love, later transformed into a god of love, who, in Hesiod's words, "overcomes the reason and purpose in the breasts of all gods and all men"
- Erebus, the realm of darkness associated with bleak Tartarus
- Nyx, the female personification of night

Bursting with powerful life-force, the primal goddess Gaia, or Earth, is "broad-breasted, the secure foundation of all forever" as she gives birth to Uranus, the "star-studded heaven" and the divine personification of the sky. Free from the taboo of incest, as were other ancient gods, Uranus becomes his mother's consort and "beds" her. The notion that sky and earth were once beings united in a sexual embrace is a common ancient idea, as in the Egyptian tale of earth god Geb and sky goddess Nut, or the Sumerian deities An and Ki.

Fertile Gaia next bears the mountains, the seas, and the nymphs, who were associated with the trees, springs, rivers, and forests. Gaia and Uranus then produce a terrible trio of sons called Hecatonchires ("the hundred-handed"), monsters who each have three heads. In the next of their curious litters are three more children known as the one-eyed Cyclopes.*

Gaia and Uranus also gave birth to a dozen children known as the Titans, the first generation of gods who preceded the later gods of Olympus. Of monstrous size and strength, they provide the source of the word "titanic." They were:

- Oceanus, a sea god whose waters encircled the earth, and his sister/mate Tethys
- Hyperion, sometimes called the sun, and his mate Theia (who together produce the sun, moon, and dawn)
- Themis (called Law) and Rhea, two more earth goddesses
- Mnemosyne, the goddess of memory

*These are not the same as the more famous literary Cyclops who appears in Homer's *Odyssey*, a one-eyed giant named Polyphemus, a son of the sea god Poseidon, who is discussed on page 250.

- Iapetus, Coeus, Crius, and Phoebe, four Titans with no specific roles
- Cronus, the youngest and craftiest, described as the "crooked schemer"

Siring so many extraordinary children was quite an achievement, but Uranus wasn't happy with his brood. He feared that these children might rise up and overthrow him—a common theme in Greek and other Near Eastern myths. So, Uranus made an interesting decision—perhaps the result of some deep, dark male-fantasy impulse—to lock himself in perpetual intercourse with Gaia so that nothing could emerge from their union. Pressed down upon Gaia, Uranus kept all of these children trapped in a cave within the earth's huge body.

Resentful and in pain, Gaia wants the children to "do in" dear old dad. But only Cronus, the youngest Titan, has the right stuff. Gaia gives Cronus a sickle with which to attack Uranus in a moment of treacherous surprise that may have left the men in Hesiod's audience feeling a bit uncomfortable: "From ambush Cronus' left hand seized the genital parts of his father; he reached out his right with the sickle, saw-toothed, deadly and sharp. Like a reaper, he sliced away the genitals of his own father."

As Hesiod tells it, these severed genitals were then carried out to the ocean, where sea foam magically mixed with Uranus's blood and semen, to create Aphrodite, goddess of sexual love, who emerged from the sea. (There is another, later and different, version of Aphrodite's birth.)

Having emasculated his father, Cronus frees his Titan siblings from their cave inside Gaia and becomes king of the gods. (Again, this parallels the Mesopotamian Creation myth, in which the primordial sea god Apsu had been overthrown by one of his offspring, Enki.) During Cronus's reign, the work of creating the world continued and hundreds more divinities were born, including more Titans, such as Atlas and Prometheus, the gods or goddesses of death, the rainbow, the rivers, and sleep—their names meticulously catalogued by Hesiod. And as we read them, we can only imagine the singer—perhaps accompanying himself on a lyre—crooning these names at a wedding feast, in celebration of the glorious divinities.

The eager crowd is now primed. The stage is set for the entrance of some of the most pivotal and familiar figures in Greek myth—the Olympians—some of whom will descend from Cronus. Knowing how he had deposed his own father, Cronus fears that the offspring of his marriage to this sister Rhea may do the same, so he swallows his first five children as soon as their mother delivers them. To save her sixth child, Rhea tricks Cronus into swallowing a stone wrapped in baby clothes, and then hides the infant in a cave on the island of Crete, where he is raised by nymphs on goat's milk and honey.* Fearful that Cronus might hear the infant's crying, Rhea orders a group of semidivine men to dance around noisily at the entrance to the cave in which he is hidden. That child, saved by Rhea, is Zeus.

The most powerful of all the Greek deities, Zeus will rise to lord over the pantheon of Greek gods. But first he must prove himself worthy. His trials begin when he returns to challenge his father's supremacy and rescue his siblings—an instant replay of Cronus's battle with his own father, Uranus. With Rhea's aid, he first tricks Cronus into drinking a liquid that makes him vomit up all five children, plus Rhea's stone. Zeus then frees the fearsome Cyclopes, still trapped within the earth, and they make magical weapons for Zeus and his two brothers, including the great three-pronged spear, or trident, for Poseidon; a helmet of invisibility for Hades; and the thunderbolts that becomes Zeus's awesome weapon and symbol of power. Zeus also frees the fearsome Hecatonchires from the depths of Tartarus, where they have been imprisoned. Though Gaia, the Mother Earth, urges the Titans to accept Zeus as the supreme god, most of them refuse, and an epic ten-year war—the Titanomachy—follows. Ultimately, Zeus and his siblings, along with their allies, prevail over the Titans, who are exiled to the depths of Tartarus.

Of the defeated Titans, only the one named Atlas receives a different fate. He is condemned by Zeus to live at the edge of the world, where he must hold up the heavens and continue the separation of sky and earth for all eternity. (The Atlas Mountains in Morocco near the Atlantic Ocean are supposedly where Atlas is forced to stand. And when a mapmaker created a collection of the maps of the known world in 1570 CE, he called it an "atlas," in his honor.)

*In Greek myths, the cornucopia, or "horn of plenty," was one of the horns of Amalthaea, the goat who nursed Zeus. The horn produced ambrosia and nectar, the food and drink of the gods. But in Roman stories, the cornucopia was the horn of a river god, which Hercules broke off. Water nymphs filled the horn with flowers and fruit and offered it to Copia, the goddess of plenty.

But Zeus's work is not done. Before he can fully assert his rule, Zeus must also defeat a race of Giants—born from the blood spilled by Uranus's castration. With the help of the half-human, half-god Heracles (Hercules to the Romans), Zeus and the gods defeat the Giants, who, according to legend, were then buried under volcanoes in various parts of Greece and Italy. When Greeks later unearthed the bones of prehistoric animals, they believed they had found remains of the Giants.

Finally, Zeus defeats Typhon (or Typhoeus), a monster with one hundred dragon heads, fire-blazing eyes, and many voices, using thunderbolts to blast him down to the Tartarus region, where he remained the source of hurricanes. (The word "typhoon" is actually a blending of this Greek name, adapted later in Arabic, with the Chinese words *toi fung*, for "big wind.") With Titans, Giants, and monsters all reduced to notches on his godly belt, Zeus is chosen to be ruler by the other gods and goddesses, who agree to live with him on Mount Olympus. The highest mountain in Greece, Olympus rises 9,570 feet (2,917 meters) in northern Greece, and divides the region of Thessaly from Macedonia. The summit is usually covered with snow and hidden in clouds, adding to its mystery as the traditional home of the gods. (The first recorded climb to the summit was not made until 1913.)

The origins of Hesiod's Greek Creation tale have been a source of debate. Historically, it is believed to be part of a much older oral tradition before Hesiod set it down in poetic form. However, symbolically, it has been thought that the story of Zeus gaining supremacy was an allegory of the gradual ascent of male power over female, with the warlike male Zeus supplanting a more primitive earth goddess. Part of this suggestion was that the Mycenaeans brought their macho mythology with them when they invaded Greece and replaced the kinder, gentler goddess worship of the Minoans. But many scholars believe that these Greek stories are actually rooted in other ancient Near Eastern myths, such as the Babylonian *Enuma Elish* or the myths of the Indo-European Hittites who ruled central Anatolia (now Turkey) and were in place before the Mycenaeans arrived. Either way, the fact remains that the Greek myths, like many other mythic systems, were drawn from earlier sources and beliefs, but were ultimately crafted into their own unique, classic Greek form.

WHO'S WHO OF THE OLYMPIANS

This list of the central gods of Mount Olympus shows their Greek name, followed in parentheses by the name used by the Romans, who later adapted much of Greek mythology as their own. Traditionally, twelve gods are called Olympians, but that list was not always the same, as some gods became more or less important at different times in Greek history. (The gods listed with a bullet are the twelve who appeared on a frieze on the Parthenon, the great temple to Athena in Athens.)

While Hesiod and Homer provided much of the earliest written source for Greek mythology, other later playwrights and poets added immensely to the traditions and stories of the gods. The famous trio of Greek tragedians—Aeschylus, Euripides, and Sophocles—took the old tales and transformed them into dramatic works of timeless power. During the Roman era, Roman poets took the Greek traditions and added new layers of complexity. Chief among these sources is Ovid (43 BCE–17 CE?), best known for his witty and sophisticated love poems, including *The Art of Love*, which is a verse "how-to" manual on finding and keeping a lover. More significant from the perspective of mythology was the *Metamorphoses*, which Ovid believed was his greatest work. In this narrative poem, largely filled with stories of mythical and magical "transformations," Ovid moves from the creation of the world to his own time. The poem describes the adventures and love affairs of deities and heroes, with more than two hundred tales taken from Greek and Roman legends and myths.

The Roman poet Virgil added a Roman dimension to Greek myth by connecting the fall of Troy to the foundation of Rome (see below, *Was Homer on the Romans' reading list?*). Other important sources for these Greek (and, later, Roman) myths were later Roman poets and playwrights, and an Alexandrian Greek Apollodorus, who collected many of the myths in his library (usually dated to the first and second centuries CE).

In addition to these classical literary sources, recent archaeological and linguistics studies have greatly contributed to our understanding of the "historical" origins of these deities.

•**Aphrodite (Venus)** Goddess of love and beauty, "Golden Aphrodite" supposedly emerges fully formed from the sea foam when the genitals of the castrated Uranus are cut off by Cronus and thrown into the sea. Her birth is a popular theme in art, and is perhaps most famously depicted in Sandro Botticelli's Italian Renaissance masterpiece *The Birth of Venus*, in which she is seen standing on a half-shell. That was Hesiod's version of the story. In Homeric versions, she is born of the union between Zeus and a goddess named Dione, again reflecting different regional traditions.

An ancient goddess who embodies overpowering sexuality and reproduction capability, Aphrodite may have been connected to other ancient Eastern fertility goddesses of Mesopotamia and Canaan, such as Inanna, Ishtar, and Astarte. In many ancient cities, Greek girls about to be married made a sacrifice to Aphrodite in the hopes that their first sexual experience would be productive. Aphrodite was also worshipped by prostitutes, of which there were two classes in Greece. The hetaerae were the "courtesans," or call girls, who entertained at the drinking-and-sex parties known as symposia, enjoyed by aristocratic Greek men; and *porne* were the common prostitutes. (The original Greek meaning of the word "pornography" was, literally, to "write about prostitutes.") Aphrodite was apparently highly revered in Corinth, a city of merchants famed for its prostitutes. Corinthian prostitutes were said to be especially beautiful and lived in luxury, and the city, as Thomas Cahill notes in *Sailing the Wine-Dark Sea*, became a "byword for sybaritic self-indulgence."*

In Greek myth, Aphrodite is always accompanied by **Eros (Cupid)**, the god of carnal desire. According to Hesiod, Eros is a much older deity, who emerges from Chaos at the same time as Gaia. But in later accounts, Eros is viewed as Aphrodite's son, always armed with a bow and quiver of arrows that cause anyone struck by

*A taste of first-century CE Corinth can be found in two of the most famous biblical letters, or "Epistles," of St. Paul—written to the early Christian Church there. In the first, he writes, "It is actually reported that there is a sexual immorality among you, and of a kind that is not found even among pagans; for a man is living with his father's wife." (I Corinthians 5:1)

them to fall in love with whatever he or she sees. Greek (and, later, Roman) myths are filled with stories of Eros shooting his arrows at random, without concern for the consequences of the sexual passion he arouses.

Among her many lovers, Aphrodite counts the other gods Ares, Poseidon, Dionysus, and Zeus. She also sleeps with Hermes in return for one of her sandals, which had been stolen by Zeus's eagle. The result of this union with the messenger god is Hermaphroditus, a boy of remarkable beauty. In a story made famous in Ovid's *Metamorphoses*, a water nymph sees Hermaphroditus walking in the woods and falls in love with him. As he bathes in a spring, the nymph jumps into the water and clings to the boy, praying that they will never be separated. The nymph's prayers are answered as they are joined into a single being with a woman's breasts and a man's genitals—source of the word "hermaphrodite."

Another tale from Ovid involving Aphrodite is the famous story of Pygmalion, the legendary king of Cyprus. Pygmalion has grown so disenchanted with the women of his land that he carves a statue of a perfect maiden. He is so taken by it that he falls in love with the statue and prays that it might become real. Aphrodite hears his prayer and grants his wish. This story is the inspiration for George Bernard Shaw's play *Pygmalion*, about a linguistics professor who teaches a working-class girl to behave like royalty. Shaw's play, in turn, inspired the musical *My Fair Lady*.

Priapus, an ancient Near Eastern god who appears in other mythologies, is another of Aphrodite's children. Because he is a very old fertility symbol and a popular god of procreation, statues of Priapus were often placed in Greek and Roman gardens. Although Priapus is dwarflike, these statues always depict him with an enormous, erect phallus, and he is also supposedly a good-luck god who could ward off the "evil eye." Priapus was even more popular among the Romans, who liked to suspend obscene poetry from the prominent phallus on his statues. ("Priapic" now means "relating to the phallus," and a disease called "priapism" is a persistent and usually painful erection, not related to sexual arousal. With Viagra, Levitra, and Cialis crowding the market, it's a wonder no drug company has decided to market "Priapus.")

•**Apollo (Apollo)** The son of Zeus and the Titan Leto, Apollo (who is also known as Phoebus, which means "bright" or "radiant") is worshipped as the god of light. Although not the true sun god, Apollo is later identified with the sun and is also seen as the civilizing god of music, poetry, and prophecy, as well as the protector of flocks and herds.*

Apollo's origins in Greek mythology are mysterious, and he may have been introduced to Greece as late as during the Dark Ages. But he was well known to Homer and Hesiod, and becomes one of the greatest gods in the Greek pantheon. Associated with the healing arts and medicine, he is also the god of disease whose arrows bring plagues. Apollo's role in prophecy was especially important, and his cult shrine at Delphi was one of the most significant in Greece. (See below, *What was the Delphic Oracle?*)

By the classical period in Greece, Apollo represented the Greek ideal of vigorous manhood, but he was not especially lucky in love. In one myth, he falls in love with one of his priestesses, the Sibyl at Cumae, one of the mythical women gifted with the power of prophecy. Taken by her beauty, Apollo offers to give her as many years in her life as grains of sand she can scoop in her hand. She accepts his offer, but then refuses to sleep with Apollo. Keeping his word, Apollo grants her long life, but denies her everlasting youth, so she becomes a shriveled old crone.

In another myth, Apollo falls in love with Daphne, a nymph. Unimpressed by Apollo's come-on, she prays for help to her father, a river god, and is changed into a laurel tree. The laurel became Apollo's sacred plant, and the crown of laurels a symbol of victory in Greece, adorning the heads of winners at the Olympic games.

Apollo also has a taste for young men. One of these is Hyacinth, a beautiful boy. While he and Apollo are practicing the discus, a gust

*The original sun god of ancient Greece was known as **Helios**, who drove his chariot across the sky each day, journeying back each night in a chariot. Although regarded with reverence, Helios was not a major cult figure. In one legend, his half-mortal son, **Phaeton**, once asked if he could drive the sun chariot for a day, and Helios agreed. But the boy could not control the horses, which threatened to set the world on fire, so Zeus killed Phaeton with a thunderbolt.

of wind causes the discus to hit Hyacinth in the head and kill him. From the dead youth's blood, Apollo creates the flower called a hyacinth—a white flower with splashes of red.

•**Ares (Mars)** A son of Zeus and Hera, Ares is the god of battle, blood lust, and war—in its destructive sense, as opposed to Athena, who represents the orderly use of war to defend the community. Disliked by Zeus, and less popular among the gods, Ares was not widely worshipped by the Greeks, but was highly admired and honored by the more militaristic Romans as Mars. Although he had no wife, Ares did have a very steamy affair with Aphrodite. Among their children were **Phobos (Panic)** and **Deimos (Fear)**, who accompanied Ares on the battlefield. They also provide the names of the two moons that orbit the planet Mars. (While Ares ruled the battlefield, the honored title of goddess of victory in battle went to **Nike**, rewarded by Zeus because she fought with the gods against the Titans. Otherwise unimportant in mythology, Nike is also the goddess of athletic victory—hence her connection to the footwear with the Olympian price tag.)

In the *Odyssey*, Homer tells a humorous story of the adulterous couple being "caught in the act" by the cuckolded Hephaestus, the lame blacksmith god to whom Aphrodite was married. When the sun god Helios sees the lovers in bed as he crosses the sky, he snitches on them to Hephaestus, who has fashioned a net hidden in the bed that catches Aphrodite and Ares in flagrante. Suspended in midair, Aphrodite and Ares become an Olympian spectacle when Hephaestus summons all of the other gods to see the netted lovers in this awkward, compromising position. In Homer's words, the gathered Olympians see "the lovebirds, snuggled so sweetly together."

•**Artemis (Diana)** With ancient origins as a mother goddess and patroness of animals, to the classical Greeks, Artemis is the daughter of Zeus and Leto, and Apollo's twin sister. She is the "virgin goddess," the patron of the hunt, the untamed protector of wild animals. She presides over the rites of passage in which Greek women changed from "wild" *parthenos* (virgin) to fully "tamed" *gyne* (woman). She is also a merciless judge of anyone who breaks her laws, and when any woman dies suddenly, it is believed that she has been struck by the arrows of Artemis.

One of the most famous stories demonstrating Artemis's swift and cruel justice involves Actaeon, the handsome hunter who accidentally comes upon the goddess as she is bathing naked in a spring. Offended, Artemis turns Actaeon into a stag and then sends his own hunting hounds out to tear their master to pieces. Told by Ovid, this myth has been a favorite subject of artists throughout history—proving, if nothing else, that there has always been a market for naked women and violence.

• **Athena (Minerva)** The patron of Athens—from whom her name is taken—Athena is the virgin goddess of war and wisdom, as well as the patron of arts and crafts, including building and carpentry. The daughter of Zeus and Metis (a goddess whose name meant "cleverness"), Athena is said to be born from Zeus's head. According to Hesiod, Zeus fears that one of his children will depose him, as he had done to his own father. To avoid this fate, Zeus swallows the pregnant Metis, hoping to absorb her cleverness and wisdom. When he complains of a severe headache, one of the other gods strikes his head with an ax, and out springs Athena, fully formed and armed, screaming a war cry. In this way, the child who might depose him is never truly "born."

Ever virginal and masculine in behavior, Athena is almost always depicted in full armor, holding a shield and spear. As patroness of Athens, she represents everything that Greek culture later idealizes—wisdom, the power of intelligence, and reason over unbridled love or passion—making her, in many respects, the opposite of Aphrodite.

She is not perfect, however. One myth illustrates that she can be swift to anger if her supremacy is questioned—especially by a mortal. A young woman named Arachne challenges Athena to a weaving competition. Taking the guise of an old woman, Athena tries to dissuade Arachne from the contest, but the mortal Arachne dismisses the warning. As the two work at their looms, Athena sees that Arachne's weaving has illustrations that seem to mock the gods by showing all of their deceptions and love affairs. She can also see that the mortal girl's weaving is better than her own. Snatching the tapestry from the loom, Athena starts to beat the poor mortal girl with a shuttle. In fear, Arachne tries to hang herself with a noose made of thread. As she hangs there, Athena sprinkles the mortal girl with poison and Arachne becomes a spider, which is why, of course, spiders spin webs and are called arachnids.

On a more noble note, the great temple dedicated to Athena in Athens was the Parthenon ("Temple to the Virgin"), which stands on a hill called the Acropolis overlooking the city. Probably the greatest example of classic Greek architecture, the Parthenon was built between 447 and 432 BCE as a celebration of Athenian pride during the Golden Age of Pericles, when Athens reigned supreme.*

- **Demeter (Ceres)** The mother goddess of crops, Demeter plays a featured role in one of the central myths of Greece, the tale of Persephone, her daughter with Zeus. When Persephone is carried off to the underworld by Hades, Demeter is enraged and prevents the crops from growing. To restore the natural order, Zeus arranges his daughter's release by negotiating a settlement between Demeter and Hades. But Hades had already given Persephone a pomegranate seed, and since she has eaten the food of the underworld, she is compelled to spend one-third of the year there with Hades and the other two-thirds in the world above. (The Greeks thought of the year in terms of only three seasons: spring, summer, and winter.)

  This "deal with the devil" was always thought to explain the arrival of spring, which is when Persephone returns to earth. Her subsequent return to the underworld means the end of the growing season and the coming of winter, seen as the time of death. While sim-

*The famed temple's sacred status changed with history. The Parthenon was converted into a Christian church about 500 CE, and then a mosque, after the Ottoman Empire captured Athens in the mid-1400s. A sculptural frieze from the front of the Parthenon remains an artistic controversy. Known as the Elgin Marbles after the English Lord Elgin, who removed the sculptures from the Parthenon in 1816, they are now held by the British Museum. But the Greek government believes that the Elgin Marbles are a Greek treasure that should be returned to their rightful place. Their unsuccessful plea was repeated during the 2004 Olympics.

ple and appealing, this explanation does not accurately fit the Greek growing season, some scholars note.* Instead, they view the tale of Persephone's abduction as an allegory explaining the fate of Greek girls who were often turned over to much older men in arranged marriages. Demeter's grief over the loss of Persephone was typical of the experiences of Greek mothers who gave up their daughters in arranged marriages, usually to an older stranger.

- **Dionysus** (**Bacchus**) One of the most widely celebrated gods of Greece (and, later, of Rome), Dionysus is not only the god of wine and ecstasy, but also the male life-force, a masculine fertility god. Unmentioned by Hesiod and little-mentioned by Homer, Dionysus is another "foreign import" who arrived in Greece much later than the other gods, a transplant from the ancient Near East. (References to him date to about 1250 BCE, and there is no evidence that he was worshipped before the Archaic Age.) But as god of wine and the sexual life-force, he was clearly a hit with the Greeks, and eventually supplanted Hestia (see below) as one of the twelve Olympians. The festivals in his honor—Dionysia in Greece and Bacchanalia in Rome—were probably the original "toga parties." And followers of Dionysus might have been some of the ancient world's biggest "party animals." These festivals became the occasion for wild dancing in the streets and ecstatic behavior by his devoted followers. Later, in Rome, they acquired even greater notoriety, forcing the Roman Senate to ban the feasts and apparently execute some of the "Dionysians" as a

*There is another myth thought to explain the seasons. Another "foreign import," **Adonis** was the child of an incestuous relationship, an exceedingly handsome young man who originated in the ancient Near East and was adapted by the Greeks. The name "Adonis" derives from the Semitic word for "lord," and he is connected with Dumuzi/Tammuz of Mesopotamian fame. Taken by his beauty, Aphrodite finds the infant Adonis and places him in a chest and gives him to Persephone, the queen of the underworld, for safekeeping. Enchanted with the youth, Persephone refuses to return him. Zeus ruled that Adonis would spend part of the year with Aphrodite and part of the year with Persephone. When Adonis stayed with Aphrodite on earth, plants and crops flourished. During his time in the underworld, vegetation died; Adonis supposedly fits into the "dying-rising" vegetation-god theory.

threat to civil order (see below, *What were the Bacchanalia and the Saturnalia?*).

But in Greek myth, Dionysus is depicted as the son of Zeus and a mortal woman Semele, who is burned to a crisp when she asks Zeus to appear to her as he really is. She unfortunately gets what she asks for—she's zapped by a thunderbolt. The dying Semele's fetus is saved by the messenger god Hermes, and Zeus sews the unborn child into his right thigh. A few months later, Dionysus is born. Ripped from his mother's womb and then from Zeus's leg, Dionysus would be described as "twice born."

In spite of his "multiple births," Dionysus is still on the hit list of Zeus's wife Hera. To save him from Hera's jealous vengeance, Zeus disguises the infant as a girl and takes him to be raised by his mortal aunt and uncle. Not fooled, Hera makes the child's mortal guardians go mad. They kill their own children and then commit suicide. But again Dionysus is spared, and Zeus transforms him into a young goat.

The vindictive Hera is not yet done with Zeus's "love child." After Dionysus returns to human form, Hera makes him go mad, and Dionysus wanders the Eastern world until he meets a goddess known as Cybele, from Asia Minor, a mother goddess (related to the Mesopotamian Inanna/Ishtar) whose cult followers indulged in ritual orgies and self-castration. Cybele cures Dionysus of his madness and introduces him to all of her secret fertility rites.

One of the most complex gods, Dionysus was sometimes perceived as both man and animal, had male and effeminate qualities, and was seen at times as both young and old. The ancient Greeks associated Dionysus with violent and unpredictable behavior, especially actions caused by drinking too much wine, and many stories about this god of intoxication involve epic sessions of drunken merrymaking. At one of these sessions, Dionysus grants the legendary King Midas his wish that everything he touches turns to gold. In another of those "be careful what you wish for" stories, Midas unfortunately discovers that Dionysus has made his wish literally come true, as his food turns to gold, and even his daughter is turned into a golden statue when he touches her. Dionysus reverses the golden curse by telling Midas to dive into a river, which accounts for the gold that was found in that area for generations. Dionysus's followers at these epic carousing sessions included nymphs, creatures called satyrs that were half-man and half-horse or goat, and women attendants called maenads.

Dionysus was also at the center of Greek drama, which had its roots in religious celebrations that incorporated song and dance. By the sixth century BCE, the rural celebration of Dionysus as an agricultural god who had brought farming, winemaking, and herding techniques to mankind was transformed in Athens into the Dionysia, a festival in which dancing choruses competed for prizes. At some point, a poet introduced the concept of a masked actor interacting with the chorus.

The playwright Aeschylus (525–456 BCE) took this idea further by adding two actors, each playing different parts. This soon evolved into full-scale plays featuring many actors and a chorus, allowing for more complex plots. Following the defeat of the Persians in 479 BCE, Athens emerged as the Greek superpower, and the annual drama festival, or Dionysia, became both a celebration and a spectacle, lasting four or five days. Thousands of Athenians watched plays in an enormous outdoor theater that could seat 17,000 spectators. At the end of the festival, prizes were awarded to the tragedians. The word "tragedy" comes from the Greek word *tragos*, meaning "goat," the sacred symbolic animal of Dionysus. Much of the lore of Dionysus is based on *The Bacchae* (c. 407 BCE), a play by Euripides (c. 480–406 BCE), one of the three great writers of Greek tragedy. In *The Bacchae*, he writes,

> Mankind . . . possesses two supreme blessings. First of these is the goddess Demeter, or Earth—whichever name you choose to call her by. It was she who gave to man his nourishment of grain. But after her there came the son of Semele, who matched her present by inventing liquid wine as his gift to man. For filled with that good gift, suffering mankind forgets its grief; from it comes sleep, with it oblivion of the troubles of the day. There is no other medicine for misery.

The other significant role of Dionysus is as a resurrected god. In one legend, Dionysus is ripped into seven parts by the Titans at

Hera's request. They throw the parts into a cauldron, cook them, and eat them. But Dionysus is immortal and returns to life—though the exact method of his resurrection is unclear. His return from death connects Dionysus to the earlier resurrection gods, such as the Egyptian Osiris, as well as early Christian worship when it eventually spread to Greece.

Hades (**Pluto**) Son of Cronus and Rhea, the ruler of the underworld, Hades did not live on Mount Olympus, and he is not usually counted among the twelve Olympians. But he shared in ruling the universe with his two brothers, Zeus and Poseidon. His name, which originally meant "invisible" or "unseen," was considered unlucky, and the Greeks often referred to him instead as Pluton ("the rich one") or by other honorific names.

Although a grim figure, Hades is not considered evil, and his underworld realm, also called Hades, is not a hellish place, but a kingdom where Hades administers justice. Nor is he actually death, which the Greeks personified in the god Thanatos, a child of the goddess of night. The Greeks believed that the dead arrived in the underworld domain after being brought by Hermes to the banks of the River Styx (which meant "hateful"). The arrivals were expected to give the boatman, Charon, a coin to ferry them across the river—ancient Greeks buried their dead with a coin in their mouth as payment to Charon. Those who did not receive proper funeral rites were forced to wander along the riverbank for one hundred years before obtaining passage from Charon. The entrance to the underworld was guarded by the terrible three-headed dog Cerberus, who wagged his tail to welcome new arrivals but devoured those who tried to leave and return to the land of the living.

Unlike the later Christian version of hell, Hades was not originally a place of terror, but a hilly landscape, dotted with trees and flowing with rivers. One of these was the River Lethe, or Oblivion, where the events of life could be forgotten. In later Greek traditions, some of the dead went to the Elysian Fields, a paradise reserved for the distinguished, and to the Fields of Asphodel, where most souls wandered in the gloom, looking for flowers.

But then there was also Erebus, one of the original elements of the Creation, which was a region of the deep, dark Tartarus, reserved for the grossest sinners who had violated some divine law or otherwise crossed Zeus. One of these was a king named Tantalus, who commited the cardinal sin of talking about having once dined with the gods, or, in another version of the myth, having cooked his own son and served him to the gods to see if they could detect this forbidden food. This mythic moment may mark the rejection of both cannibalism and human sacrifice. For his crime, Tantalus was sentenced to stand in a pool of water, which drained away when he tried to drink, and with fruit dangling before his eyes, which was whisked away as soon as he reached to eat it. Tantalus was, in other words, eternally "tantalized."

The other famed denizen of Tartarus was Sisyphus, a clever king and founder of Corinth, who saw Zeus seduce a nymph and made the mistake of talking about it. Angry at Sisyphus for revealing his secret, Zeus told Thanatos to capture Sisyphus and place him in chains. But Sisyphus pulled a very old trick by convincing Thanatos to demonstrate how to put the chains on himself. With Thanatos out of action, "death takes a holiday," and no one could die in the land of mortals. Upset that nobody was dying in battle, the war god Ares stepped in, killing Sisyphus and freeing Thanatos.

But Sisyphus had one more trick up his sleeve. He had earlier instructed his wife not to bury him if he died. Since he hadn't been buried, he convinced Persephone that he shouldn't be in Hades, and she freed him, supposedly to attend his own funeral. Realizing that the gods had been tricked once more, Hades dragged Sisyphus back to the underworld, where three judges of the dead ordered his punishment. He was forced to push a boulder up a hill. Every time he reached the top, the stone would roll back down again, and Sisyphus had to start again—pushing the same stone up the hill for all eternity. The story of Sisyphus was converted into one of the great twentieth-century allegories of existentialism by Albert Camus, who saw the plight of modern man in *The Myth of Sisyphus* (1942). Camus wrote, "The struggle to reach the top is itself enough to fulfill the heart of man. One must believe that Sisyphus is happy."

The only mortal to defeat Hades and death was the fabled singer Orpheus, whose songs had supernatural powers. When his beloved wife, Eurydice, dies of a snakebite, Orpheus descends to the underworld and enchants Hades and Persephone with his singing. They allow Eurydice to leave, only Orpheus is instructed not to look back at her before leaving the underworld. But Orpheus can't resist a backward glance at his beloved, and she is lost forever.

•**Hephaestus (Vulcan)** God of fire, blacksmiths, and metalwork, Hephaestus is the son of Hera, and something of a trickster god, a typical role for gods of smiths and crafts in other mythologies as well. The identity of his father is a mystery, and at birth, he is dwarfish and disfigured with a limp, so his mother, Hera, throws him from Mount Olympus, and he falls into the sea. Another version blames Zeus, angry at apparently being cuckolded, for throwing Hephaestus into the sea, which cripples him.

Raised for years in a cave by nymphs who teach him the arts of metalwork, Hephaestus creates a magical golden throne as a gift for his mother. As soon as Hera sits in it, she is trapped in a fine golden mesh. Hephaestus only agrees to release her when Dionysus, the god of wine, gets him drunk and brings him back to Olympus. Hephaestus releases Hera on the promise that he can marry the beautiful Aphrodite, which makes them the odd couple of Olympus: beautiful, sexy Aphrodite and the crippled dwarf Hephaestus.

The divine craftsman, he also builds the palaces of the gods and plays a key role in a very important myth—the creation of the first woman, Pandora (see below, *What was in Pandora's "box"?*).

•**Hera (Juno)** Hera, the queen of the gods, is presented in Hesiod's poems as the daughter of Cronus, but she may have originated as a pre-Greek earth goddess, and in the view of some modern scholars, may have been a widely worshipped deity in Greece before Zeus arrived with the Mycenaean invaders. Chiefly a goddess of marriage, women's sexuality, and fertility, like the Egyptian Hathor, Hera is associated with cattle and was often called "cow-eyed."

As Zeus's jealous wife, Hera is usually most preoccupied with his constant sexual misadventures. In wooing Hera, Zeus had disguised himself as a cuckoo bird in a rainstorm to win her sympathy. When she picks up the pitiful, wet bird, Zeus drops his disguise and rapes her. It was the beginning of their frequently stormy relationship, which is at the center of so many of the Greek myths. Yet, in the face of Zeus's many affairs, Hera never wavers in her commitment to her husband, and her fidelity has been taken to represent the ideal Greek wife, upholding monogamy—at least on the part of wives—and the orderly inheritance of property and rank in Greek culture. She has three children with Zeus: Ares, one of the Olympians; Eileithyia, a patron of midwives and childbirth; and Hebe, the embodiment of youth.

Hera and Zeus are also supposed to be the parents of the smith god Hephaestus. But in Hesiod's account of his origins, Hera conceives Hephaestus on her own, in an act of jealous revenge, so typical of the motivating force at the heart of many of the myths about Hera. Frequently betrayed by Zeus, Hera often turns her anger toward his lovers and many offspring. One lover, Semele, was burned up. Another, the young princess Io, whom Zeus had turned into a heifer to conceal her from Hera, is tormented by a gadfly and gallops all over the world with a perpetual itch. Hera may have reserved her greatest anger for the hero Heracles (see below, *What kind of hero kills his wife and children?*), the son of Zeus and the mortal Alcmene, princess of Thebes.

• **Hermes** (**Mercury**) Messenger of the gods, famed for his winged feet and helmet, Hermes is also the patron of travelers. His name has been thought to derive from the ancient word *herm*, for "stone heap," as it was a common practice for travelers to mark their trails by piling up stones, not as a guide to return but simply as a symbol of having passed by. (A common worldwide tradition, such stone piles, or cairns, appear in the Bible, ancient America, the British Isles, and elsewhere in the ancient world. Such markers are still traditionally left by modern hikers who "mark trail" by adding to stone piles.) But he is also a trickster, and from the moment of his birth, Hermes gets into mischief, starting by stealing the sacred cattle of his brother Apollo, an act which also cast him as the patron of thieves.

Hermes is the messenger of the gods, as well, and the protector of human messengers—men who had the dangerous but important job of traveling between hostile communities, delivering diplomatic messages. In ancient times, messengers were not supposed to be harmed, just as modern diplomats are supposed to travel with immunity.

Additionally, as god of travel, Hermes has the job of escorting the dead in their journey to Hades.

His most notable offspring is **Pan (Faunus)**, the pastoral god of woods and pastures and the protector of shepherds and their flocks. Half-man and half-goat, Pan is one of the few Greek deities who is not all human, but is also one of the most popular gods. Believed to have a wild, unpredictable—and especially lusty—nature, he can fill humans and animals with sudden, unreasoning terror, which is why the word "panic" comes from his name. Pan has many love affairs with nymphs and other minor deities, but when he pursues the nymph Syrinx, she runs away from him in terror and begs the gods to help her. The gods change her into a bed of reeds, from which Pan makes a musical instrument called the panpipe (though another myth credits Pan's father, Hermes, with that invention).

Eventually Hermes became the protector of merchants and was considered an important god as the Greeks moved from farming to more commercial pursuits.

Hestia (Vesta) The eldest child of Cronus and Rhea, Hestia is the sister of Zeus and the first child swallowed by Cronus. Her name means "hearth," and she is the traditional Greek protectress of the home and also guards the hearth fire, one of the crucial duties of a woman in a Greek home. Although she was worshipped in the home of every Greek, Hestia is perhaps the least significant of the Olympians, and there are few stories about her. She is, in essence, the first "stay at home" woman. As a result, she has little chance for adventure—or mischief. In later times, Hestia's place in Olympus among the twelve is taken by Dionysus.

Hestia was considered far more important in Rome, where she had been adopted as Vesta, and served as symbol of the city. Residing at her shrine in a temple in the Roman Forum were six vestal virgins, who tended an eternal flame. Chosen when they were between six and ten years old, they served for thirty years, and the punishment for losing their virginity was severe. A vestal virgin who broke the taboo was whipped and buried alive in a small chamber with only a bed. Over a thousand years, about twenty vestals were known to have been punished this way.

•**Poseidon** (**Neptune**) One of the three sons of Cronus, Poseidon becomes ruler of the sea and is, in Homer's words, the "shaker of earth"—literally responsible for earthquakes, which were frequent and violent in Greece and the Aegean Sea region. Almost always depicted carrying his three-pronged spear, the trident, and driving a chariot, he is one of the most widely—and anciently—worshipped of the Greek gods.

Some scholars believe that Poseidon may have been an older god already worshipped in Greece when the Mycenaeans arrived, perhaps as a fertility god, associated with the water. But by the time of Homer and Hesiod, he is thought of as Lord of the Deep. A powerful figure who often resists his brother Zeus, he becomes one of the most significant figures in Homer's *Odyssey*, as the god who is most hostile to the hero Odysseus.

•**Zeus** (**Jupiter**) King of the gods, god of thunder and weather, and originally bearing a name that meant "shining sky," Zeus is the son of the Titan Cronus, who had toppled his own father, Uranus. Similarly, Zeus brings his father down and supplants him as the chief deity. The only major Greek god whose Indo-European origins are undisputed, Zeus is connected with older gods who probably arrived in Greece with the people later known as Mycenaeans. Some scholars see parallels between his story and the Mesopotamian god-feud in which Enki killed Apsu. (See chapter 3.) There are also similarities between Zeus and Marduk, hinting that the Greeks may have been influenced by the earlier Mesopotamian myths. In Greek myth, this old god of the bright sky is transformed into Zeus, the weather god. After the great war with the Titans, Zeus draws lots with his brothers, and divides the world. He is lord of the sky; one of his brothers, Hades, becomes lord of the underworld; and the other, Poseidon, gets dominion over the sea.

Zeus's first wife is Metis, a sea nymph known for her wisdom, but Zeus is most famously married to Hera. Still, he is a notorious adulterer and has many lovers, both divine and human, and Hera deeply resents all of his many offspring. Some scholars believe that this was another vestige of the early rivalry between the male-dominated Zeus cult that arrived with the Mycenaeans and Hera's goddess/earth mother-religion, which may have predated the Mycenaean era. Whatever the sexual politics may have been, Zeus became ruler of the world, presiding over law and justice—which essentially meant Greek customs. In that role, he often metes out justice to those attempting to defy the right order of the world through hubris. A word commonly misidentified today as "excessive pride"—as in, "The Yankees lost to the Red Sox because of their hubris"—this Greek concept actually meant a form of insolence, or intentionally dishonorable behavior, a powerful term of condemnation in ancient Greece.

Dispensing justice for dishonorable behavior seems a strange notion coming from a god best known for being a "serial adulterer." His many notorious sexual escapades include both divine and mortal women. This makes him not only king of the gods, but father of quite a few of them as well. Among his lovers are the Titan Themis, with whom he has the three Horae (Seasons) and the Moirae (Fates); the goddess Mnemosyne (Memory), who produces the Nine Muses, who inspire poetry, dancing, music, and the other arts;* the grain goddess Demeter, who gives birth to Persephone; and the Titan Leto, mother of two of the greatest gods, Apollo and Artemis. Scholars believe that all of these affairs were allegorical tales meant to explain how the great "father" was responsible for creating the order of the world as it existed in the Greek mind.

But the tales that describe his exploits grew very colorful over the centuries. To seduce his many conquests, Zeus—a master of disguise—overcomes resistance by taking many different forms and shapes, perhaps most famously as a swan. That is how he appears to

*The Muses are Clio (history), Euterpe (flute playing), Thalia (comedy), Melpomene (tragedy), Erato (love lyrics), Terpsichore (dance), Polyhymnia (pantomime), Urania (astrology), and Calliope (epic poetry).

the queen of Sparta, Leda. Zeus mates with Leda in the form of a swan, and they conceive two children, one of them famed as Helen of Troy and the other Polydeuces (or Pollux). When Leda sleeps with her husband on the same night, she also conceives the mortals Castor (twin of Polydeuces) and Clytemnestra, who becomes the wife of King Agamemnon, commander in chief of the Greeks against Troy.

Among Zeus's many mortal lovers are young boys—which strikes the modern mind as unnatural, but was not unusual among elite Greeks of the Classical Period. One of his most famous male lovers is Ganymede, a prince of the royal Trojan house and the most beautiful of mortals. In one legend, Zeus, in the form of an eagle, abducts Ganymede and carries him to Olympus, where he serves as a cup bearer to the gods. This was how handsome young boys functioned in the Greek drinking-and-sex parties called symposia, where older men initiated young boys into sexual knowledge, a practice known as pederasty, and the subject of Plato's dialogue, *Symposium*.

MYTHIC VOICES

He bound Prometheus the schemer in inescapable fetters
a torment to bear, and through them he drove a mighty
stone pylon,
and sent a long-winged eagle to gnaw his incorruptible
liver.
By day the bird fed upon it, but each night as much was
replenished
as was lost on the day before.

—HESIOD, Theogony

The beautiful fables of the Greeks, being proper creations of the imagination and not of the fancy, are universal verities. What a range of meanings and what perpetual pertinence has the story of Prometheus! Besides its primary value as the first chapter of the history of Europe . . . it gives the history of religion with some closeness to the faith of later ages. Prometheus is the Jesus of the old mythology. He

is a friend of man; stands between the unjust "justice" of the Eternal Father and the race of mortals, and readily suffers all things on their account.

—RALPH WALDO EMERSON *on Prometheus*

How did man get fire?

In Hesiod's Creation story, not all of the Titans fought against Zeus. Prometheus—a Titan who was a fire god, master craftsman, and trickster whose name connoted "forethought"—joined Zeus in the war against the other Titans. But as time went by, Prometheus rebelled. He was offended when Zeus took a dislike to the first humans, whom Prometheus had molded out of clay. In an argument over sacrifices to the gods, Prometheus balked when Zeus decided to deprive men of fire.

Taking the side of man, Prometheus tricked Zeus into receiving only the bones and fat of sacrificed animals instead of their meat. He had disguised the bones under a layer of glistening fat that Zeus chose instead of a plate of meat hidden under the animal's stomach. (If that turns your stomach, don't order haggis, a traditional Scottish delicacy featuring the stomach of a sheep.) Outraged at the deceit, Zeus decided that man can have meat but not the fire to cook it. When Prometheus hid fire in the hollow of a dried stalk of fennel and gave it to mankind, Zeus retaliated by chaining the Titan to a mountain peak in the Caucasus Mountains, where each day an eagle pecked at his liver—and each night the liver grew back.

Destined to suffer this torture for eternity, Prometheus was only freed when he used his gift of forethought to assure Zeus he had nothing to fear from a seemingly threatening prophecy. Zeus let Prometheus go. But he had one more "trick up his toga" to spring on mankind. The Lord of Olympus instructed Hephaestus to sculpt a lovely girl from earth and water. When the craft god was finished, all the gods then contributed other gifts to the first woman. As in a scene from a fairy-tale "finishing school," Athena gave this creation beautiful clothes and taught her to weave on the loom. Aphrodite endowed her with beauty and charm, but the heartbreak and the sorrow of love as well. Finally Hermes—at Zeus's urging—instilled in her the ability to lie persuasively. (Hermes was instructed to give to the woman "thievish morals and to add the soul of a bitch"—in Hesiod's less than loving words.) So, Hermes "filled her with lies, with swindles, all sorts of thievish behavior," and she was named Pandora, which in Greek means "all gifts."

Though Prometheus warned his brother Epimetheus—a not-so-sharp tool whose name means "afterthought"—not to accept this gift from Zeus, Epimetheus was enchanted and married Pandora, who arrived bearing a package.

What was in Pandora's "box"?

First of all, it wasn't a "box," but a covered jar. But that's another story, and we'll come to it in a bit.

The Greek equivalent of the biblical Eve, Pandora was the first woman, and according to the myth, created by Zeus as a punishment for men. Just as Zeus had been tricked by Prometheus with skin and bones that had been concealed under some enticingly glistening fat—an offering that looked good on the outside— Zeus returned the favor by sending Pandora, a "package" that seemed well wrapped but concealed trouble. Zeus also sent along a somewhat mysterious jar.

When Hermes delivered Pandora to Prometheus's brother, Epimetheus, he was smitten by her, even though she was a "curse to men who must live by bread," in Hesiod's woman-hating words. In spite of Prometheus's warnings against accepting anything from Zeus, Epimetheus welcomed Pandora. Hesiod never says that Pandora was told not to open the jar. But plagued by insatiable curiosity, Pandora opened the jar given to Epimetheus by Zeus. Out flew all the ills that torment mankind—hard work, pain, and dreadful diseases that bring death. They all escaped from the jar to plague humanity.

The curious twist to Hesiod's story is that only hope did not escape from the jar. Pandora put the lid back on the jar before hope could escape. But there is some ambiguity in that. Does it mean man has hope because it has not flown away? Or is it trapped within the jar? Hesiod does not explain. What is clear is that he takes a dim view of women, much like the authors of the biblical folktale in Genesis who blame the suffering of the world on Eve. Of Pandora, Hesiod says in *Theogony*, "From her descends the ruinous race and tribe of women."

Commenting on Pandora and the Greek view of women, classicist Barry Powell wrote in *Classical Mythology*, "Among the Greeks, misogyny seems to be based not so much on primitive magical terror, or economic resentment as . . . on a male resentment of the institution of monogamy itself. Greek myth is obsessed with hostile relations between the sexes, especially between married couples. . . . We need to remember that . . . ancient literature, and myth, was composed by men for men in an environment ruled by men."

As for the common expression "Pandora's box," it has a long history. In 1508, the Dutch author Desiderius Erasmus first used the phrase "Pandora's box" instead of the original *pithos* in Greek, a traditional jar for storing grain. And since then, "Pandora's box" has come to symbolize any object or situation that seems harmless on the outside but has a great potential for discord, evil, and unlimited harm.

MYTHIC VOICES

He hastily stored away the thunderbolts, forged by
Cyclopes,
and conceived a different design, of opening dark heavy
rain clouds,
In every quarter of heaven, and drowning mankind in the
waters.

—OVID, Metamorphoses

Why does Zeus send a great flood to destroy man?

Prometheus plays a supporting role in another Greek story, which may be less familiar than that of Pandora, but has important biblical parallels.

In *Works and Days*, Hesiod described the creation of humanity in five separate ages. First came a golden race of mortals, during the time of Cronus, which disappeared without explanation. Next, Zeus created a race made of the precious metal silver, but they refused to make sacrifices to the gods and were wiped out. A third age was made of bronze, but they proved to be so warlike that they wiped themselves out. The fourth age was the Heroic Age, populated by a race of demigods created by Zeus. When they died, many of these heroes either were placed in the heavens as constellations, became companions of the gods, or went to live on the mythical Island of the Blessed, which was ruled by Cronus.

It is later, in the age of iron, that Zeus finally created the present generation of humans. But, according to Roman poet Ovid's *Metamorphoses*, when Zeus (Jupiter to Ovid) walked among these humans, he was disgusted, especially by a king who practiced cannibalism and human sacrifice. Zeus decided to destroy them. With the help of Poseidon, Zeus unleashed a tremendous flood and nearly all of humanity was killed. Two good souls, however, were saved. Deucalion, the son of Prometheus, and his wife, Pyrrha, who was the daughter of Pandora, had been warned by the prescient Prometheus of the imminent flood. Deucalion built a boat, sent out a bird—a dove, in his case—and, after the floodwaters subsided, the boat came to rest on a mountaintop.

All of these details, of course, echo both the Mesopotamian flood accounts and the biblical story. Like Noah, Deucalion and Pyrrha were allowed to live. But they were sad and lonely in an empty world. The voice of a goddess from a nearby cave told them to throw their "mighty mother's bones" over their shoulders.

Puzzled at first, Deucalion realized that the command referred not to his own mother, but to Mother Earth—whose bones are rocks. Picking up stones, Deucalion and Pyrrha threw them over their shoulders, and they were turned into people, and Deucalion and Pyrrha were responsible for repopulating the earth. Among the "children" they created was their son Hellen, who gave his name to the entire Greek race, later known as the "Hellenes."

Which mythical monster has the worst "bad hair day"?

First among the heroes of the Heroic Age was Perseus, the son of Zeus and his mortal lover Danaë. When Danaë's father, a king, learns from an oracle that his own grandson will someday kill him, he sets Danaë and the infant Perseus adrift in a chest. They are saved by a fisherman, whose brother, Polydectes, rules the isle of Seriphus. Over time Polydectes falls in love with Danaë and wants to marry her, but she is unwilling. To prevent the marriage, the grown Perseus agrees to slay the Medusa, one of three monstrous sisters known as the Gorgons, whose ugliness turns men to stone. Once beautiful, Medusa had boasted of her beauty to Athena, who became jealous and changed her into a hideous monster with living snakes for hair. The Greeks carved images of Medusa's head on their armor to frighten their enemies, and images of Medusa's head were also used as charms to protect them from evil spells.

Aided by Hermes and Athena, Perseus sets off on his quest. He is given a curved sword, a cloak to render him invisible, Hermes' winged shoes, and a leather bag to carry Medusa's head. In the most familiar version of the myth, Perseus slays Medusa by looking at her reflection in his mirrorlike shield, although other accounts say that Athena guides his hand while he looks away. After he decapitates Medusa, who has been made pregnant by Poseidon, Perseus places the deadly head in the leather bag. As Medusa dies, the winged horse Pegasus springs from her body, and poisonous snakes rise from the blood that drips from her head. Athena saves the blood from Medusa's body and later gives it to Asclepius, the god of healing (see below, *Which Argonaut was a god of healing?*). Although the blood from Medusa's left side is deadly poison, that from her right side has the power to revive the dead.

On his way home, Perseus rescues a beautiful maiden, Andromeda, from a giant sea monster and marries her. Once back in Seriphus, he turns Polydectes to stone by showing him the head of the Medusa. Unfortunately, fulfilling this prophecy, Perseus accidentally kills his grandfather with a discus. Although he is entitled to become king of Argos after that, Perseus chooses instead to rule Tiryns, where he and Andromeda found a great dynasty. Among his descendants is the great hero Heracles.

What kind of hero kills his wife and children?

If you only heard of one Greek god or hero when you were a kid, it was probably Hercules, whose name in Greek legends was Heracles. The leading character in many a B-movie featuring brawny, bad actors, Heracles was a legendary figure of the Heroic Age, who probably was just as popular in ancient Greece as he is today. Heracles was born in Thebes, the son of the mortal princess Alcmene and the philandering Zeus. Because he is another of Zeus's illegitimate offspring, Heracles incurs the wrath of Hera, who has it in for any child born from Zeus's cheating ways.

Hera's spite takes on some creative forms. First, she causes the birth of Heracles to be delayed so that he is not the firstborn child, cannot wear the crown, and, in fact, is made a slave. Hera then sends two snakes to kill Heracles as he sleeps, but the baby boy amazes everyone by strangling them with his bare hands. Fond of the boy, Zeus intervenes to try to put a stop to Hera's sabotage. The Olympian places the infant Heracles at the sleeping Hera's breast so that he will receive the mother's milk of the gods. But Heracles bites down so hard that Hera wakes up and pushes the baby away—denying Heracles complete immortality. When her breast milk spills, it spreads across the sky as the Milky Way.

The semidivine Heracles goes on to become a warrior of great strength and skill. After helping the Thebans defeat an enemy, Heracles marries the Theban king's daughter Megara, and has three children, who become Hera's new targets. Seeing an opportunity to do harm, she causes Heracles to suffer a fit of madness, in which he lets fly his arrows, killing his whole family. Seeking to purify himself and atone for this crime, Heracles goes to the Oracle at Delphi and learns he must serve his cousin, Eurystheus, king of Mycenae. Over the course of twelve years, he performs twelve labors—which in the original Greek were conveyed by the word *athoi*. It meant "contest" and was the source of the word "athletics."

Here are the labors of Heracles, which have been described over the centuries with many variations:

1. *The Nemean Lion*

Heracles takes on the fierce lion of Nemea, which has been killing all the flocks near Mycenae. At first, his arrows simply bounce off the animal's skin, so Heracles chases the lion and kills it with his bare hands.* He keeps the lion's impenetrable skin as a trophy and is often depicted in art with the lion's jaws covering his head like a helmet.

2. *The Lernean Hydra*
Heracles takes on a many-headed snake with the body of a hound, whose mere breath can kill and whose heads grow back as soon as they are cut off. The deadly Hydra lives in the swamps of Lerna, also near Mycenae, where it kills livestock. At first, Heracles makes no progress in his fight with the beast, and Hera even sends a giant crab to bite him as he fights. But Heracles' nephew Iolaüs pitches in. As Heracles cuts off a head, Iolaüs seals each neck with fire to prevent it from growing back. After killing the Hydra, Heracles dips his arrows in the beast's blood to make them even deadlier. Hera later raises both the Hydra and the crab into the sky, where they are known as the constellations Hydra and Cancer.

3. *The Erymanthian Boar*
Heracles captures this huge boar who lives in the central Peloponnesus region of Arcadia with little difficulty. But later, when he stops to eat and drink with some centaurs—the half-man, half-horse beasts—they end up in a drunken brawl. During the fight, Heracles kills several of the centaurs with his poison arrows, including his friend, the centaur Pholus, who dies accidentally when an arrow falls on his foot. The centaur Nessus, who escapes this free-for-all, will reappear in Heracles' story with disastrous consequences.

4. *The Ceryneian Deer (also called the Arcadian Stag)*
Ordered to capture a deer, famed for its golden antlers and metal hoofs, Heracles succeeds after tracking the animal for a year. But when Heracles meets Apollo, the god claims the deer is sacred to his sister, Artemis. Heracles apologizes and later releases the animal.

*This is one of several parallels between the feats of Heracles and those of the biblical Samson, another ancient strongman, who also killed a lion and was renowned for his hefty sexual appetites.

5. *The Stymphalian Birds*
   Near a lake in Arcadia lives a flock of vicious birds with wings that fire arrows, beaks that can pierce armor, and whose droppings are lethal to crops. After Heracles startles the birds by banging some metal castanets, they fly into the sky, and he kills them with his poisoned arrows.

6. *The Augean Stables*
   Ordered to clean the enormous stables of King Augeas, Heracles finds himself knee-deep in dung. Heracles cleverly punches holes in the sides of the stable and diverts a river to flow through the stables, cleaning them overnight.

7. *The Cretan Bull*
   Told to capture the sacred bull of King Minos, Heracles goes to the island of Crete. "Seizing the bull by its horns," he tosses it into the sea and then rides it, rodeo-style, back to Mycenae, where he releases the bull, which is later killed by the Athenian hero Theseus. (This version of the story conflicts with the more famous tale of Theseus and the Minotaur. See below, *Was Atlantis ever discussed in Greek myth?*)

8. *The Horse of Diomede*
   Heracles captures Diomede, Ares's son and the barbarous king of Thrace who owns four deadly horses that feed on human flesh. Heracles feeds the wicked king to his horses, whom he tames and sets loose. They are later killed by wolves.

9. *The Girdle of Hippolyta*
   Heracles is asked to obtain the girdle—a sash or belt—of Hippolyta, queen of the Amazon women warriors who, according to myth, severed a breast so that it would not interfere with drawing a bow. Expecting a fierce battle, Heracles gathers a small army. But smitten by the hunky hero, Hippolyta simply agrees to give Heracles the belt. In some interpretations, taking the girdle would have been viewed as a metaphor for rape, while surrendering it was seen as consensual sex.

Infuriated at this turn of events, Hera takes the guise of an Amazon warrior and leads an attack. Heracles strangles Hippolyta in battle, thinking that she has betrayed him.

10. *The Cattle of Geryon*
Heracles is sent to get a flock of magical cattle belonging to Geryon, a three-headed monster who lives at the western edge of the known world (modern Spain). The great hero trudges across northern Africa until he reaches the spot where the cattle are kept—the point where the Mediterranean Sea meets the Atlantic Ocean. He erects two great columns of stone—the Rock of Ceuta in Tangiers and the Rock of Gibraltar, afterwards known as the Pillars of Heracles. Killing the herders who keep the cattle, he drives them all the way across Europe, to Mycenae, where he sacrifices them to Hera. (This is the first of the labors set outside Greece, and it is thought that these "foreign" adventures were told as a way of describing the wider world as the Greeks sailed to the farthest reaches of the Mediterranean.)

11. *The Apples of the Hesperides*
While looking for the golden apples that grew on a magical tree of life, Heracles finds Prometheus nailed to the rock. He kills the eagle that torments the Titan and sets Prometheus free. In gratitude, Prometheus tells Heracles how to get the apples from the Hesperides, the daughters of Atlas, who possess them. Heracles offers to hold up the sky while Atlas gets the apples. Freed from his backbreaking and onerous task, Atlas decides to leave Heracles where he is, stuck with his job. Heracles outsmarts the rather dimwitted Titan by asking him to hold the sky for just a moment. Atlas obliges and Heracles takes back the apples.

12. *Cerberus, the Hound of Hades*
In his most daunting feat, Heracles must descend into Hades to steal the three-headed dog Cerberus, who guards the gates of the underworld. There are different versions of how Heracles does this. In one account, he fights with the lord of the underworld himself and wounds him. While Hades is off getting his wound healed, Heracles captures the dog, and brings it back to the upper world. In "defeat-

ing" death, Heracles supposedly gains immortality. But in another version, Hades is more compliant, and allows Heracles to take the dog as long as he uses no weapons. Protected by his lion skin, Heracles wrestles the dog, chains it, and drags it to the land of the living. Having accomplished this, he returns Cerberus to Hades.

After completing the twelve labors, Heracles marries the princess Deianira—a name that means "man-killer." As they travel together, they come to a river where they meet the centaur Nessus—one of the group that Heracles had fought with during the labors. For a small fee, Nessus ferries travelers across the river. While carrying Heracles' bride, Nessus tries to rape Deianira, and Heracles shoots him with a poisoned arrow. As he lies dying, the centaur convinces Deianira to take some of his now-poisoned blood and semen and smear it on Heracles' robe if she ever wants a love potion to keep her husband faithful.

After Heracles falls in love with another princess, Deianira follows the centaur's advice. But when Heracles puts on the robe, now poisoned with the centaur's tainted blood, it burns him so terribly that he begs to be placed on a funeral pyre. Heracles then leaps into the flames. His grief-stricken wife also kills herself by jumping into the funeral pyre when she realizes what she has done.

Ascending to the home of the gods, Heracles resolves his differences with Hera, marries Hera's daughter Hebe ("Youth"), and enjoys immortality among the gods on Olympus.

Which great hero gets "fleeced"?

Heracles plays a bit part in a tale of family feuding and power-grabbing that takes to the high seas in the story of Jason and the Golden Fleece, one of the first nautical adventures in Western literature. This much-loved tale has parts recounted by Homer, the playwrights Aeschylus and Euripides (whose *Medea* covers Jason's later years), as well as the philosopher Socrates. But the most familiar version was compiled in *Argonautica* by Apollonius of Rhodes, a third-century BCE account of the legend of the young prince Jason, who is forced to flee his home in the city of Iolcus after the throne there is unlawfully seized by his uncle.

Fearing for Jason's life, his mother tucks him away in the cave of the wise centaur Chiron, who has tutored some of the greatest heroes in Greek myth.

When Jason returns to his home sometime later, his wicked uncle Pelias is in power and poised to kill his young rival. But there is one small problem—it's a feast day, and the ancient laws of hospitality are in force. Ever resourceful, Pelias tries another tactic. He tells Jason he will step down if the young man can bring back the Golden Fleece, which hangs from a tree in Colchis and is guarded by a dragon that never sleeps.

The fearless Jason recruits a crew of fifty heroes—including Heracles—who become known as the Argonauts, after their ship, the *Argo* ("Swift"). The largest vessel ever made, it is outfitted with a magical talking beam cut from Zeus's sacred oak at Dodona and given to Jason by Athena. Sailing from Iolcus in Thessaly, the Argonauts reach Colchis, but only after surviving a series of dangerous adventures, including a battle with the Harpies, winged monsters with hooked beaks and claws that swoop down and take the food from the table of a king. The grateful king tells the Argonauts how to defeat the next danger, the "clashing rocks" that smash together to crush any ship entering the Black Sea. Sending a dove ahead of them as a decoy, the wily Argonauts pass safely through the deadly rocks by rowing hard as the dove flies through.

Before winning the Golden Fleece, Jason discovers that he has two more obstacles—he must yoke together a pair of fire-breathing oxen and plow a large field where armed warriors spring up out of the dragon's teeth that have been sown there. The attractive Jason and his plight draw the attention of the king's daughter, Medea, a sorceress who gives the Argonauts' leader magic ointments to spread on his sword, shield, and body, which will protect him from the monstrous dragon guarding the fleece. When Jason's mission has been accomplished, Medea sails home with him on the *Argo*. The fiendish Medea then does what few ordinary sisters will do—she cuts up her brother Aspyrtus into little pieces and scatters these in the water so that her father, who is in hot pursuit, must stop to recover his son's body for proper burial.

But it was not to be happily ever after for these lovers.

When Jason unexpectedly returns with the fleece, King Pelias refuses to honor the bargain. Once again, it is Medea to the rescue. Pretending she has a magic charm to make the king young again, she tricks his daughters into killing him. Outraged at this "regicide," the people of Iolcus force Jason and Medea to flee to Corinth, where they live happily for ten years and have two children. As fate would have it, though, the couple's life unravels when Jason falls in love with the king of Corinth's daughter. Not one to take such a betrayal lying down, Medea kills her two sons and flees to Athens, where she has a son named Medus with the king of Athens. Broken, sick, and old, Jason is sitting beneath the prow of the *Argo*, when a piece of it breaks off and kills him. Medea is later banished back to Colchis. But she lives on as a central character in the tragedies of the playwright Euripides.

MYTHIC VOICES

. . . Let no one think of me
As humble or weak or passive; let them understand
I am of a different kind: dangerous to my enemies,
Loyal to my friends. To such a life glory belongs.

—EURIPIDES, Medea *(431 BCE)*

Which Argonaut was a god of healing?

If you've ever been to a doctor's office, a pharmacy, or a hospital, you've probably seen it and wondered—why is there a symbol of a double snake entwined around a staff? This emblem of the medical profession is actually a mistake of sorts, and originates with Asclepius, depicted by Homer as a tribal wound-healer, and also one of Jason's Argonauts, who was no doubt brought along on the trip for his healing skills.*

Like Jason, Asclepius is raised by Chiron, the wise centaur. As a baby, Asclepius is sent to live with the mythical creature, after Asclepius's

*Asclepius was identified with a single snake. Perhaps because of its ability to shed its skin, it was viewed in ancient times as a sign of immortality. The staff with two snakes, chosen by the American Medical Association as an emblem, is actually the wand of Hermes, called a caduceus, which he used to conduct the dead to Hades.

divine father, Apollo, discovers that his lover, Coronis, had been unfaithful to him while pregnant with their child. Miffed at the betrayal, Apollo does what any jilted Greek god might do—he strikes this woman with a bolt of lightning. Before she dies, however, Apollo suffers remorse. As Ovid tells it:

> *But Phoebus flatly refused to allow the child of his loins*
> *to crumble to ashes, cremated in the funeral pyre of its*
> *mother,*
> *seizing the child from the womb, he bore it off to a cavern*
> *where dwelt the double-formed Chiron, the Centaur . . .*

Apollo rescues his unborn child, who goes on to become an important Greek god, revered as the inventor of medicine throughout both ancient Greece and Rome, where he was known as Aesculapius.

From ancient records, we know that the Greeks held Asclepius in very high esteem. During plagues and in times of illness, the Greeks prayed to him for help and relief, setting up special temples where they went to communicate with him. Epidaurus, the site of several ancient Greek ruins including a famous outdoor theater built in the 300s BCE, was a special gathering place of the first physicians, who were known as the Asclepians. The ruins of an ancient temple honoring their patron was also found near the Epidaurus theater, apparently a place where many sick people went in hope that Asclepius would cure them through their dreams while they slept in a nearby guesthouse. By 200 BCE, according to Roy Porter's history of medicine, *Blood and Guts*, every Greek city-state had its temple to the god, where sick pilgrims slept overnight in special incubation chambers before an image of the healer god.

The admiration for Asclepius in Rome was equal to that of Greece. Not only did the Romans build a major shrine to the healing god after their city was delivered from a plague, they equipped the Asclepian temples with baths to capitalize on the healing power of water. The priests of Asclepius supposedly had extensive knowledge of herbal cures and other natural remedies—what we might call "alternative" treatments today—and crowds flocked to the "spas" of the ancient experts for these remedies as people today seek out spa treatments around the world.

Interestingly, the revered Asclepius gets into trouble in Greek myth when he oversteps his bounds. This happens when he uses his healing gifts to try and revive a dead man. Offended by this, Hades complains to Zeus, who delivers to Asclepius the same fate as Apollo had dealt his mother—Asclepius is killed with a thunderbolt and sent to the underworld. When Apollo discovers what has happened, he grants Asclepius divine status as the god of medicine.

MYTHIC VOICES

I swear by Apollo Physician and Asclepius and Hygieia and Panacea and all the gods and goddesses, making them my witnesses, that I will fulfill according to my ability and judgment this oath and this covenant:

I will follow that system of regimen which, according to my ability and judgment, I consider for the benefit of my patients, and abstain from whatever is deleterious and mischievous. I will give no deadly medicine to any one if asked, nor suggest any such counsel; and in like manner I will not give to a woman a pessary to produce abortion. With purity and with holiness I will pass my life and practice my Art. I will not cut persons laboring under the stone, but will leave this to be done by men who are practitioners of this work. Into whatever houses I enter, I will go into them for the benefit of the sick, and will abstain from every voluntary act of mischief and corruption; and further from the seduction of females or males, of freemen and slaves. Whatever, in connection with my professional practice or not in connection with it, I see or hear in the life of men, which ought not to be spoken of abroad, I will not divulge, as reckoning that all such should be kept secret. While I continue to keep this Oath unviolated, may it be granted to me to enjoy life and the practice of the art, respected by all men, in all times! But should I trespass and violate this Oath, may the reverse be my lot!

—from the Hippocratic Oath (original version)

Was Hippocrates a man or myth?

In contrast to the mythical Asclepius, there is a historical basis to the life of the other most famous doctor of ancient Greece, Hippocrates. (430?-380? BCE). Often called the father of medicine, Hippocrates was a well-known ancient physician who practiced medicine on the Greek island of Cos. Hippocrates challenged the notion of using magic, myth, and witchcraft to treat disease. Taking the fairly radical step of dismissing "root-gatherers, diviners and others whom they dismissed as ignoramuses and quacks," as medical historian Roy Porter writes, Hippocrates and his followers believed that diseases had natural causes and could therefore be studied and possibly cured according to the workings of nature. As Porter puts it, "No longer pretending to be an intercessor with the gods, the true doctor would be the wise and trusty bedside friend of the sick."

While there is no evidence that he actually wrote the texts attributed to him, which probably derive from a variety of hands over time, Hippocrates is still credited with teaching his followers, the first physicians, to view the patient as a whole; accept that much healing takes place naturally; follow a simple diet to achieve good health; and regard the first duty of the doctor as to his patients rather than to himself. The profound maxim that permeates medicine today, "First do no harm," is attributed to his *Epidemics*, but is not actually part of the Hippocratic Oath, a modern version of which is still recited by many new doctors.

Mythic Voices

At a later time, there occurred portentous earthquakes and floods, and one grievous day and night befell them, when the whole body of your warriors was swallowed up by the Earth and the island of Atlantis in like manner was swallowed up by the sea and vanished; wherefore also the ocean at that spot has now become impassable and unsearchable, being blocked up by the shoal mud which the island created as it settled down.

—Plato

Was Atlantis ever discussed in Greek myth?

As fantastical places go, the so-called Lost Continent of Atlantis has had a long and intriguing history, peppered by inspiring stories, theories, bad science-fiction movies, a recent Disney animated feature, and even a sixties rock song by the pop singer Donovan ("Way down below the ocean/Where I want to be"). In the seventeenth century, a Jesuit writer published *Underwater World*, placing Atlantis in the Atlantic Ocean. Jules Verne included a description of Atlantis in his nineteenth-century adventure classic *20,000 Leagues Under the Sea*. But contrary to popular belief, the story of this ancient but highly advanced civilization that disappeared beneath the ocean is nowhere to be found in ancient myth—either in Hesiod or Homer. What we know of Atlantis—"the island of Atlas"—actually comes from a rather unlikely source—*Timaeus* and *Critias*. These two "dialogues" were composed by the philosopher Plato (428–348 BCE), the pupil of Socrates, who founded the Academy—later called the School of Athens—which flourished for more than nine hundred years.*

Plato readily acknowledged that what he knew of Atlantis had been handed down through a long series of storytellers and was possibly first spoken of by Egyptian priests—giving the impression of a long round of the game of "telephone," played out in ancient times. According to Plato's version, a brilliant and highly superior, wealthy and powerful civilization once existed on the isle of Atlantis, supposedly located beyond the Pillars of Heracles. This would have been the Strait of Gibraltar, which would place Atlantis in the ocean named after the legendary island, the Atlantic. But recently, others have argued that these pillars are actually the Bosporus Strait, which separates the Black Sea from the Mediterranean Sea, and that Atlantis truly existed in the Mediterranean.

In this legendary civilization, which supposedly flourished more than ten thousand years ago, "the most civilized men," as Plato described them, were descended from the sea god Poseidon and had created an earthly paradise. Food was plentiful, and the buildings and temples were

*The Academy started in a grove of olive trees shared with a public *gymnasium*, or place of nude exercise. The Academy was shut down by the Byzantine Roman emperor Justinian, who thought he was stamping out paganism.

magnificent. One of these temples, according to Plato's description, was "coated with silver save only the pinnacles and these were coated with gold. As to the exterior, they made the roof all of ivory in appearance, variegated with gold and silver. . . ."

As Plato described it, Atlantis was a great military power that could muster an army of more than a million men. But its people turned corrupt and greedy, so the gods punished them. During one day and night, great explosions shook Atlantis, and the continent sank into the sea. Plato's apocalyptic tale of Atlantis has fascinated people ever since, providing both serious archaeologists and plenty of more imaginative "occult" theorists with an appealing target for their investigations and theories. Over the years, numerous expeditions have attempted to locate the remains of Atlantis, but so far none has discovered a "lost island" beneath the Atlantic Ocean. Among the most popular of these theories was that of Edgar Cayce, a famed American clairvoyant and psychic healer who died in 1945. In best-selling books that have attracted millions of readers over the years, Cayce claimed that Atlantis was a highly advanced society that possessed the equal of modern technology, and he prophesied that Atlantis would rise again in the latter part of the twentieth century. Needless to say, that prophecy has not been fulfilled.

Aristotle, Plato's student, had what may be a sounder theory. He suggested that Plato had made up the story in order to illustrate his own philosophy of ideal government, thoroughly summarized in *The Republic*. In this utopia, an intellectual elite ruled. Drawn from the ablest people of all backgrounds and sexes, these educated, qualified people were to rule as "philosopher kings." In this ideal society, they would live communally, share food, lodging, spouses, and own no property. Ruled by knowledge, they would govern for the benefit of all the other classes in a virtuous society embodying the ideals of wisdom, courage, temperance, and justice.

While Plato was speaking allegorically, there may still be some historical basis to the Atlantean legend. The consensus among many archaeologists and historians is that the myth of Atlantis is probably based on the first major civilization in the region of Greece, which arose on Crete, an island that separates the Aegean Sea from the Mediterranean. Occupying a central position in the eastern Mediterranean, with proximity to Egypt, the Near East, and mainland Greece, Crete developed the first great seagoing power of the ancient world, beginning about 3000 BCE. It was this island culture, which produced lavishly decorated palaces, indoor plumbing, elegant pottery, and jewelry, that may have been the source of the legend of Atlantis.

Today, many scholars believe the cause of the cataclysmic destruction in the Atlantis legend was actually a volcano on the island of Thera in the Aegean Sea, about 70 miles (110 kilometers) north of Crete. Volcanic eruptions destroyed most of Thera about 1550 BCE, largely wiping out the Minoan civilization, which had flourished on both Thera and Crete. "Minoan" got its name from King Minos, the legendary ruler of Crete and central character in one of the most significant Greek myths—the story of Theseus and the Minotaur.

Is Theseus and the Minotaur just another "bull" story?

If you've ever been lost in a maze, played the game Labyrinth, or been accused of telling a "bull" story, you've been connecting with a famous Greek myth. What's the whole story of Theseus and the Minotaur?

According to the myth, Crete's king Minos asks the sea god Poseidon for a sign of favor. A beautiful white bull emerges from the sea, and Minos is then expected to sacrifice this wondrous animal to Poseidon. Instead, Minos keeps the white bull and substitutes a lesser animal. As anyone remotely familiar with the Greek gods knows, holding out on an Olympian is never a good idea. Poseidon angrily curses Minos by causing his wife, Pasiphaë, to fall in love with the white bull.

This is where the story takes a kinky turn. To satisfy her lust for the bull, Pasiphaë has Daedalus, the Athenian statue-maker, make her a wooden cow. While hiding inside of it, Pasiphaë is impregnated by the white bull and gives birth to a monster—a bull with a human head called the Minotaur. (Ovid relates this story in a poem in *Art of Love*, which concludes: "Well the lord of the harem, deceived by a wooden plush covered dummy/Got Pasiphaë pregnant. The child looked just like his dad.") In order to keep this grotesque reminder of his wife's bestial infidelity hidden from view, Minos orders Daedalus to build beneath his palace an escape-proof secret maze called the Labyrinth, to house the Minotaur.

Enter Theseus, one of the legendary men of the Heroic Age and the greatest hero of Athens. The son of King Aegeus of Athens, Theseus had been raised away from his home, unaware of his royal blood. But his father, the king, had once buried a sword and a pair of sandals beneath a rock. He told the mother of Theseus that when her son was strong enough to move the rock, he could claim his inheritance—an ancient inspiration for King Arthur's later "sword in the stone." At sixteen, Theseus found rock, sandals, and sword, and went to Athens to reclaim his place as heir.

When Theseus reaches Athens, his father does not recognize him. But the sorceress Medea, now married to the king, knows exactly who he is. She tries to have Theseus done in, but he survives all of her tricks.

Theseus's most dangerous adventure now lies ahead of him. Ever since some Athenians had killed the son of Minos, the city of Athens has been compelled to send seven youths and seven maidens to Crete each year to be eaten by the Minotaur. To end this tragic deal, the heroic Theseus announces that he will go as one of the youths to be sacrificed, and kill the Minotaur. The Athenian victims always sail for Crete aboard a black-sailed ship. Before departing Athens, Theseus promises his father that if all goes well, he will return in a ship flying white sails.

In Crete, Theseus encounters Ariadne, the daughter of King Minos, who immediately falls in love with the young hero and decides to help him kill the Minotaur. Ariadne gives Theseus a ball of thread she has received from Daedalus and tells the young man to trail it behind him as he descends into the Minotaur's lair. In one of the most memorable moments in myth, Theseus kills the Minotaur and retraces his steps through the maze's twisting passages by following the string. With the Minotaur dead, Theseus sails back for Athens. (There are different versions of Ariadne's fate. In the happy account, she marries the god Dionysus. But another says she died of a broken heart.)

What should be a triumphal moment for Theseus turns painfully tragic. First, in Crete, when Minos learns that the inventor Daedalus helped Ariadne in the plot to kill the Minotaur, he throws the inventor and his son Icarus into prison. While imprisoned, Daedalus constructs two sets of wings made from feathers held together by wax. According to Ovid, Daedalus tells his son where to fly—halfway between sun and water. But Icarus is daring and wants to fly higher. When he flies too close to the sun, the wax melts, and he plunges into the sea. According to Barry Powell, the story of Icarus is a mythical illustration of the Greek maxim "Nothing too much," one of the proverbs inscribed over the temple of Apollo at Delphi. Powell comments in *Classical Mythology*: "Doubtless because overdoing things was a common weakness of the Greeks, their sages were fond of preaching the virtue of the 'Golden Mean.'"

The second tragedy involves the heroic Theseus, who has forgotten his assurance to raise white sails if he should come back alive. In his hurry to return home, Theseus forgets to take down the black sails, and when Aegeus sees the returning ship, he jumps into the sea, thinking Theseus is dead. (The Aegean Sea is named for the dead king of Athens.)

With his father dead, Theseus became the king of Athens, and his rule is marked as the legendary beginnings of Athenian democracy. Theseus supposedly abolished the monarchy, minted the first coins, and created a unified state. Aristotle later viewed the story of Theseus and the Minotaur as an allegory for the victory of democracy over tyranny, and the story of Theseus became a national myth, Athenian propaganda. Historically speaking, the myths of Theseus are just that—myths. But as classicist Barry Powell puts it, "History and myth are a perennial tangle; humans are mythmaking animals, retelling ancient stories to fulfill present needs."

In actual history, Athenian democracy had its beginnings under the lawmaker Solon (639?-559? BCE), who led Athenian government until his retirement. One of his accomplishments was to reform the harsh Athenian laws drawn up earlier by Draco—a code so harsh it inspired the word "draconian." After Solon's retirement, Athenian democracy backpedaled under Solon's cousin, Pisistratus, and did not fully arrive until the thirty-year period under Pericles. Beginning about 460 BCE, Athenian democracy—while far from perfect—began to flourish. As historian Charles Freeman points out in *Egypt, Greece, and Rome*, "It remains unique as the world's only example of a successfully functioning and sustained direct democracy. It lasted for nearly 140 years—a remarkable achievement in a period of history where instability was the norm. It involved its citizens as officials, legislators and law enforcers in a way few modern democracies would dare to do and it is remarkable for breaking the traditional connection between political power and wealth. And all this when the city was also acting as a major and innovative cultural center."

Mythic Voices

What did they believe, these Greeks? Were the gods real to them or just metaphors? Certainly they did not have creeds or dogmas, confessional or doctrinal positions such as we have come to expect from religions. And just as certainly, there was a graduated spectrum of interpretation, as there must always be in things religious, that spanned classes and communities and that shifted in emphasis from one period to another. What is so striking about the Homeric gods—as opposed to the One that most of us are familiar with (though familiar is surely the wrong word)—is their lack of godliness. Oh sure, they have power beyond the dreams of the world's most powerful king, but they exercise this power just the way we would—heavy handedly, often mercilessly, even spitefully. And they are taken up with their own predictable domestic crises—who's sleeping with whom, who's getting back at whom, who's belittling whom. Could anyone actually believe in such gods?

—Thomas Cahill, Sailing the Wine-Dark Sea

What was the Delphic Oracle?

Of the many sacred places in ancient Greece, none was more significant than Delphi, home of the oldest and most influential religious sanctuary in ancient Greece. It was not just an important center—it was the center, literally. Delphi had come to be regarded as the omphalos, or navel, of the world, and the site was marked with a large conical stone. The sacred stone at Delphi was supposedly the very stone Rhea tricked Cronus into swallowing at the time of the Creation. After eating his first five children, the father god had swallowed this stone, wrapped in swaddling clothes, instead of the sixth child, Zeus. When Zeus later forced him to vomit forth the other children, this stone came out, too.

Delphi is near the Gulf of Corinth, on the slopes of Mount Parnassus, and a religious shrine was founded there sometime before 1200 BCE. Originally a shrine to Gaia, the earth goddess, the temple at Delphi by the eighth century BCE was dedicated to Apollo, the god of prophecy. For at least twelve centuries, the oracle at Delphi spoke on behalf of the gods, advising rulers, citizens, and philosophers on everything from their sex lives to affairs of state. The oracle spoke out, often deliriously, exerting wide influence.

As part of the ritual at Delphi, a petitioner brought an offering of sacred cake, a goat, or a sheep, before consulting the Pythia, the priestess of the shrine. After careful purification, Pythia sat on a tripod and fell into a trancelike state in which she received messages and prophecies from Apollo. In this trance, and sometimes in a frenzy, she would answer questions, give orders, and make predictions. Some scholars say her divine communications were then interpreted and written down by male priests, often in ambiguous verse. But others say the oracle communicated directly with petitioners.

For years, modern scholars have dismissed the theory that vapors rising from beneath the temple floor were responsible for the "inspiration," which is how the Greeks explained it. Despite many efforts, no underlying fissure or source of intoxicating fumes was ever found, and the vapors were assumed to be mythical, like much else about the site. But recent scientific work at the site is shaking that view. As reported in the *Scientific American* in August 2003, a geologist, an archaeologist, a chemist, and a toxicologist have uncovered a wealth of evidence that suggests the ancients had it exactly right. They have solid evidence that petrochemical fumes from the region's underlying rocks could rise to the surface to help induce visions. Specifically, the team found that the oracle probably came under the influence of ethylene—a sweet-smelling gas once used as an anesthetic, and which, in light doses, produces feelings of euphoria.

With the rise of Christianity, the temple eventually decayed and fell from favor. Around 361 CE, the Roman emperor Julian the Apostate tried to restore the temple, but the oracle wailed that her powers had vanished. In 390 CE, the Christian Roman emperor Theodosius permanently closed the temple as part of his drive to stamp out any vestiges of pagan worship.

Modern science may be reopening its lost secrets.

Mythic Voices

But do not worry about marriage with your mother;
No end of males have dreamed of sleeping with theirs.

—Sophocles, Oedipus the King

Do all little boys want to kill their father and sleep with their mother?

The oracle plays a central role in a myth that was made a household name by Sigmund Freud. An "Oedipal complex" is—in Freud's view—a boy's desire to compete with his father and sleep with his mother. But what is the myth behind the psychology?

Oedipus was born the son of Laius, king of Thebes, and his wife, Jocasta. An oracle said Laius would die at the hands of his own son—a story with echoes back to the beginnings of Greek mythic Creation—who would then marry his mother. To protect himself, Laius places the three-year-old Oedipus on a mountainside to die. The boy is discovered while still alive by a shepherd who gives him to Polybus, the childless king of Corinth, and his wife, Merope. The couple rear Oedipus as their own, and he grows up unaware of his mysterious past. But when he goes to Delphi and hears the same grim prophecy that had troubled Laius, Oedipus leaves home, believing that he is sparing his true father and mother from harm.

That is when fate strikes. As he heads toward Thebes, Oedipus is run off the road by a chariot and fights with the driver and passenger, killing them both in a case of ancient Greek "road rage." What Oedipus could not have known was that one of the men he has killed is his real father, King Laius. Part one of the prophecy is thus fulfilled.

But the Delphic Oracle had also predicted that the man who solved "the riddle of the Sphinx" will be king of Thebes and marry the queen.

On his way to Thebes, Oedipus next encounters the Sphinx, a creature with the head of a woman, the body of a lion, a serpent tail, and wings. Sent to plague the city after Laius had apparently disrespected the gods, the Sphinx lived on a high rock outside the city of Thebes, and would ask anyone who passed by to solve a riddle: "What has one voice and becomes four-footed, two-footed, and three-footed?"

A wrong answer results in death, and the Sphinx has been devouring Thebans one by one. Confronted by the Sphinx, Oedipus replies, "Man, who crawls on all fours as a baby, then walks on two legs, and finally needs a cane in old age."

Furious because Oedipus has solved the riddle, the Sphinx jumps off the rocky perch to her death. Having solved the riddle, Oedipus arrives in Thebes, where he is made king and marries the queen. Jocasta, of course, is unaware that her new husband is really her son. Part two of the prophecy has been fulfilled. They have two sons, Polynices and Eteocles, and two daughters, Antigone and Ismene.

Oedipus the King by Sophocles (c. 496–406 BCE), the second of the three great Greek tragedians, is the most famous play to treat this extraordinary story of confused identities. As the play opens, Oedipus is already the king, and is trying to discover why the city is suffering from a plague, not realizing that his own actions are the cause. Through a turn of events, Jocasta realizes what has happened and rushes to her bedroom. Then the truth is revealed to Oedipus as well. He goes to the bedroom and finds that Jocasta has hung herself. Cursing himself, Oedipus then puts out his own eyes. Blinded and bloodied, he returns to the stage and asks to be sent into exile. Oedipus ends his days near Athens, where he is secretly buried.

On its own, as myth and tragic drama, the story of Oedipus is powerful stuff. But it gained a completely different currency when the term "Oedipus complex" was first used by the Austrian psychiatrist Sigmund Freud in 1900. Freud used the Greek tragedy as support for his claim that every boy fantasized about killing his father and having incestuous sex with his mother, a desire that must be repressed. Freud described the "Oedipus conflict," or "Oedipus complex," as a state of psychosexual development and awareness that first occurs around three and a half years of age. Freud similarly claimed that all girls wanted to have sex with their fathers, what he called the "Electra complex."

Today, many psychoanalytic researchers and anthropologists have largely dismissed this idea. They believe that if such a complex develops, it is a result of personal factors and social environment, and is certainly not part of a universal mind-set.

Another in-vogue psychiatric term also gets its origins from Greek myth. Narcissism, often described as excessive or malignant self-love, comes from the brief tale of Narcissus. Son of a river god and a nymph, Narcissus was a boy of transcendent beauty. When his parents asked a seer if Narcissus would have a long life, they were told he would, as long as he did not see his own face. When he was grown, Narcissus loved no one until he saw his reflection in a pond. He stared at himself and reached to touch his face, falling into the pond and drowning. Ovid told a slightly different version of the myth, in which Narcissus was actually punished for his self-absorption. When he rejected the love of Echo, a nymph, she was so overcome by grief that she wasted away until only her voice remained. For his cruelty to the nymph, Narcissus was punished by drowning. After his death, Narcissus was changed into the flower that bears his name.

Mythic Voices

Whoever obeys the gods, to him they particularly listen. . . .

The Olympian is a difficult foe to oppose. . . .

The glorious gifts of the gods are not to be cast aside. . . .

Not at all similar are the race of the immortal gods and the race of men who walk upon the earth. . . .

Thus have the gods spun the thread for wretched mortals; that they live in grief while they themselves are without cares; for two jars stand on the floor of Zeus of the gifts which he gives, one of evils and another of blessings.

—*From the* Iliad

Is Homer just a guy from *The Simpsons*?

Two long poems. One is a bloody, blow-by-blow account of men hacking at each other. The other is a tale of a lost wanderer trying to get home. Homer's *Iliad* and *Odyssey* are still considered touchstones of Western culture. Yet we certainly don't know much about the guy who supposedly wrote them almost three thousand years ago.

The man we call Homer remains a mystery, for the most part, and scholars know almost nothing about the poet who has influenced our language and literature so thoroughly and significantly. Traditionally considered a blind Greek poet, Homer is thought by some scholars to have lived in a Greek-speaking city on the eastern shore of the Aegean Sea or on the island of Khios. But that is it. Beyond these meager clues, there is only the speculation of generations of readers and scholars.

For thousands of years people have argued over whether an actual Homer existed, and whether he wrote the *Iliad* and the *Odyssey*. New research into writing in Greece at that time, along with extensive studies of how oral poetry was composed and preserved, have changed the debate. There are several schools of thought concerning Homer. While one school contends that Homer actually composed and wrote down the poems himself just as writing emerged in Greek history, others say he was an illiterate bard who only sang the poems until writing emerged near the end of his life. At that point, literate scribes came to Homer's assistance and took dictation. Still a third school of thought contends that Homer's poems were memorized by a guild of public reciters called "rhapsodes"—the ancient Greek version of "wedding singers"—who carried on Homer's oral tradition until writing appeared in Athens much later.

During the twentieth century, researchers in the Balkan regions, where bards once sang, found living bards who still recite epics the length of Homer's and even longer. Accustomed as we are to instant news—with short attention spans and memories completely reliant on our Palm Pilots or Blackberries to keep track of a few phone numbers—to most of us that sort of expansive storytelling ability seems astonishing. But the Homeric epics originated in centuries long before Homer's time, when the bards were improvising and improving and even adding to older story lines. Most likely, the bards created a series of poems that told the entire story of the Trojan War, and it was Homer who may have given these stories their characteristic individual genius.

In *Sailing the Wine-Dark Sea*, Thomas Cahill persuasively makes the case that Homer not only existed but had firsthand experience with what he wrote. "Homer was thought to have been a wandering blind bard, but this is almost certainly due to Homer's description of a blind bard who performs in the *Odyssey*, later taken to be a self-description of the poet. Whatever the case, he must have been sighted, at least earlier in life, for there is too much in the *Iliad* of gritty reportage for us to think that the poet never saw battle. It would, in fact, be most unlikely if Homer did not serve as a soldier. . . . There is scarcely a Greek figure of any consequence who did not serve in the military as a young man or did not afterwards take a keen interest in warfare."

Blind or not, real or not, the man we call Homer transformed the way people experienced myth. And finally it all comes down to the poems anyway. When you compare the words, emotions, and action of his two epics to the earlier literature of mythology—in Egypt and Mesopotamia, for instance—you see how Homer humanized the myths. Certainly, his gods could be remote and powerful. But they were also powerfully human—with all the flaws that implies. They raged, they lusted, they envied, and, like Hera, they sought vengeance. And it was that sense of making the divine human that may lie at the heart of what Homer and the rest of the "Greek Miracle" was all about.

How did Homer fit a ten-year war into a poem?

First of all, the *Iliad*—which means a poem about Ilium (Troy)—is not the history of the Trojan War. Rather it describes events in the final year of the Trojan War, fought between armies of the kingdoms of Mycenaean Greece and the city of Troy, located on the coast of what is now Turkey. According to legend, the Trojan War lasted ten years, until Greece defeated Troy—all because Helen, the young wife of Sparta's King Menelaus, had run off with the handsome Paris, prince of Troy. But the story of the *Iliad*—divided into twenty-four books and consisting of more than 15,600 lines—covers only fifty-four days. And much of it describes only four days of fighting, separated by two days of truce. When it ends, Achilles is still alive and Troy not yet taken.

So, what's the story?

Believe it or not, stubborn men fighting over a beautiful woman. The war itself, of course, is ostensibly fought over Helen. But as the epic opens, an angry quarrel has broken out between Agamemnon, brother of Menelaus, king of the Mycenaeans and leader of the Greek forces, and Achilles, the greatest warrior among the Greeks. Agamemnon demands that a captured Trojan girl be given to him as war booty. But she is the daughter of a priest of Apollo, and the Greeks are advised to return the girl to her father. When Agamemnon refuses, Apollo strikes the Greek forces with plague, wiping out hundreds of warriors, until Agamemnon relents. In exchange for the girl he gives up, however, the stubborn Greek king demands another girl, who has already been given to Achilles as war booty.

This seemingly minor incident, and the conflict between Achilles and Agamemnon, is at the heart of the poem. More like a sulking child than the greatest warrior of all time, Achilles withdraws into his tent and refuses to fight. Without Achilles, the Greeks are driven back to their ships by the Trojan forces and their leader Hector, son of Troy's king Priam, and Troy's greatest champion.

Wearing the armor of Achilles, Patroclus—who is Achilles' closest friend, tent mate, and, many scholars contend, his lover—tries to lead the Greeks into battle. But he is no match for Hector, who kills Patroclus. With the death of his friend and comrade, Achilles is aroused to seek revenge. Given a new suit of armor made for him by the smith god Hephaestus, Achilles returns to the battle and, after slaughtering many Trojans, kills Hector outside Troy. Lashing the body of the fallen Trojan hero to his chariot, Achilles drags Hector around the walls of Troy and finally back to his own tent. He keeps Hector's body, executes some Trojan captives, and threatens to cut Hector to pieces until King Priam comes to plead with him. Achilles is commanded by the gods to grant Priam's request, but it is Achilles' own sense of human pity that makes him yield to the broken old man. He gives Priam the body for proper burial, and the story ends with the funeral of Hector.

For nearly 3,000 years, readers have found the *Iliad* a moving expression of the heroism, idealism, and tragedy of war. In addition to the battle scenes, the *Iliad* tells about life within Troy. It describes the emotional farewell between Hector and his wife, Andromache, who foresees his death. A great soldier, Hector is also a family man, who is called on to defend his country and, in so doing, loses his life. A reluctant warrior, he berates his brother Paris for causing the war but is also loyal to him. In many ways the truest "hero" of *Iliad*, Hector embodies Homer's themes of honor, loyalty, and social obligation.

Is the *Iliad* all there is to go on when it comes to the Trojan War?

In a word, no.

Many of the events that lead up to the *Iliad* are not actually described in the poem, which is sharply focused on the war itself, with vivid, pulsing descriptions of battle and the conduct of both men and gods. One of the most gruesome moments is told in the play *Agamemnon* by Aeschylus, and describes the sacrifice of Iphigenia, Agamemnon's daughter. Ready to sail, the Greeks cannot get a favorable wind, because Agamemnon had once slighted the goddess Artemis. To save the expedition, Agamemnon is advised to sacrifice his own daughter, Iphigenia. Deceived into believing that she is going to marry Achilles, the young girl is dressed in a wedding dress and brought to the altar only to learn that she is going to be sacrificed. Once Iphigenia is dead, the winds blow fair. (In another version, Artemis intervenes at the last second and sends a substitute animal, just as God gave Abraham a substitute ram in Genesis when he was about to sacrifice his son Isaac.)

Another scene not in the *Iliad*—this one about Achilles—comes from myth. As Ovid tells it, Achilles had been dipped in the River Styx as a baby. Therefore, he could never be wounded, except at the spot where his mother held him by the heel. Achilles dies when he is shot in the heel by Paris. This was, of course, the origin of our phrase "Achilles' heel," which means a person's weakness or vulnerable point.

Continuing the list of what is not in the *Iliad* is the actual fall of Troy.

This scene is described in the *Aeneid*, an epic poem by the Roman poet Virgil. The *Aeneid* tells how the Greeks build a huge wooden horse, "the Trojan horse," and place it outside the walls of Troy. Odysseus and other warriors hide inside the horse while the rest of the Greek army sails away. Although the prophetess Cassandra* and the priest Laocoon warn the Trojans against taking the horse into their city, they are ignored. But a Greek named Sinon, left behind to provide "disinformation," persuades the Trojans that the horse is a sacred offering, which will bring them the protection of the gods. The Trojans then pull the horse into Troy, and in the night, as the Trojans "sleep off" their victory celebrations, Odysseus and his companions creep out of the horse. The gates of the impregnable Troy are opened, and the Greek army storms the city, having returned from a nearby island where their ships had been hidden. The Greeks wipe out almost all the Trojans, burn Troy, and take back Helen.

And finally, the cause of the war itself comes from ancient myth, not Homer. The real troubles begin with an incident at a divine wedding feast. All the gods and goddesses have been invited except Eris, the goddess of discord. Eris is offended and tries to stir up trouble. She sends a golden apple to the feast, inscribed with the words "For the most beautiful." Three goddesses—Hera, Athena, and Aphrodite—each claim the apple for herself. Finally the handsome Paris, the son of Troy's King Priam, is brought in to judge the dispute. While all three goddesses try to bribe him, he awards the apple to Aphrodite, because she promises him Helen, the most beautiful woman in the world, the semidivine daughter of Leda and Zeus.

Helen is already married to King Menelaus of Sparta. But when Paris visits her, Aphrodite causes her to fall in love with the Trojan prince, and

*The daughter of Priam, Cassandra is so beautiful that Apollo falls in love with her and gives her the power of prophecy. But she rejects him, and Apollo punishes Cassandra by ordering that no one will ever believe her prophecies, including her own Trojan people when she advises them to return Helen. After Troy falls, Agamemnon takes Cassandra to Mycenae as a slave and ignores her prophecy of his death. After Agamemnon's wife, Clytemnestra, has the king killed, Cassandra is also murdered. Her name is now used to describe any prophet of doom.

she flees back to Troy with him. Paris has not only stolen his host's wife, he has broken a sacred code of being a proper guest. Menelaus and his brother, Agamemnon, organize a large Greek expedition against Troy to win back Helen—and for this she is, in the words of playwright Christopher Marlowe, "the face that launched a thousand ships/And burnt the topless towers of Ilium."

Was there really a Trojan War?

Did it happen? Was Troy real? Did Agamemnon, Helen, and Hector live and breathe? Or did Homer, like Shakespeare in his plays about real kings, embroider a tale that made the mortal immortal?

Since Heinrich Schliemann's nineteenth-century discovery of Troy, much digging has been done to try and get to the bottom of the Troy question. What Schliemann thought was Troy turned out to be actually a much earlier city, and after more than one hundred years of archaeology, scholars still don't agree on Troy and the legends of the war. While some think Homer's epic is an outright fiction, others believe it exaggerates small conflicts involving the Greeks from about 1500 to 1200 BCE. Still others say the legend of Troy is based on one great war between the Mycenaean Greeks and the city of Troy in the mid-1200s BCE. Archaeology and recent scholarship have combined to paint a portrait of this ancient face-off between two regional "superpowers." Archaeologists have found strong historical evidence in the ruins of Troy and other places that confirms certain events described in the epics.

In an article for the *Archaeology Institute of America* (May 2004), Manfred Korfmann, a director of excavations at Troy and a professor of archaeology at the University of Tübingen, had this answer to the question of the "real" Trojan war:

"According to the archaeological and historical findings of the past decade especially, it is now more likely than not that there were several armed conflicts in and around Troy at the end of the Late Bronze Age. At present we do not know whether all or some of these conflicts were distilled in later memory into the 'Trojan War' or whether among them there was an especially memorable, single 'Trojan War.' However, everything currently suggests that Homer should be taken seriously, that his story of a military conflict between Greeks and the inhabitants of Troy is based on a memory of historical events—whatever these may have been. If someone came up to me at the excavation one day and expressed his or her belief that the Trojan War did indeed happen here, my response as an archaeologist working at Troy would be: Why not?"

MYTHIC VOICES

Look now how mortals are blaming the gods, for they say that evils come from us, but in fact they themselves have woes beyond their share because of their own follies. . . .

All men have need of the gods. . . .

Olympus, where they say there is an abode of the gods, is ever unchanging; it is neither shaken by winds nor ever wet with rain, nor does snow come near it, but clear weather spreads cloudless about it, and a white radiance stretches above it. . . .

The gods, likening themselves to all kinds of strangers, go in various disguises from city to city, observing the wrongdoing and righteousness of men. . . .

So it is that the gods do not give all men gifts of grace—neither good looks nor intelligence nor eloquence. . . .

—*From the* Odyssey

Which crafty Greek hero can't wait to get home?

Home is a powerful idea—as anyone who has seen *E.T.: The Extra-Terrestrial* knows. But the little alien had it easy compared to the most famous homebound traveler in literature, Odysseus, the cunning king of

Ithaca. He is the star of the *Odyssey*, one of the most influential works in Western history and among the greatest adventure stories ever told. Scholars still fight over its origins, but, traditionally, the *Odyssey* is thought to have been composed by Homer, probably in the 700s BCE. The poem describes Odysseus's long journey home to Ithaca, an island off the northwest coast of Greece, after he fights against Troy. One of the heroes of the *Iliad*, Odysseus (changed in Latin to "Ulixes" and translated into English as "Ulysses"), is credited with the idea of the Trojan horse, and just as he used trickery to end the ten years of fighting, he relies on his wits to defy even greater odds in the *Odyssey*.

Like the *Iliad*, the *Odyssey* consists of twenty-four books, but it is considerably shorter, running some 12,000 lines long, and takes place over a period of about ten years. Unlike the *Iliad*, which is more of a tragedy, the *Odyssey* is an adventure tale, and in many ways more "fun." It has been called a "comedy," in the original sense of the word, which meant order was restored with the reuniting of a family. The good guys and bad guys are easily identifiable. Very different from Achilles or Hector, Odysseus is the crafty hero—resolute, curious, but mostly devoted, like E.T. or Dorothy Gale of Kansas, to getting back home after nearly twenty years away—ten of them fighting at Troy, three lost at sea, and seven more on the island of Calypso, where his tale begins.

Odysseus has been the prisoner of the sea nymph Calypso (whose name means "concealer"), when the gods of Mount Olympus decide that the time has come for him to return to Ithaca and his loyal wife, Penelope. During his long absence, she has been under pressure to accept that her husband is dead, and marry again so that Ithaca has a new king. Telemachus, the son of Odysseus, resents his mother's noblemen suitors, and the goddess Athena suggests that he go to seek news of his father. Telemachus sets off in search of him.

Meanwhile, the tale returns to Odysseus's adventures. When Calypso releases Odysseus, he sails away on a raft, but Poseidon—angry at Odysseus for reasons that will emerge—sends a storm that shipwrecks him. Washed ashore on a beach, he is discovered by Nausicaa, the beautiful daughter of the Phaeacian king. Sheltered by the Phaeacians, he recounts for them his years of wandering since the Trojan War when he set out for home with twelve ships carrying fifty men each.

First, he tells of his escape from the lotus-eaters, who consume a drug that makes men forget home and purpose. Next, he recounts his blinding of the Cyclops Polyphemus with a hot wooden stake. Odysseus had cleverly told Polyphemus that his name was "Nobody," so that the other Cyclopes would be befuddled when the wounded Cyclops roars that "Nobody" is trying to kill him. Concealing his crewmen under some sheep so they can pass by the blinded Cyclops, Odysseus and the crewmen eventually escape—but Odysseus then makes the mistake of taunting the giant and reveals his true name. Cyclops then prays to his father, Poseidon, who avenges the creature by vowing to make Odysseus's homecoming a nightmare come true.

After being blown off course, Odysseus sails on to the island of the enchantress Circe, who changes all of the crewmen into pigs but wants Odysseus for a lover. Protected from the spell of Circe by a magical herb, Odysseus beds Circe and subsequently learns how to return his crewmen to human form and sail past the sea monsters Scylla and Charybdis. The sorceress also tells Odysseus how to navigate past the Sirens, sea nymphs who use their beautiful singing to lure sailors to death on a magic island. Finally, she warns the men not to eat the sacred cattle of Helius (the sun).

Odysseus's ship survives most of these dangers and seems ready to reach Ithaca without further trouble until some of his men ignore Circe's warnings and eat the sacred cattle of the sun. As punishment, the ship is destroyed by a thunderbolt. All the men drown, except Odysseus, and he is washed up on the island of the beautiful nymph Calypso, who promises Odysseus eternal life if he marries her. After seven years on Calypso's island, Odysseus goes to the shore one day and weeps for his beloved wife, Penelope. Seeing this, Athena takes pity on him and asks Zeus to release Odysseus from his suffering. Odysseus builds a raft and lands on the island of the Phaeacians, where the young princess Nausicaa discovers him, naked, save for a strategically placed tree branch. The princess takes Odysseus to her father's court, where he begins to recount his adventures.

After Odysseus finishes his story, he returns home. Reunited with his son, Telemachus, Odysseus goes to the palace, dressed in beggar's rags. Penelope has spent years tricking her suitors by promising that she will choose one of them when she finishes a weaving, a project she unravels each night. The exasperated suitors demand that she finally choose among them, and she finally agrees to marry the man who can string Odysseus's great bow and shoot an arrow through twelve axes. Taking the bow himself, the disguised Odysseus wins the contest, then kills all 108 of the unarmed young suitors and is reunited with Penelope.

Did the Romans take all their myths from the Greeks?

The "Greek Miracle" in Athens—highlighted in the works of the three tragedians who based most of their works on the myths—soon came crashing down. A series of wars with rival Sparta began in 431 BCE. A great plague struck Athens, killing Pericles, among many others, in 429 BCE. In 404 BCE, Athens surrendered to Sparta, concluding the disastrous Peloponnesian Wars that had split Greece. Oligarchy, the rule of a few wealthy aristocrats, returned to Athens.

The doom of the Golden Age was sealed in 338 BCE, when King Philip from the northern province of Macedonia united all Greece under his rule. An era had ended. The curtains and lights had gone down on the glorious age of the city-state and all its remarkable accomplishments. But a new act was about to open in the drama of Greek glory when Philip was assassinated and replaced by his ambitious son, Alexander, a student of Aristotle, who, as Alexander the Great, spread Greek culture, language, and ideas throughout the eastern Mediterranean. Establishing his namesake city, Alexandria in Egypt, Alexander made it the center of Greek culture, a position it held for the next three hundred years. In the Hall of the Muses there, the classics of Greek literature were gathered, and science flourished as scholars took up the serious study of mathematics, astronomy, and medicine. In Alexandria, Greek-speaking Jews translated the ancient Hebrew writings into the first Greek version of the Bible, known as the Septuagint, and Apollodorus collected his library—the most complete and straightforward accounting of Greek myths from the creation of the world to the death of Odysseus. Alexander's massive effort to "Hellenize" his empire continued even beyond his death in the city of Babylon in 323 BCE.

But a new star was rising in the Mediterranean. A small tribe of Indo-European speakers* on the Tiber River, living near the future city of Rome, had begun to build an unparalleled empire that would, over the course of the next three hundred years, dominate and control the entire Mediterranean world and well beyond. These warriors first entered Greece in 229 BCE, and, in 146 BCE, sacked Corinth, and soon all of Greece became a Roman province. Instead of forcing their own myths and gods on the people they conquered, the Romans quickly absorbed the ideas and cultures of the conquered, especially the Greeks, whose glorious legends and stories they adopted as their own.

Roman mythology, in fact, largely seems a copy of Greek mythology. As Thomas Cahill put it, "Of the many people of Earth, the Romans may have had the most boring religion of all. . . . Contact with the impressive stories of Greek mythology and the thrilling art that accompanied them—a contact that began as a result of the Greek colonization of southern Italy—encouraged the Romans to dress up their own religion in Greek fashions."

From ancient times, the earliest Romans did possess a mythology of their own. In fact, many of the basic similarities between Roman and Greek mythology can be traced to the common Indo-European heritage shared by Rome and Greece. Before the Romans came into contact with Greek culture, they worshipped the gods of their direct ancestors, the Latini, who may have arrived on the Italian peninsula around 1500 BCE and were on the future site of Rome by about 1200 BCE. The native Romans had many of their own gods, including three major deities—Jupiter, Mars, and Quirinus—who are known as the "archaic triad." Jupiter ruled as god of the heavens and came to be identified with Zeus. Mars was god of war and occupied a much more important place in Roman mythology than did Ares, the war god of the Greeks. Quirinus, an agricultural god, eventually faded from prominence, absorbed by the Greek gods.

*"Indo-European" generally refers to people who lived in the area north of the Black Sea, in southeastern Europe. This culture worshipped a warrior god who ruled the sky. One group of Indo-Europeans migrated westward to what is now Greece and Rome. Another group migrated southward into northern India. Called the Aryans, they developed the warlike sky god Indra and are discussed in chapter 6.

By the late 500s BCE, the Romans replaced the archaic triad with the "Capitoline triad"—Jupiter, Juno, and Minerva—a name that came from the Capitoline Hill in Rome, where the main temple of Jupiter stood. In this new triad, Jupiter remained the Romans' chief god. They identified Juno with the Greek Hera, and Minerva with Athena. It was during the 300s BCE, as the Romans came into increasing contact with Greek ideas, that they began to worship Greek gods and goddesses, gave them Roman names, and built temples and shrines in their honor.

Between the 500s and 100s BCE, additional Roman mythological figures appeared, nearly all of them based on Greek divinities. Besides Greek-inspired divinities, the Romans worshipped many native gods and goddesses, including Faunus, a nature spirit later connected to Pan; Pomona, goddess of fruits and trees; Terminus, god of boundaries; and Tiberinus, god of the Tiber River.

The earliest Romans had believed that gods and goddesses had power over agriculture and all aspects of daily life. For example, Ceres was the goddess of the harvest and became associated with the Greek Demeter. Her festival was the Cerealia, a ceremony held in April (and the source of the word "cereal"). Her daughter Persephone became the Roman Proserpina. The goddess Vesta guarded the hearth fire and was associated with Hestia. The god Janus stood watch at doors and gates. As such, Janus looked both ways and controlled beginnings, which is how his name gets connected with the first month in the Roman calendar, January. Jupiter, later the supreme Roman god, was first worshipped as a sky god with power over the weather, which, obviously, connected him with Zeus. (Their names are also connected, according to most linguists, by the same Indo-European root words for "sky.") Liber, the ancient Roman god of wine, became associated with Dionysus and was also called Bacchus.

Despite the connections to Greek myths and deities, as Rome grew into a republic and then an empire, its religion was very different from that of the Greeks. It is true that, like the Greeks and other ancients, average Romans frequented temples, made sacrifices, embraced superstition, believed in the power of "augury," or divination, became fixated on astrology, and honored household deities. But the Romans were far less interested in myth or theology than they were in raw power, order, and Roman glory—enforced through military superiority and the rule of law known as Pax Romana. This is what made Rome tick, as the empire came to dominate the European and Mediterranean world. The Romans were far more concerned with building good roads on which their legions could travel than imposing their mythic traditions on the people they conquered. In fact, historian Charles Freeman notes, "Roman tolerance to local cults and even their readiness to join them was one important way in which the empire was cemented." When Julius Caesar and Augustus were both deified after their deaths, ushering in an era of emperor worship, it was a consolidation of political power, not a new theology. But it was one which the Roman citizen was wise to acknowledge.

Who were Romulus and Remus?

Unlike the Greeks, the Romans considered their divinities historical persons and used the myths to explain the founding and history of their nation. The best example of this historical emphasis is found in the story of Romulus and Remus, the legendary founders of Rome.

Romulus and his brother, Remus, were the twin sons of the war god Mars (Greek Ares) who had raped Rhea Silvia, the first of the vestal virgins, as she was bathing. For breaking her vow of chastity, Rhea Silvia was imprisoned and her babies taken from her, set afloat on the Tiber River in a small boat.* When the boat came to rest, the infant boys were found and rescued by a woodpecker and a she-wolf—the sacred animals of Mars. The she-wolf cared for the pair until a shepherd discovered the twins and raised them.

The pair became hunters and warriors who were so respected that men agreed to live under their rule in a new city. Romulus and Remus decided to build a city at the spot on the Tiber where the she-wolf had found them. But at the founding, a bitter quarrel erupted between the brothers, and they fought. Romulus killed his brother and wept over his

*In some versions it is a basket, an interesting parallel to the stories of Moses, the Mesopotamian Sargon, and the Persian Cyrus, great leaders who were all abandoned in baskets on rivers.

corpse. Recovering from his grief, Romulus built the new city of Rome, supposedly in 753 BCE.

In the city at first settled only by runaway slaves, bandits, and murderers, and with a dangerous shortage of women, the Romans realized that they needed wives. When a nearby group called Sabines came to a religious festival, the Romans rushd through the crowds, seizing the young Sabine women as captive brides, an incident frequently depicted in classical art as *The Rape of the Sabine Women*. This episode was followed by a fight between the Sabine tribes and Rome. At the request of Jupiter, the Sabine women stood between the opposing armies and demanded peace. The Sabines eventually joined Rome.

The Romans believed that Romulus became the city's first king, and, according to Roman mythology, he ruled for forty years before vanishing in a thundercloud. Romulus was supposedly the first of seven legendary kings who ruled Rome from its founding until the early 500s BCE. There is little evidence that these seven kings actually existed or that any of the events connected with their reigns ever took place. But it made for a good ending to the story of Rome's epic foundation.

Mythic Voices

Arms, and the man I sing, who, forc'd by fate,
And haughty Juno's unrelenting hate,
Expell'd and exil'd, left the Trojan shore.
Long labors, both by sea and land, he bore,
And in the doubtful war, before he won
The Latian realm, and built the destin'd town;
His banish'd gods restor'd to rites divine,
And settled sure succession in his line,
From whence the race of Alban fathers come,
And the long glories of majestic Rome.

—Virgil, The Aeneid
(c.19 BCE, translated by John Dryden)

Was Homer on the Romans' reading list?

Apparently so. Those Romans knew a good thing when they saw it. The national epic of ancient Rome, the *Aeneid,* is largely modeled on the great Greek epics, the *Iliad* and the *Odyssey*.

A complex poem celebrating Roman virtues and giving the new empire a glorious past, the *Aeneid* was written by the Roman poet Virgil (also sometimes spelled Vergil) between 30 and 19 BCE. Virgil chose the mythical Trojan hero Aeneas, son of the goddess Aphrodite and Anchises, a prince related to the royal house of Troy, as a way of expressing Rome's ancient moral and religious values. Composed to honor Augustus, the first emperor, who was later believed to be a descendant of Aeneas, the *Aeneid* comprises twelve books. The first six of these books imitate the *Odyssey* by describing Aeneas's adventures at sea following the capture of Troy by the Greeks.

As the *Aeneid* begins, Aeneas and his Trojan followers have survived a shipwreck and reach Carthage, a city actually founded by the Phoenicians in North Africa about 800 BCE—hundreds of years after the *Iliad*'s Troy might have fallen. Once ashore, Aeneas meets and falls in love with Carthage's Queen Dido, and recounts for her court the fall of Troy: the well-loved story of the wooden horse, the tales of Sinon and Laocoon; and his own escape. Then, just as Odysseus had regaled the Phaeacians with his tales in the *Odyssey*, Aeneas spins the long history of his adventures.

Dido and Aeneas are soon caught up in a steamy romance, but the gods have Roman destiny to worry about. They order Aeneas—the soul of that destiny—to leave Dido. In despair and anger, Dido commits suicide, cursing Aeneas and his descendants with her dying words. Later, after reaching Italy, Aeneas goes down to the underworld—where he encounters Dido and his dead father—and learns about his future descendants, the Romans. He returns to the upper world and, with his followers, lands at the mouth of the Tiber River in Latium.

Virgil based the last six books of the *Aeneid* on the *Iliad*, and these begin as Aeneas arrives near the future site of Rome. There, the local king, Latinus, offers him land for his people and marriage to his daughter, Lavinia, who had already been promised to a local king. War erupts between the locals and the Trojan survivors. The battle is hotly contested, and finally Aeneas and the rival king agree to settle the conflict by single battle. Aeneas wounds his opponent, and is about to show him mercy when he sees a reminder of a friend he had lost—as Achilles had lost Patroclus. He plunges his sword into the breast of the warrior king.

Aeneas founds a town called Lavium, after his wife, Lavinia, before he dies in battle. Aeneas's son, Ascanius, later moves the town to Alba Longa, where twelve generations—or 450 years—later the twins Romulus and Remus are born.

Mythic Voices

While Paul was waiting for them in Athens, he was deeply distressed to see that the city was full of idols. So he argued in the synagogue with the Jews and the devout persons, and also in the marketplace every day with those who happened to be there. Also some Epicurean and Stoic philosophers debated with him. Some said, "What does this babbler want to say?" Others said, "He seems to be a proclaimer of foreign divinities." (This was because he was telling the good news about Jesus and the resurrection.) So they took him and brought him to the Areopagus [a hill west of the Acropolis] and asked him, "May we know what this new teaching is that you are presenting? It sounds rather strange to us, so we would like to know what it means." Now all the Athenians and the foreigners living there would spend their time in nothing but telling or hearing something new.

Then Paul stood in front of the Areopagus and said, "Athenians, I see how extremely religious you are in every way. For as I went through the city and looked carefully at the objects of your worship, I found among them an altar with the inscription, 'To an unknown god.' What therefore you worship as unknown, this I proclaim to you. The God who made the world and everything in it, he who is Lord of heaven and earth, does not live in shrines made by human hands, nor is he served by human hands, as though he needed anything, since he himself gives to all mortals life and breath and all things. From one ancestor he made all

nations to inhabit the whole earth, and he allotted the times of their existence and the boundaries of the places where they would live, so that they would search for God and perhaps grope for him and find him—though indeed he is not far from each one of us. For 'In him we live and move and have our being'; as even some your own poets have said."

—*The Acts of the Apostles, 17:22–28*

What were the Bacchanalia and the Saturnalia?

Following a miraculous conversion, the Apostle Paul, a Jew from Tarsus (in what is now Turkey) who had once persecuted followers of Jesus, spent years traveling the Greco-Roman world of the first century, preaching the gospel, or "good news," of Jesus Christ. This biblical passage described his experience in Athens, where he tried to convince first-century Athenians that Jesus was the one god.

This scene was followed by another interesting episode, in which Paul caused a riot. During a trip to Ephesus, home of the temple to Artemis, known as one of the Seven Wonders of the Ancient World, Paul continued to preach against idols. But the silversmiths and other craftsmen, who made a good living crafting idols and statues in that Greek city, were not happy with a man preaching a religion that said, "Get rid of your false idols". The silversmiths started a riot and captured two of Paul's traveling companions. A reasonable town clerk stepped in and quieted the crowd, ultimately giving the tradesmen some very modern advice: If they wanted to do something about Paul and the other Christians, they should sue!

The Western world had reached another crossroads: the introduction of the Apostle Paul and the New Testament. Although the Romans crucified Jesus in Jerusalem for treason c. CE 30, his followers spread Christianity throughout the empire. Paul, a Roman citizen, would eventually go to Rome, where he was imprisoned and, legend has it, killed. Peter, one of the original twelve disciples of Jesus, also supposedly died in Rome during the persecution of the early Christian Church. But Rome was about be transformed. And when it was, some ancient practices would collide head-on with Christianity.

During the time of the Roman Empire (roughly from 27 BCE to 476 CE), Roman religion in the empire increasingly centered on the imperial house, and Emperor Augustus himself was deified after his death, as his uncle Julius Caesar had been deified after his assassination. Yet, as Thomas Cahill writes in *Sailing the Wine-Dark Sea*, "Roman religion was basically a businessman's religion of contractual obligations. . . . Not only were there few Roman myths, there was virtually no theology . . . the very enigmas that sparked the speculations of the earliest Greek philosophers."

The exception to the "boring" Roman religion might have been the Bacchanalia, wild and mystic festivals celebrating the Roman (and Greek) wine god Bacchus. Introduced into Italy around 200 BCE, the Bacchanalia were held in secret and attended first by women only. Admission to the rites was later extended to men, and the notoriety of these festivals, which from earlier Greek times had an air of drunken revelry and probably sexual liberty attached to them, came to be viewed as a threat. In Rome, the cult grew to the point that it was thought that crimes and political conspiracies were being hatched at the Bacchanalia. That led in 186 BCE to a decree of the Senate that severely restricted the festival. In spite of the harsh punishment inflicted on those found in violation of this decree, the Bacchanalia were not stamped out, particularly in the south of Italy, for a very long time.

Another popular Roman festival was the ancient celebration of Saturnalia, a thanksgiving holiday marking the winter solstice and honoring Saturn, the god of agriculture. The Saturnalia began on December 17, and while it only lasted two days at first, it was eventually extended into a weeklong period that lost its agricultural significance and simply became a time of general merriment. Even slaves were given temporary freedom to do as they pleased, while the Romans feasted, visited one another, lit candles, and gave gifts.

All of the similarities between Saturnalia and Christmas are no accident. Christians in the fourth century assigned December 25 as Christ's birthday because pagans already observed the day as a holiday. This would sidestep the problem of eliminating an already-popular holiday while Christianizing the population. In 350, Pope Julius I declared that

Christ's birth would be celebrated on December 25. There is little doubt that he was trying to make it as painless as possible for pagan Romans to convert to Christianity; the new religion went down a bit easier with them when they realized that their feasts would not be taken away from them. (Another mythical connection to this special Christian date is the birth of Attis, a vegetation god from Asia Minor who was the consort of a goddess known as Cybele, another "foreign" goddess that the Romans were drawn to worship. Her temple in Rome, appropriated by Christians in the fourth century, was on the site of the Vatican.)

From the time Rome had conquered Greece, even more exotic religions were finding their way into the empire, including the worship of the Egyptian goddess Isis and Mithraism, a Persian mystery religion of male initiates that flourished in the Roman Empire in the second and third centuries CE. Roman soldiers may have brought this cult of the Persian god Mithra back to Rome, one of a whole crowd of mystery religions competing for converts in the empire. The historian Plutarch (46–125 CE) reported that the worship of Mithra was introduced to Rome by captive pirates brought back from Cilicia. By around 100 CE, it had become widely popular among Roman bureaucrats, soldiers, and slaves. Among the legions, this was especially so, with Mithraism's strong emphasis on honor and courage, the brotherhood of the Good combating Evil. It had several similarities to Christianity, including a holy day celebrated on December 25, and was popular enough to warrant suppression by the Christian fathers by the fourth century.

It was in this rather fertile ground of competing cults that Christianity made its debut in the Roman world. Despite persecutions, usually at times of civic tensions beginning with Nero—who was, according to many biblical authorities, the "Beast" with the infamous number 666 in the Book of Revelation—Christianity steadily gained converts. Things changed permanently with the reign of Constantine I, who was named emperor of Rome's western provinces in 306 CE. In 312, Constantine defeated his major rival after having a vision promising victory if he fought under the sign of the Christian cross. In 313, Constantine and Licinius, emperor of the eastern provinces, granted Christians freedom of worship. And after Constantine defeated his coemperor in 324, he moved his capital to Byzantium in 330, renamed the city Constantinople (modern Istanbul, Turkey), and made Christianity the officially supported religion in the Roman state.

After Constantine died in 337, his three sons and two of his nephews fought for control of the Roman Empire. One of the nephews, Julian—later called the Apostate—became emperor in 361. A student of the Greek classics, Julian had been drawn to the Greek gods and underwent a "pagan conversion." As emperor, he tried to check the spread of Christianity and restore the traditional Roman religion. In 363 CE, Julian was killed in an attempt to invade Persia. By the late 300s, Christianity was well established as the official religion of the empire, and Rome was becoming Christianity's central city. All cults, save Christianity, were prohibited in 391 CE by an edict of Emperor Theodosius I. The empire was permanently split into the Western Roman Empire and the Eastern Roman Empire after Emperor Theodosius I died in 395.

The Western Roman Empire grew steadily weaker. The Vandals, Visigoths, and other Germanic peoples invaded Spain, Gaul, and northern Africa. In 410, the Visigoths looted Rome, and the empire "fell" in 476, the year that the Germanic chieftain Odoacer forced Romulus Augustulus, the last ruler of the empire, from the throne. The Eastern Roman Empire survived as the Byzantine Empire until 1453, when the Turks captured Constantinople.

CHAPTER FIVE

AN AGE OF AXES, AN AGE OF SWORDS

The Myths of the Celts and Norse

In this great carnage on Murtheimme Plain Cuchulainn slew one hundred and thirty kings, as well as an uncountable horde of dogs and horses, women and boys and children and rabble of all kinds. Not one man in three escaped without his thighbone or his head or his eye being smashed, or without some blemish for the rest of his life. And when the battle was over Cuchulainn left without a scratch or a stain on himself, his helper or either of his horses.

—from the *Táin*,
translated by Thomas Kinsella

Break no more my heart today—
I will reach my grave soon enough,
Sorrow is stronger than the sea . . .

—"The Poem of Derdriu,"
from *The Exile of the Sons of Uisliu*

An age of axes, an age of swords, shattered shields
An age of tempests, an age of wolfs, before the age of men
crashes down.

—Poetic Eddas

The Romans, in their first encounters with these exposed, insane warriors, were shocked and frightened. Not only were the men naked, they were howling and, it seemed, possessed, so outrageous were their strength and verve. Urged on by the infernal skirl of pipers, they presented to the unaccustomed and throbbing Roman sensorium a multimedia event featuring all the terrors of hell itself.

—THOMAS CAHILL,
How the Irish Saved Civilization (1995)

How do we know what the Celts believed?

Did the Druids practice human sacrifice?

What did Druids have to do with Stonehenge?

Who's Who of the Celtic Gods

What was *The Cattle Raid of Cooley*?

How does eating a mythical fish make you really smart?

What do the Celts have to do with Halloween?

What is the *Mabinogion*?

What mythology besides Celtic came storming out of northern Europe?

How do a giant's armpit and a cow help create the Norse world?

Who's Who of the Norse Gods

Who is the most important hero in Norse myth?

MYTHIC MILESTONES

Celtic and Northern Europe
Before the Common Era

3500–3200 Stone circles and alignments and rows of standing stones are built throughout northern and western Europe.

Stonehenge begun in southern England (completed about 1500); its alignment with the sunrise on the summer solstice seems connected to its purpose. Sacrifice of some kind may have taken place there as well. The quarrying, mining, and transportation of these large stones over long distances suggests a sophisticated social organization, but no written records have been found.

c. 3000 Elaborate passage graves are constructed in Ireland.

c. 2300 European Bronze Age; bronze objects begin to appear in tombs.

c. 1200 Urnfield culture emerges in Danube area, so named because cremated ashes are placed in large urns in communal burial fields.

c. 1000 Earliest fortified hilltop sites in western Europe.

c. 800 Celtic Iron Age begins in Hallstatt (Austria).

753 Rome founded.

c. 500 Graves in France show Greek and Etruscan imports—indications of trade between Celts and Mediterranean civilizations; burials include chariots and weapons.

450 Celtic La Tène culture emerges in west and central Europe and a distinctive art style arises. The La Tène style emphasized elaborate patterns of interwoven curves and spirals and featured highly stylized plants and animals that had little resemblance to nature.

c. 400 Celts expand into British Isles.

Greece's Golden Age flowers in Athens.

390 Celtic tribes burn Rome.

c. 350 Celtic tribes cross to Ireland.

272 Celtic invaders sack Delphi in Greece.

228 Celts settle Galatia in Asia Minor (modern Turkey).

c. 100 Fortified Celtic settlements are built in western Europe.

70 Rome's Golden Age: Cicero, Ovid, Virgil.

58–50 Julius Caesar completes conquest of Gaul.

31 Octavian becomes Emperor Caesar Augustus.

Common Era

9 Three Roman legions are destroyed by German tribes on the Rhine.

47 Britain invaded by Romans.

100 Legendary Queen Medb (Maeve) of Connacht reigns in Ireland.

122 Emperor Hadrian builds defensive walls and towers to fortify the northern boundary of Roman Britain.

166 German tribes invade northern Italy.

253 Germanic invasion into Gaul cripples the prosperous northwestern provinces.

378 Mistreatment of the Visigoths by Roman officials causes uprising; the emperor Valens is killed and his army wiped out.

401 Patricius, a Briton, is taken into slavery in Ireland. He will later become known as St. Patrick.

406 German tribesmen invade the Roman Empire.

410 Final withdrawal of Roman Legions from Britain.

Alaric the Goth sacks the city of Rome.

431 Council of Ephesus declares that the Virgin Mary is the Mother of God.

432 Bishop Patrick arrives in Ireland; converts Irish Celts to Christianity.

441 Anglo-Saxons start to colonize England.

451 Attila the Hun defeated at Troyes.

455 In sea attack launched from Africa, Vandals sack Rome.

476 The last Roman emperor, Romulus Augustulus, is deposed; he is replaced by Odoacer, "king of Italy," which marks the end of the Roman Empire in the West.

c. 500 Brigid (later St. Brigid) founds an abbey at Kildare, Ireland.

597 St. Augustine converts Anglo-Saxons to Christianity.

636 Lindisfarne Monastery founded.

789 First recorded Viking raid on England at Weymouth.

793 Vikings plunder Lindisfarne Monastery off British coast.

866 Vikings occupy British city of York.

870 Vikings settle Iceland.

902 Vikings establish a permanent base at Dublin.

911 Vikings found Duchy of Normandy.

982 Vikings settle Greenland.

986 Vikings reach North America and establish settlements.

999–1000 Christianity accepted in Iceland.

1016 Danish king Canute crowned king of England.

1066 Battle of Hastings: Normans—descendants of the Vikings—invade and conquer England.

c. 1220 Prose Edda, Norse myths compiled by Snorri Sturluson.

Picture this. It is about fifty years before the birth of Jesus, a typical day in the ancient world. In Greece, philosophers and their students stroll the streets of Athens, thinking Big Thoughts as they walk past centuries-old temples and statues gracefully carved from elegant marble. In Egypt—where the pyramids are already more than two thousand years old!—the Library of Alexandria is filled with scholars reading great works of classic literature, contemplating philosophy, drawing maps of the world, and studying higher mathematics and astronomy. In Rome, a Classic Age of poets and writers has begun to flourish and, before long, Imperial Rome will spread its language, law, martial order, and carefully constructed roads across the Mediterranean and European world.

But on a remote battlefield somewhere in Europe, the Roman general Julius Caesar leads his well-ordered legions against a howling band of naked warriors. These barbarians rush into battle with weird musical instruments—shrieking pipes made out of animal skins and strange, curved trumpets. If they win the day, these "savages" will surely take their Roman enemies' heads as trophies and sacrifice hundreds of captives in ceremonies led by priestly magicians called Druids. These Druid priests don't worship gods in majestic temples in city centers. Their gods are everywhere around them, a host of mythical spirits that fill every forest, field, mountain, lake, and spring. Even in the strange and mysterious circles of stones that dot the European landscape.

Descended from an ancient people of Indo-European origins, these savage warriors are called Celtae or Galli by the Romans, and Keltoi and Galatatae by the Greeks. Today they are known by a catchall word as the Celts.*

While great and glorious civilizations rose and fell in the Mediterranean worlds of Egypt, Mesopotamia, Greece, and Rome, the rampaging Celts eked out a seminomadic existence in a world of savage cruelty and nearly constant warring that was far removed from their contempo-

*The NBA's famed Boston franchise is commonly pronounced *sehl-tics*, but the word "Celts" is more accurately pronounced *kelts*, though some authorities prefer *shelts*.

raries in Alexandria, Babylon, Athens, or Rome. Migrating across Europe over a thousand years of history, the Celts had settled uneasily on the fringes of the Roman Empire—and of its eventual successor, the Church of Rome. By the start of the first century, their principal outposts were in Ireland and the British Isles, and Brittany, in northwest France.

What little we know of these wild people stretches much further back. According to hints from history and archaeological clues, the Celts first settled in northern Europe before occupying a wide swath of territory that spread across most of western Europe. Based on what we learned from digs in Austria and Germany, they are first known to have lived in Hallstatt, near Salzburg, where hundreds of Celtic graves have been unearthed, dating from about 700 BCE. At such sites as Hochdorf in Germany, other sets of Celtic graves revealed bodies buried with entire horse-drawn wagons filled with luxury goods—obviously meant for people who thought that they were going somewhere else in the next life. Unfortunately, they did not leave a "Swiss Alps" version of the Egyptian Book of the Dead to help succeeding generations discern just what it was they were thinking.

But around 500 BCE, something happens. Just as Athens entered its Golden Age and the Roman Republic was born, the Celts began to spread across western Europe. Around the same time, or possibly around 350 BCE, groups of them crossed the seas to the British Isles and Ireland, where they established their most enduring societies. The reasons for this mass migration are still unclear—climate changes, famine, and overpopulation are all likely suspects. But the Celts were on the move. And they were fierce, as the Romans and Greeks would learn. In 387 BCE, a group of Celts attacked and burned Rome in its early days. Another group of Celtic raiders ransacked the sacred Greek Temple at Delphi in 279 BCE.*

*Among the many Celtic offshoots were the people addressed by St. Paul in his Epistle to Galatia, a region around what is now Ankara, Turkey. This significant letter made clear that, in Paul's view, Gentiles (non-Jews) did not have to become Jews before becoming Christians. And Paul included a very specific catalog of vices that may have been commonplace in Galatia, among them fornication, impurity, licentiousness, idolatry, sorcery, envy, drunkenness, and carousing. But the Galatians—and other Celts—surely had no monopoly on this sort of behavior in the first-century world.

The Celts were also on a collision course with destiny. While terrifying and not easily subdued, they never achieved true "nation" status, remaining loose collections of tribes led by warlord kings. Plagued by constant warring among themselves, the Celts began falling to the onslaught of more "civilized" opponents between 300 BCE and about 100 BCE. During this time, the Romans conquered much of Celtic Europe, basically wiping out most vestiges of Celtic society on the continent, absorbing some bits and pieces of their myths into Roman worship or merging Roman beliefs and gods with the local deities. When the Celtic leader Vercingetorix managed to unite many of the Celtic tribes in Gaul, it was a last gasp. In 52 BCE, Caesar obliterated them after a hard-fought, eight-year-long campaign. Two thousand survivors of one battle were spared, but Caesar had all of the warriors' hands cut off. Their leader, Vercingetorix, was later executed in Rome. The only Celts who preserved their own culture for any length of time were those sheltered by the sea, on the British Isles and in Ireland—a Celtic stronghold that never succumbed to the Roman Empire but finally did submit to the Roman Church when St. Patrick converted the Irish Celts to Christianity. That is why so many elements of Celtic myth, belief, and worship are associated with the Irish, Welsh, Scottish, and British branches of the Celtic tree.

When we think of the Celts today, the image is one of a fraternity house gone really bad. Loud, boisterous, lots of feasting and drinking—especially before a battle. That impression would be largely correct. According to historian William K. Klingaman in *The First Century*, "Nothing terrified the common Roman soldier of this age more than the nightmarish prospect of capture, torture and mutilation by the Druidic priests. . . . Facing civilized Greeks or even the ferocious Parthians was one thing: battling barely human enemies, who according to rumor, drank human blood and roasted human flesh, was quite another."

But that is only part of the story. Free-spirited, clannish, and primitive, the Celts had a softer side. They could also be poetic, artistic, even romantic—and deeply religious. Although their ancient spiritual practices might leave much to be desired today, the Celts were powerfully connected to the gods of the natural world. Theirs was a religion of sacred groves and hilltops, pools and springs. They believed in the healing power of water; and sacred plants—like the evergreen mistletoe—

were used to cure diseases, promote fertility in women, and celebrate life in the midst of winter.

From what little has survived of the earliest Celtic myths, we know they found their gods all around them—in earth, water, woods, and in the animals they prized, especially horses. While their sun god was important, he was not an overpowering deity, as in Egypt. Perhaps that made sense in a colder, often darker part of the world where the sun didn't shine as often or as brightly. But just as the myths of Egypt, Mesopotamia, Greece, and Rome illuminated their cultures, so did the legends of the Celts shed light on a people who would find a unique place in Western civilization.

MYTHIC VOICES

> **As a nation they are extremely superstitious. People suffering from diseases, as well as those who are exposed to danger in battle, offer human sacrifices at ceremonies conducted by the Druids. They believe that the only way of preserving one man's life is to let another man die in his place. Regular tribal sacrifices are held, at which colossal figures made of wickerwork are filled with living men, and then set alight so that the victims burn to death. They think that the gods prefer the sacrifice of thieves and bandits, but whenever there is a shortage of criminals, they do not hesitate to make up the number with innocent men.**
>
> —JULIUS CAESAR, The Battle for Gaul

How do we know what the Celts believed?

In dealing with the Celts—and especially their myths and beliefs—we are a bit like the proverbial six blind men touching an elephant: each feels a different part of the animal and makes a very different assumption about the creature he is touching. When it comes to understanding the Celts, there are lots of disconnected parts, but it is hard to see the whole picture.

Unlike the great civilizations before them, the Celts left very few indelible marks. They were mostly a nonliterate people who produced no lasting writings in their earliest known periods—no *Gilgamesh*, no Book of the Dead, no *Iliad*, no Holy Bible. Although they went from being nomadic wanderers to settled farmers, the Celts never built large cities and left no records or bureaucracy to provide insights into their habits and customs. Some of their Druid priests did have a rudimentary form of writing, but if they recorded any religious writings, myths, poetry, or hymns, none survive. An identifiable Celtic Creation story has never been found.

That leaves us with a handful of other sources, including writers from the Classical Period in Rome, chief among them Julius Caesar—bane of generations of Latin students. Caesar and other Roman reporters often recount a Celtic fascination with rituals that the "civilized" Romans found barbaric, including human sacrifice, headhunting, strange forms of divination, and an attitude toward life after death that the Romans found curious. But because these writers were looking down their prominent noses at a people they considered well beneath them, Roman views of the Celts must be taken with a healthy grain of salt.

Archaeology also offers some clues to who the early Celts were and how they lived, but here, too, there are large gaps in the record. The Celts did not leave behind pyramids and temples, libraries filled with cuneiform tablets, and ancient cities waiting to be unearthed, such as Knossos, Troy, or Nineveh. Sacred spaces of Celtic worship often consisted of open-air enclosures, like a grove of sacred oaks, or holy lakes and springs. The Celts dug deep pits or shafts in order to communicate with the mysterious powers of the underworld. But the more enduring places that survive from Celtic settlements were often tainted by later conquerors. For instance, the Celts considered the famed mineral waters found at Bath, England, to be sacred healing waters associated with Sulis, an otherwise obscure local goddess of these thermal springs. After the Roman conquest of Britain, the site was transformed by the Romans into Aquae Sulis ("Waters of Sulis"), with a temple to a goddess the Romans called Sulis Minerva, simply attaching the name of one of their familiar deities to that of the existing local goddess. Later generations of British royalty turned the waters of Bath into a regal spa, and it finally became a Victorian-era resort where the English aristocracy could "take the waters."*

As for the early Celtic burial sites uncovered in Alpine Germany, these, too, have yielded some clues to their myths, religious practices, and beliefs. But even some of these recent finds date from the post-Roman era and are sometimes tainted by Roman influence. There are a few surviving images of Celtic gods from the pre-Roman period, which depict a god with the horns of a stag. And some stone figurines show three seated women, presumably representing a three-person mother goddess as a maiden, a mother, and an older woman. But museum shelves aren't exactly groaning with impressive collections of Celtic statuary and decorated pottery. Getting a visual impression of early Celtic culture is, ultimately, slippery business.

There is one shining bright spot in this otherwise dimly lit room of the Celtic past. One branch of the Celtic family tree deserves a laurel wreath for record-keeping. Fortunately, the rich oral traditions, foundation stories, and tales of gods and legendary heroes of the Celts who settled in Ireland, Wales, and southwest England were preserved. And several important collections of Irish and Welsh myths capture the voice and spirit of this pre-Christian Celtic world.

Granted, these sources come with a big red warning label attached. Most of the surviving tales were not written down until the eleventh and twelfth centuries, long after Ireland and the British Isles were Christianized in the fifth century. In Irish and British monasteries, the literate monks—the same ones who are largely responsible for preserving the Bible during Europe's Dark Ages—recorded many of the traditional Irish and Welsh Celtic stories, but probably laundered the Celtic originals, layering them with biblical or Christian sentiments. But beggars can't be choosers. These Irish and Welsh tales are the best we have—and they have made an enormous contribution to Irish and British literature.

*The sacredness of water to Celts is also attested to by the thousands of coins found in the springs of Bath. We know the Celtic practice of throwing an object into the water was widespread, based on discoveries of many objects—including swords and shields—in lakes and wells near Celtic sites. This vestige of Celtic belief lives on in the commonplace practice of throwing coins in wishing wells and fountains.

Of these later sources, three from Ireland are most significant and entertaining. The Book of Invasions (*Leabhar Gabhala*); the Ulster Cycle—which includes a masterful Irish epic called the *Táin Bó Cúailnge* (pronounced *toyn boe kool-ee*), or *The Cattle Raid of Cooley*; and the Fionn (Fenian) Cycle were all written down in Irish monasteries that would be crucial to preserving the written word during the Middle Ages. A fourth collection, the *Mabinogion*, was written in Wales, although exactly how it found its way into print is a mystery. The oldest known fragments date to 1225, but the oldest complete *Mabinogion* is dated to around 1400.

The first of these collections—the Irish Book of Invasions (*Leabhar Gabhala*)—is a twelfth-century attempt to compile a "history" of Ireland. Certainly derived from a much older oral tradition—just as *Gilgamesh* or Hesiod's *Theogony* had been—it describes a series of five successive mythical occupations of Ireland, including a generation said to be descendants of the biblical Noah. Such a biblical flourish was typical of the medieval Christian attempt to add a touch of religious "legitimacy" to these old pagan myths. It concludes with the arrival of the ancestors of the Celts in Ireland.

At the center of this account is the story of the last race of gods in Ireland, the Tuatha ("tribe," or "people"), told in a foundation myth known as the Tuatha Dé Danaan. The Tuatha—the "people of the Goddess Danu"—were the fourth of five races that invaded Ireland and fought two battles for supremacy. In the first, they defeated the clumsy Firblogs. The second was against the Fomorians, a race of misshapen, violent, and evil beings who controlled the country. But after defeating the Fomorians, the Tuatha gave them the province of Connacht. Because this account provides a list of most of the divinities that the Irish Celts worshipped before they were Christianized after 400 CE, it is a valuable resource for piecing together the rudiments of late Celtic mythology. The Tuatha were ultimately replaced with the arrival of the Celts, who were said to come from Spain (perhaps Celtic Galicia, hence the derivation of the word "Gaels," for Irish). Following their defeat, the Tuatha retreated to the underground mounds called *sídh*, where they continued to play a major role in Irish legend as the "little people," aka leprechauns.

The second collection of tales is called the Ulster Cycle, and the most important of these is the *Táin Bó Cúailnge*, or *The Cattle Raid of Cooley*—and often referred to simply as the *Táin*. Combining ancient myth with legends of early Irish heroes, the *Táin* is Ireland's *Iliad* and *Aeneid* all wrapped into one, a story that describes the conflicts between two of Ireland's northern provinces, Ulster and Connacht. Steeped in the supernatural, the *Táin* features a goddess-queen Medb (Maeve), who may well be based on an actual historical figure, and Ireland's greatest national hero, Cuchulainn (*koo-hool-n*), an Irish version of Gilgamesh, Hercules, and Achilles.

The third group of significant Irish stories is found in the Fionn (Fenian) Cycle, also compiled in the twelfth century, which chronicles the adventures of another Irish folk hero, Finn MacCool, and his band of warriors, called the Fianna, who are famed for their great size and strength. Again, these characters are legendary figures, probably based on real people—just as the *Iliad* may have been—although they also interact with true mythic deities. The events in the Fionn Cycle are believed to hint at the actual political and social conditions in Ireland around the year 200 CE.

Finally, the *Mabinogion* is a collection of Welsh tales that was also compiled sometime in the twelfth century CE. These stories describe the mythical history of parts of Britain, though many of the gods who appear in Welsh mythology largely resemble the Tuatha Dé Danaan in Irish mythology, possibly because Irish Celts migrated to Britain and took their myths with them. These stories are significant not only because they offer a view of Welsh Celtic myths, but also because they introduce the first references and early tales of a figure who would evolve over centuries into the legendary King Arthur.

MYTHIC VOICES

(Druids) concern themselves with questions of ethics in addition to their study of natural phenomena. And because they are considered the most just of all, they possess the power to decide judicial matters, both those dealing with individuals and those involving the common good. They have been known to control the course of wars, and to check armies about to join battle, and especially to judge

cases of homicide. . . . And both they and others maintain that the soul and the cosmos are immortal, though at some time in the future fire and water will prevail over them.

—STRABO *(63 BCE-24 CE?),* Geography
(translated by Timothy Gantz)

Did the Druids practice human sacrifice?

When they weren't storming around on horseback, sacking villages, and plundering their enemies, the ancient Celts had time to gather for worship ceremonies in natural, outdoor settings, like forests, where the oak was considered especially sacred. But before you conjure up some pastoral image from a Walt Disney film in which the birds, rabbits, and other forest creatures join forces to gently drape daisy chains around the neck of some benevolent Merlin-like character, consider this—human sacrifice was clearly part of the deal for the Celts. Clubbing, a sliced jugular, garroting—being strangled with a knotted cord—and drowning were all among the usual methods. While the Romans were antagonistic toward the Druids, and some of their reports may be exaggerated, sacrificial victims may have also been burned in giant wicker baskets wrought in the shape of a human figure, as Julius Caesar reported. The first-century Roman writer Tacitus recorded that Druids analyzed the death throes and blood flows of sacrificial victims to divine the future. Then the body might be tossed in a bog.

In 1984, the mummified remains of a man were dug out of a peat bog in Lindow Moss, near Manchester, England. Peat is an excellent natural preservative and the fellow in the bog—since known as Lindow Man—was exceptionally well preserved. Hands uncalloused, indicating he was probably highborn and not a laborer, Lindow Man might have been an Irish Druid prince. We even know what Lindow Man ate before his ritual death—bits of a blackened hearth cake that included traces of mistletoe. Then his skull was flattened with three blows of an ax; he was strangled by a cord knotted three times; and his blood was emptied with a slice through his jugular. According to authorities on the Celtic world who studied his remains, Lindow Man may have offered himself as a sacrifice to the gods in order to aid in the defeat of the Romans then assaulting Britain, in about 60 CE. He was a willing victim—a sacrifice for the good of his people.*

But, to be fair, early Celtic worship was not just about human sacrifice. Archaeological evidence from burial sites suggests that the Celts believed in the afterlife as well as the immortality of the soul. They provided their dead with weapons and other necessities to carry along on their journey. Sometimes, with a buried body, they placed small wheels that were intended to be emblems of the sun, to provide light in the afterlife.

The Celts were also pantheists who revered a range of nature deities, including the gods of thunder, light, water, and sun, as well as stags and horses. Concerned about having a continuous food supply, they looked to gods like Sucellos—the "Good Striker"—who made sure the plants woke up in the spring. Sucellos did this by striking the winter-hardened earth with the long-handled hammer he always carried.

Perhaps the least understood—and, recently, most romanticized—aspect of Celtic belief was the class of hereditary priests called Druids. Skilled in magic and fortune-telling, they advised kings and chieftains, served as judges in trials, and oversaw religious ceremonies—including sacrifices—often in groves of oak trees. (Linguists suggest a connection between the words "druid" and "oak.") In Celtic Ireland, Druids were also "knowledge-keepers," who memorized the tribe's history—as opposed to the bards, who sang the legends, and seers called *filidh*, who kept the sacred traditions and managed, unlike Druids, to survive into the Christian era. Though few historical reports exist, one memorable and oft-cited passage by a Roman writer describes how the Druids dressed in white robes and used a golden sickle to cut down mistletoe. The sacred plant they called "all-heal," mistletoe was thought to possess the miraculous power to cure disease, promote fertility in women, make poisons harmless, protect against witchcraft, and generally bring bless-

*Perhaps this helps explain why the Irish Celts so readily accepted the teachings of St. Patrick (c. 389-461 CE) when he explained that Jesus had also sacrificed his life to save his people. That concept may have appealed to the Celts, along with the three-leafed shamrock that Patrick supposedly used to illustrate the idea of a Holy Trinity, as three was a number sacred to all Celts.

ings and good luck. It was also baked into the cake eaten by Lindow Man before his ritual death.

In fact, mistletoe was considered so sacred that even enemies who happened to meet beneath it in the forest would lay down their arms, exchange a friendly greeting, and keep a truce until the following day. From this old custom grew the practice of suspending mistletoe over a doorway or in a room as a token of peace. The use of this once-powerful Druidic plant in modern Christmas festivities is just one example of the crossover of Celtic and other pagan customs to Christian practices. But when Britain was converted to Christianity, the bishops did not allow the mistletoe to be used in churches, because it was considered the central symbol of a pagan religion.

What did Druids have to do with Stonehenge?

As one famous newspaper's slogan suggests, "Inquiring minds want to know." And inquiring minds have been wondering for centuries—did ancient Celtic religion have anything to do with the megalithic monument called Stonehenge?

Located in southwestern England—not too far from the waters of Bath—Stonehenge is one of the world's most recognizable sites and inspiration for many theories, both serious and pseudoscientific. It has attracted the curious, the superstitious, and the scientific for hundreds of years, yet remains shrouded in mystery. Were these huge stones—weighing tons and moved from hundreds of miles away—set in a circle on an open landscape as an ancient calendar or "clock" that helped primitive Britons measure the seasons? Or were they another landing pad for alien visitors who needed a terrestrial parking spot? Or was Merlin, the famed magician from the legend of Arthur, behind the Stonehenge mystery?

That last idea, introduced by the early "historian" Geoffrey of Monmouth, had Merlin magically construct the monument as the "Giant's Dance" to commemorate a battle victory. It is an idea that ties in with one popular theory in New Age circles—that Stonehenge was some sort of gigantic altar where the Druid priests made sacrifices, since Merlin had "Druid priest" written all over him. It is certainly conceivable that Druids found Stonehenge to be a prime spot for their own worship ceremonies—though what those ceremonies were remains a matter of conjecture. We don't have a neat set of hieroglyphics describing a Druid-led dawntime observance of the summer solstice with the first rays of the sun breaking through the gaps between these giant stone plinths. Or an etching of a ceremony on a midsummer day with the famed "Heel Stone" of Stonehenge casting a long phallic shadow into the center of the stone ring, in a symbolic "Midsummer Marriage" of Father Sky coupling with Mother Earth.

Lacking solid, authoritative evidence of Stonehenge's original purpose, people will keep speculating. As they do, it is important to keep one fact in mind: according to most authorities, Stonehenge existed long before the Celts arrived in Britain. Once arrived, Celtic Druids may have appropriated Stonehenge for their religious ceremonies. But they most likely didn't build it. According to recent archaeological findings, this ancient monument was erected in three main phases that may date back to around 3300 BCE and continued for nearly two thousand years, until about 1500 BCE. The monument's famous ring of large stones is thought to have been built between 1800 and 1700 BCE, but the Celts probably did not arrive in the British Isles until 350 BCE. And while some may argue for a much earlier date of around 700 BCE, that is still centuries removed from the construction of Stonehenge.

WHO'S WHO OF THE CELTIC GODS

This list is divided into two parts. Part I comprises the chief gods as they would have been known to early Celts in Europe before they fell to the Romans and Druidism was suppressed. Part II focuses on the chief gods and mythical characters of the Irish Celts, as preserved in the later written collections.

Part I: Early Celtic Gods Worshipped Across Europe

Belenus The Celtic god of agriculture, Belenus also represents the life-giving and healing power of the sun and was associated with Apollo by the Romans, who created their own "Apollo Belenus." The great festival of Belenus, called Beltane ("bright, or goodly, fire"), was celebrated on May 1 of the Roman calendar with bonfires lit to rekindle the earth's warmth. Animals were led past these fires to be purified and protected against disease, and some scholars believe that this practice may have been connected to the nursery rhyme line about "the cow jumping over the moon."

Cernunnos (pronounced *Kur-noo-nohs*) Called the "horned one," Cernunnos (his Latin name, given by the Romans) is among the most ancient of the Celtic gods and his origins are linked with the horned figures depicted in the Paleolithic or Stone Age European cave paintings found in northern and central France and Britain. With the antlers of a stag, Cernunnos was seen as lord of the beasts, a "shape shifter" who also took the form of a snake or wolf.

A pastoral and agricultural god of both fertility and abundance, Cernunnos is thought to dispense fruit, grain, and wealth. But he is also associated with the small "solar wheels" that the Celts placed in graves, presumably as emblems of the sun to provide light in the underworld.

Epona Known as the "horse goddess," Epona is also associated with the earth and fertility, and is one of the most popular Celtic goddesses. In a very ancient story, it was said that Epona was born when her father, who hated women, mated with a horse. Epona is one of the few deities to whom stone monuments were erected that still survive, most of them in France. Representations of Epona usually show her with a horse, revered in the Celtic world for its beauty, speed, bravery, and sexual vigor. Sometimes Epona was shown riding sidesaddle or standing between two ponies.

When Roman cavalry officers learned of Epona, they adopted her and held an official Roman festival in her honor each year on December 18. She is the only Celtic deity to be accorded the honor of a Roman festival.

Nantosuelta The goddess whose name meant "wandering river," Nantosuelta was once thought to be a water goddess, but is now more often viewed as a fertility goddess—water being seen as a powerful symbol of birth. The patron of hearth and home, she is the consort of Sucellus, an agricultural god, and she is usually depicted carrying a basket of apples.

Sucellus Sometimes described as the "king of the gods," Sucellus is a male fertility deity whose name meant the "good striker." Always depicted carrying a long-handled hammer, he uses this tool to wake up the plants and herald spring.

Taranis The thunder god, Taranis rides across the sky in his chariot, which emits thunder from its wheels and lightning from the sparks of his horses' hooves. A powerful Celtic war god, Taranis was equated by the Romans with Jupiter (like Zeus, the god of thunder) and sometimes with their war deity, Mars. (He is also connected to the Norse god Thor; see below.) The Roman writer Lucan singles Taranis out as the god to whom human sacrifices were made, although more recent scholarship shows sacrifices were made to several Celtic gods. Seven altars dedicated to Taranis are known to have existed in the Celtic world, all dating from Roman times.

Part II: The Celtic Gods of Ireland

Brigid Known as the "exalted one," Brigid is an Irish fertility and war goddess. Supposedly raised by a Druid, she is a divine "multitasker," responsible for healing, fire, blacksmiths, poetry, wisdom, and protecting the flocks. As the saying goes, a woman's work is never done.

Her holy day, called Imbolc, is one of the four major Celtic religious festivals of the year, an important springtime event celebrating ewes coming into milk—a powerful symbol of rebirth and fertility for Irish Celts. It was also traditionally a time during which a wife or husband could legitimately walk out of their marriage.

More intriguing than Brigid's mythical stories are the parallels between this pagan goddess and her sixth-century namesake, St. Brigid (450 CE-523 CE), who blinded herself in order to avoid an arranged marriage and become a nun. The goddess Brigid is known for her generosity, and St. Brigid became one of Ireland's patron saints, known for her miraculous ability to feed people and perform endless acts of kindness. St. Brigid also tended a fire that was said to burn continuously for hundreds of days, just as the goddess Brigid was associated with the ritual fires of purification. Finally, St. Brigid's feast day is celebrated on February 1, the same day that Imbolc, the festival of the goddess Brigid, had been celebrated.

Daghda Known as the "good god," Daghda is viewed by the Irish people as the "father of the gods," but could never be confused with a deity like Zeus. Think John Goodman: kindly, fat, and somewhat uncouth. Wearing an obscenely short tunic, Daghda drags around a gigantic weapon on wheels—a magic club with the power to kill at one end and restore life at the other. A god of magic, wisdom, and fertility, Daghda is also the "provider" god, who possesses an enormous and inexhaustible source of food that comes from the "cauldron of Daghda." His never-empty cauldron was later connected to the Holy Grail supposedly used by Jesus at the Last Supper and brought to the British Isles by Joseph of Arimathea.

The son of the great goddess Dana, Daghda freely mated with many goddesses, but his coupling with the battle goddess Morrigan was most significant, because it was thought to provide security to the Irish people. Most likely a localized version of the Celtic agricultural god Sucellos, Daghda had other names as well—Aed (fire), Ollathir (all-father), and Ruad Rofessa (lord of the great knowledge).

Dana (Danu) Mother of Daghda, Dana is the mother goddess of the entire divine race known as the Tuatha. In Irish myth, when the Tuatha are supplanted by the Celts, they retreat to underground hills and are transformed into the fairies, or "little people" of later Irish folktales. Dana finds underground residences for all of them, and these are the "fairy mounds" (*sídh*) that provide many legendary place-names around Ireland. Two famous mounds in County Kerry are known as the "paps [breasts] of Anu," another form of the great goddess's name.

Lugh Associated with sunshine and light, Lugh (pronounced *loo*) is the "shining god" as well as a fierce warrior, magician, and craftsman, related by blood to both the Tuatha Dé Danaan and the rival Fomorians. Among the many marvelous weapons he forges are a sword that cuts through anything and a spear that guarantees victory. Once the Tuatha are supplanted in Ireland and transformed into the legendary "little people," Lugh becomes the craftsman Lugh Chromain ("little stooping Lugh"), whose name was later Anglicized as the word "leprechaun."

Another vestige of his name is found, somewhat ironically, in the capital city of Ireland's colonial conqueror. The "fortress of Lugh" became Lugdunum, Latinized by the Romans into Londinium, which later became London.

Lugh's festival, called Lughnasa, was celebrated on August 1 and was one of four pivotal Celtic Irish holidays, meant to mark the beginning of the harvest. It plays a central role in Irish playwright Brian Friel's *Dancing at Lughnasa*.

Morrigan (Nemhain, Badbh, Macha) Known as the "phantom queen," Morrigan (pronounced *more-ree-an*) is a shape-shifting goddess of horses and war, who can change from human being into animal forms. Whenever Morrigan appears as a raven, death is nearby, and she is often seen to be waiting at a river ford for warriors to pass so that she can determine which will die in battle that day. Standing in the river and washing the corpses of the dead, she is also called "the washer at the ford."

One of Morrigan's most important roles comes in the great Irish story the *Táin*, when she unsuccessfully attempts to seduce the hero Cuchulainn. (Pronounced *koo-hool-n*; see below.) Intent upon making war, not love, this warrior hero rejects her advances, and in doing so, seals his fate.

Nuadu (Nudd) Supreme king of the Irish Celtic pantheon, Nuadu is the legendary ruler of the Tuatha, but loses his arm in battle and must relinquish his kingship. Later given a magical arm of silver, he is able to reclaim the throne, but he loses his courage in later wars and has to retire, giving the throne over to Lugh.

Mythic Voices

The first warp-spasm seized Cuchulainn, and made him into a monstrous thing, hideous and shapeless, unheard of. His shanks and his joints, every knuckle and angle and organ from head to foot, shook like a tree in the flood or a reed in the stream. His body made a furious twist inside his skin, so that his feet and shins and knees switched to the rear and his heels and calves switched to the front. The balled sinews of his calves switched to the front of his shins, each big knot the size of a warrior's bunched fist. On his head the temple-sinews stretched to the nape of his neck, a mighty, immense, measureless knob as big as the head of a month-old child. His face and features became a red bowl: he sucked one eye so deep into his head that a wild crane couldn't probe it onto his cheek out of the depths of his skull; the other eye fell out along his cheeks. His mouth weirdly distorted: his cheek peeled back from his jaws until the gullet appeared, his lungs and liver flapped in his mouth and throat.

—from the Táin, *translated by Thomas Kinsella*

What was *The Cattle Raid of Cooley*?

This grim description of the transformation of a handsome young Irish hero into a dreadful killing machine is the picture of Cuchulainn, the greatest warrior of Irish myth and folklore, and a central character in the Ulster Cycle and one of its central stories, *The Cattle Raid of Cooley* (*Táin Bó Cúailnge*).

First written down in the Christian era, the Ulster Cycle has an overlay of Christian culture, but the stories are about an older, darker time in Ireland, hundreds of years before the arrival of St. Patrick and Christianity. Said to have taken place about the time of Jesus Christ, the Cycle has a slim basis in fact, since its stories may be a recounting of the actual struggles among early Irish groups. But the stories in the Cycle have been layered with myth, legend, and fantastic episodes of sex, drinking, and killing—in approximately equal measures.

Although there are conflicting versions of his birth, Cuchulainn's tale begins when Lugh, the chief of the gods, impregnates Deichtine, the sister (or daughter) of Conchobor (pronounced *connor*), legendary king of Ulster, in a dream. The child she bears—Sétanta—possesses extraordinary power due to his divine parentage, and gains further strength when he is tutored by goddesses in the art of war. But the boy gets into hot water when he is attacked by the watchdog of the smith god, Culann, and kills the animal. Culann angrily demands restitution, and the boy agrees to stand in as watchdog until a new animal can be trained. As a result of this episode, Sétanta's name is changed to Cuchulainn—"the hound of Culann."

When little Cuchulainn grows up, he is a strikingly handsome man and a ferocious warrior who turns into an appalling vision of terror when a battle frenzy—usually translated as the "warp-spasm"— seizes him. Armed with a magic spear called the Gae Bulga, which can inflict only mortal wounds, and accompanied by a charioteer who makes his chariot invisible, Cuchulainn is a fierce headhunter who always takes the most heads. To help him regain his mortal shape after battle, naked maidens are paraded in front of him and he is lowered into three successive barrels of icy water until he has cooled off—clearly the ancient Celtic version of the proverbial "cold shower."

In the *Táin*, the character of Cuchulainn is equaled only by Queen Medb (Maeve), the legendary warrior queen of Ulster's rival province, Connacht. Although here a mortal queen, the mythical Medb was also a powerful goddess of fertility—headstrong, powerful, dominating, and sexually ravenous. Her name meant "she who intoxicates"—figuratively and literally—and is closely connected to the medieval drink mead. As Celtic authority Miranda Jane Green put it, "Her rampant promiscuity symbolizes Ireland's fertility, and the association of her name with an alcoholic drink is linked with the concept of the union between goddess and mortal ruler. . . ." Before a battle, she would calm the troubled warriors who knew they had to fight the next day. As Thomas Cahill writes—and this was no myth—"Insensate drunkenness was the warrior's customary prelude to sleep."

In the *Táin*, Medb only marries the older King Ailill because he has money. The *Táin* actually opens with a comic scene in which the king and queen are arguing in bed over who is the wealthier of the two. Ailill says, "It struck me today how much better off you are today than the day I married you." Medb replies that she brought him such a great dowry when they married that he is essentially a "kept man." Just as a petty argument in ancient Greece among Hera, Aphrodite, and Athena over which was most beautiful led to the Trojan War, this contentious "pillow talk" soon leads to wholesale bloodshed, destruction, and death.

When Medb's husband proves that he indeed owns more than she does—he has one more bull than she does, a special white one—the queen, determined not to be outdone, orders her men to steal a famous bull called Donn Cúailnge, the Brown Bull of Cooley, which is held in rival Ulster. But her men are thwarted by Ulster's hero, Cuchulainn, who single-handedly fights off the invaders. Frustrated by the hero of Ulster, Medb plots to kill him and employs army after army without success.

The story ends with grim irony. While all the blood is being shed by men, the Brown Bull of Cooley is off fighting with King Ailill's White Bull of the Connacht, an epic contest that rages all over Ireland. Finally, the Brown Bull—the prize first sought by Maeve—defeats the White Bull. But as it returns to Ulster, the exhausted animal dies, collapsing in blood, vomit, and excrement—not a pretty picture. All of the fighting and death have essentially been for naught, and the hero Fergus, Medb's lover and leader of the men of Connacht, offers a moral that could just as well have been applied to the *Iliad*: "We followed the rump of a misguiding woman. It is the usual thing for a herd led by a mare to be strayed and destroyed."*

As the *Táin* ends, the story is not yet finished. Other tales in the

*In legend, the author of the *Táin*, Fergus, is another Irish hero of superhuman size, strength, and sexual appetites. He was the king of Ulster before Conchobar. When one of his lovers agrees to sleep with him only if her son can be king of Ulster for a year, Fergus consents. But the son, Conchobar, proves to be so popular that Fergus is not permitted to return to the throne. Fergus and Conchobar later argue over Conchobar's cruelty to Deirdre, a young woman who throws herself to death under the wheel of a chariot rather than marry at Conchobar's command, leading Fergus to join the rival armies of Connacht and become Queen Medb's lover.

Ulster Cycle complete the legend of Cuchulainn. Medb recruits sorcerers—children of a man that Cuchulainn has earlier killed—who will do away with the supernatural hero. Finally, either killed by his own magic spear or struck by a magic spear thrown by one of these sorcerers, Cuchulainn is mortally wounded. But he secures himself to a rock, so he can die in an upright position. For three days—the Celts did love the number three—he throws back the invaders, time and again. But even his courage and superhuman strength are not enough. Finally a raven, the symbol of the war goddess Morrigan, lands on Cuchulainn's shoulder, and Ulster's great hero expires. A legendary warrior, Cuchulainn grew in Irish folk stature until he came to be treated as a defender of all Ireland. At Dublin's main post office, scene of the famous 1916 Easter Uprising, in which Irish republican fighters battled British forces, there is a statue of the mythical hero in death, almost a Christlike figure from a Pietà, with the raven of death alighting upon his shoulder.

As for Medb, she dies when her nephew, using a sling, hits her in the head with a lump of hard cheese.

How does eating a mythical fish make you really smart?

For years, mothers told children to eat fish. "Brain food," they always called it. It was advice not lost on Finn MacCool, an Irish superhero who stars in the Fenian* Cycle of tales, set in the province of Leinster around 200 CE. One popular legend tells how MacCool came to possess great wisdom by burning his thumb while cooking the Salmon of Knowledge. Yes, you read that right. MacCool is a young man working for the Druidic poet Finnegas, when he is given a fish to cook. But it is no ordinary fish. The Salmon of Knowledge possesses all the world's wis-

*"Fenian" becomes an important name later in Irish history. Beginning in the 1850s, it was used by Irish nationalists struggling to free the country from British rule. Many Fenians also belonged to a revolutionary secret society called the Irish Republican Brotherhood, founded in the United States. The Fenians had a great influence on a later generation of Irish nationalists, and after years of rebellion and guerrilla warfare, Ireland won independence in 1921—although the province of Ulster remained in British control. Political heirs of the Fenians, the nationalist party Sinn Féin ("we ourselves") began as a self-reliance movement in 1905.

dom, and the bard Finnegas has spent seven years trying to catch it. When the old poet gives the boy the fish with instructions on how to cook it, he warns young MacCool not to eat even a bite. But while cooking the fish, MacCool burns his thumb and puts it in his mouth to ease the pain. Finnegas realizes immediately that the boy will gain all the knowledge, and tells him to eat the rest of the magical salmon. From that day on, MacCool needs only put his thumb in his mouth when he has a problem, and the solution is revealed.

The Fenian Cycle includes other stories featuring MacCool. Among the most famous and popular stories is "The Pursuit of Diarmuid and Grainne," a bittersweet tale of lost love in which MacCool is about to marry Grainne, the beautiful daughter of an Irish king. But when Grainne sees one of MacCool's warriors, Diarmuid, at her wedding ceremony, she instantly falls in love, leaves her fiancé at the altar, and elopes with her new beau.

With a band of his best warriors, the Fianna, MacCool sets off in pursuit of the lovers, and much of the story describes the adventures of Diarmuid and Grainne as they flee MacCool, aided by Oenghus, the god of love. The chase goes on for sixteen years until the jilted MacCool relents and pardons the lovers, who settle down at Tara, legendary seat of Irish kings.

One day, Diarmuid is mortally wounded by a magical boar on a hunt with MacCool. MacCool has the power to save his friend's life simply by giving him water. But as he cups his hands and fills them with the water, it trickles through his fingers, and Diarmuid dies.

The tales in the Fenian Cycle also focus on MacCool's son, Oisin, and his grandson, Oscar. In one of the most prominent of these tales, Oisin, a handsome warrior-poet, is hunting when he encounters Niamh, the goddess of the Irish otherworld. The two are smitten and gallop off together to the Land of Forever Young—a place where sorrow, pain, and old age are unknown. The lovers have a child there, but Oisin is homesick for Ireland and misses his family. Niamh agrees to let him return and gives him her magic horse. But it comes with one condition: he must not dismount.

Once back in Ireland, Oisin realizes that three hundred years have passed since he left. Stopping to help some men move a boulder, he falls from his horse and immediately ages the three hundred lost years, crumbling in the dust. In another, clearly Christianized version of the story, Oisin ages horribly but does not die. Instead, he meets St. Patrick and Oisin recounts the stories of his father, compiled in another Irish collection, *The Interrogation of the Old Men* (c. 1200 CE).

As legendary Irish figures, both Finn MacCool and Oisin appear in the works of writers of generations of great Irish writers, notably in the poem "The Wanderings of Oisin" (1889) by William Butler Yeats. Perhaps most famous of all, Finn MacCool is the model for the character of Finn in James Joyce's experimental novel *Finnegans Wake* (1939).

What do the Celts have to do with Halloween?

In one of the legends of Finn MacCool, his first act as the guardian of the king's palace at Tara is to rid the court of the malicious goblin Aillen, who set fire to the palace every year at the festival of Samhain (pronounced *sow-in*). Celebrated from the night of October 31 to November 1, this New Year festival traditionally marked the end of summer and the harvest as well as the beginning of the dark, cold winter. It was a time of year often associated with death, when animals were brought in from the fields and slaughtered.

It was also considered a time of great danger. During the festival, the barriers between the worlds of the living and the dead were broken, "the curtain was drawn back," and spirits from the "other world" could walk the earth. On the night of October 31, the spirits of the dead caused mischief and damaged crops. But their presence wasn't all bad—they made it easier for the Druids to make predictions about the future.

Like many Celtic festivals, Samhain spurred the Druids to build huge sacred bonfires, sacrifice animals, and gather people together to burn crops in honor of the Celtic gods. During this fire festival, the Celts wore masks and costumes, typically consisting of animal heads and skins, and attempted to tell each other's fortunes.

Are you starting to get the picture? "Trick or treat for UNICEF" and "Elvira, Queen of the Night" got started two thousand years ago, at a pagan Irish bonfire.

When the Samhain celebration was over, the Celts relit their hearth fires from the sacred bonfire, to help protect them during the coming winter. Some scholars believe that the Lindow Man (see above) may have been a symbolic stand-in executed in a ritual slaying of the king, who was killed three times—by garroting, clubbing, and stabbing— during the feast of Samhain.

By 43 CE, the majority of Celtic territory was under Roman control. During the next four centuries, two Roman festivals were combined with the traditional Celtic celebration of Samhain. The first was Feralia, a day in late October when the Romans traditionally commemorated the passing of the dead. The second was a day to honor Pomona, the Roman goddess of fruit and trees. The symbol of Pomona is the apple, and the tradition of "bobbing" for apples that is practiced nowadays on Halloween is just one more vestige of our pagan past.

Of course, then as now, some Christians took a dim view of all this pagan frivolity. By the 800s, the influence of Christianity had spread into Celtic lands. In the seventh century, Pope Boniface IV designated November 1 as All Saints' Day, a time to honor saints and martyrs. Presumably the pope was attempting to replace the Celtic festival of the dead—masking it (get it?) with a related, but Church-sanctioned holiday. The celebration was also called All-hallows or All-hallowmas (from Middle English *Alholowmesse*, meaning all saints' day) and the night before it, the night of Samhain, came to be called All-hallows Eve and, eventually, Halloween. Even later, in 1000 CE, the church named November 2 All Souls' Day, a day to honor the dead. It was celebrated in a fashion similar to Samhain, with great bonfires, parades, and dressing up in costumes as saints, angels, and devils. Together, the three celebrations—the eve of All Saints' Eve, All Saints' Day, and All Souls' Day—were called Hallowmas. A similar convergence of native pagan beliefs and Catholicism around these dates took place during the Spanish conquest of Mexico and produces the "Hispanic Halloween," Día de los Muertos ("the Day of the Dead") (See chapter 9, What is the "Day of the Dead"?)

Another significant Celtic holiday was Beltane. Held on May 1 and heralding the arrival of summer and the planting season, Beltane was celebrated as a day of fiery purification when, the Celts believed, the fairies were especially active. In Roman Britain, Beltane was merged with a Roman festival called Floralia, which also honored the goddess of springtime, Flora. Eventually, the Celtic and Roman holidays were fused into May Day, a celebration that may date back to even older springtime festivals from ancient Egypt and India.

The modern image of May Day conjures up a merry vision of vernal innocence—children gaily dancing around a Maypole festooned with bright-colored ribbons and flowers. But originally, Beltane was a fertility festival, and the giant Maypole was an undisguised and unashamed phallic symbol. It was often the occasion for young men and women to turn their thoughts to more than just love. In a pre-Christian world, there were fewer moral constraints about sex, and lovers left the Beltane bonfires to wander off into the woods. Although the holiday was cleaned up into its G-rated version in Christian Europe, the May Day festival was not a tradition that appealed to America's Puritan Fathers, who must have had long memories of its pagan past. That is why May Day never took hold in early America while it continued to be more widely celebrated in Europe.

MYTHIC VOICES

Llenllweag the Irishman seized Caledvwlch, swung it round in a circle and killed Diwrnach the Irishman and his entire retinue; the troops of Ireland came and fought, and when these troops were put to flight Arthur and his force boarded the ship in their presence, with the cauldron filled with the treasures of Ireland.

—*"How Culhwch Won Olwen," from the* Mabinogion
(translated by Jeffrey Gantz)

What is the *Mabinogion*?

Apart from these Irish myths and legends, the other significant body of Celtic literature was preserved in Wales, where the oldest myths were not written down for centuries. Again, it is probably a case of a Christian-era writer retelling these stories from his own point of view. Nonetheless, most of what is known of Welsh mythology is contained in a collection that is called *The Four Branches of the Mabinogi*, commonly known as the *Mabinogion*, compiled sometime in the twelfth century. These stories describe the mythical history of Wales, and many of the gods who appear in the Welsh mythology resemble the Tuatha Dé Danaan in Irish mythology. The suggestion is that Irish Celts may have migrated to Britain and brought their mythology with them. The stories are significant, because they offer the only view of earlier Welsh myths and include the first early references to characters and tales that would later evolve into the legend of King Arthur.

The first of four tales in *The Four Branches of the Mabinogi* tells the story of Pwyll, his wife Rhiannon, and their son, Pryderi. The goddess Rhiannon—who is possibly a vestige of the Celtic horse goddess Epona—is betrothed against her will, and wants to marry Pwyll, a king in southwestern Wales. When she dresses in gold and rides past him on a white mare, Pwyll is smitten by her beauty. They eventually marry, and their son, Pryderi, is born. But right after his birth, the baby is stolen, and Rhiannon's six attendants, in an attempt to clear themselves of any blame, kill a dog and smear its blood on Rhiannon's lips. The queen is charged with murdering her son and is forced to sit outside her husband's door, telling strangers of her crime and offering to carry them on her back, like a horse. In truth, Pryderi was never murdered but had been snatched and left near a stable. Raised by foster parents who eventually realize who he is, Pryderi is returned to his mother, and she is released from her punishment.

Links to Arthurian legend begin to appear in another part of *The Four Branches of the Mabinogi*. The tale of "Culhwhc and Olwen," which dates to approximately 1100 CE, includes many names and places later connected to Arthur, among them a reference to a sword whose Welsh name—Caledvwlch—means "battle breach." A weapon of great power, it was later identified with Excalibur, the legendary "sword in the stone." There is also mention of Arthur's father, Uthyr Pendragon, and his wife, Gwenhwyfar—later Anglicized as Guinevere. A reference to a cauldron, which, in some stories, acquired magical properties, is thought to be an old connection to the later idea of Arthur's search for the Holy Grail and may hark back to the Irish cauldron of the god Daghda, which provided a never-ending source of food.

The first references to Arthur found in the *Mabinogion* probably emerged from even earlier Irish myths. Traditional Irish hero stories may have been merged with those of Wales, resulting in the first legends of Arthur, a character who was probably based on a powerful Celtic chief who lived in Wales during the 500s CE and led the battle against the invading Saxons. (Others have made a case that he lived during Roman times and led the revolt against Roman rule around 400 CE. The Romans withdrew from Britain in 410 CE.) In any case, the stories of Arthur were exported to Brittany, another Celtic bastion in France, around 1000, where the renowned Breton minstrels then helped spread the tales all over Europe.

The legend of Arthur that endures today is mostly derived from the traditions set down by Sir Thomas Malory (d. 1471), the English author who created the familiar Arthurian legend. No effete intellectual writer, Malory was a violent criminal who had committed robbery and murder. From 1451, he spent much of his life in prison, where he probably did most of his writing. Drawing from a variety of earlier legends and stories—such as an ancient "history" of Britain by Geoffrey of Monmouth, and a variety of other sources, including a group of eight romances originally called *The Book of King Arthur and His Noble Knights of the Round Table*—Malory's legend of Arthur was printed with the more grandiose title *Morte d'Arthur*.

One of the central characters in Malory's *Arthur* not found in the *Mabinogion* is Merlin, the wizard in Arthurian legend who brings about the king's birth. In fact, Merlin's origins go even deeper into a Celtic past, to a Welsh wizard named Myrddin. Many authorities believe the roots of Merlin's character can be traced back to the Druidic tradition. Various traditions attributed great magical feats of power to Merlin, from overcoming dragons to the construction of Stonehenge. But his role in the Arthurian story—the magical bringing together of Arthur's parents, the raising of Arthur, the placement of Excalibur in the stone—was first recorded in the twelfth century. Nor does the *Mabinogion* relate anything of the half-sister of Arthur, Morgan le Fay (or Morgaine, Morgana), who was presented as a healer and shape-shifter by Geoffrey of Monmouth. By the time of Thomas Malory, she is the cause of Arthur's downfall. To round out the circle of Celtic connections, many scholars believe that Morgan is a version of the earlier Morrigan, the Celtic war goddess, who brought about the fall of Ireland's great hero, Cuchulainn.

Mythic Voices

Valhalla stands nearby, vast and gold-bright. Odin presides there, and day by day he chooses slain men to join him. Every morning they arm themselves and fight in the great courtyard and kill one another; every evening they rise again, ride back to the hall, and feast. That hall is easily recognized: its roof is made of shields and its rafters are spears. Breast-plates litter the benches. A wolf lurks at the western door, and an eagle hovers over it.

—*from* The Norse Myths, Kevin Crossley-Holland

It is nearly 350 years that we and our fathers have inhabited this most lovely land, and never before has such a terror appeared in Britain as we have now suffered from a pagan race, nor was it thought that such an inroad from the sea could be made.

—*English scholar* Alcuin *(793 CE)*

What mythology besides Celtic came storming out of northern Europe?

Maybe your first taste came from Looney Tunes, when Elmer Fudd put on a horned helmet and sang "Kill the Wabbit, Kill the Wabbit" to music from Wagner's *Ring Cycle*. Or maybe it was the scene in *Apocalypse Now*, when American helicopters attacked a Vietnamese village as loudspeakers blared "The Ride of the Valkyries." Perhaps the Marvel Comics character Thor was your introduction. Or the video loop of the burning "Yule log" shown on television every Christmas. Or the Minnesota football team called the Vikings. Or the magical world of giants, dwarves, runes, magical swords, and powerful rings created by John Ronald Reuel Tolkien (1892–1973) in *Lord of the Rings*.*

Powerful and popular, all of these images are based on the Norse and German myths of the Vikings.

Think "Viking," and perhaps you envision burly, bearded men with broadswords, horned helmets, and dragon boats, accompanied by outsized women with names like Brunhilde. If so, you would be right. Each of these rich images represents the fierce Vikings, or Norsemen, who terrorized, raped, and pillaged their way across Europe for some three hundred years, from about 800 until 1100 CE, when they were Christianized and started to cut back on their hell-raising.

As we see from the English scholar Alcuin (above), who got his first taste of Viking handiwork when raiders sailed out of the fjords of Norway in June 793, the Norsemen were a force to be reckoned with. After looting a monastery off the northeastern coast of England, where monks had been serenely copying religious manuscripts on the island of Lindisfarne, the Viking raiders spent the next few centuries scorching other parts of England, Ireland, and Scotland. In 841, they established Dublin as a winter base and began to strike farther from home, looting and burning towns in France, Italy, and Spain, and spreading fear wherever they went.†

If the Celts were frat-house boys gone bad, the Vikings were a gang of lawless bikers—"bad to the bone"—until they finally settled down and became the respectable, civilized Scandinavians they are today.

But was it all about pillage, rape, and destruction? Or was there a kinder, gentler Viking?

The answer is—not really. For most of their history, the Vikings were

*A scholar of medieval poetry and myth, Tolkien taught Norse and Germanic literature and mythology at Oxford University for more than thirty years while writing *The Hobbit* and the *Ring Trilogy*. His books are steeped in these Norse and Germanic myths, and the name of his wizard, Gandalf—sometimes likened to the Norse god Odin—comes from Norse poetry. The scene of the *Ring Trilogy*, Middle Earth, is also drawn from Norse myth, in which the world of men is called Midgard.

†They even reached North America, establishing settlements in Canada five hundred years before Columbus arrived. But their stay there was temporary and left no permanent impact on the Americas.

fierce pirates and warriors who descended from the Germanic peoples who had settled in northwestern Europe. Going as far back as 2000 BCE, some of these Germanic tribes had migrated to modern Denmark, Norway, and Sweden, where they put down roots as farmers and fishermen until overpopulation and a harsh northern environment led them to turn their considerable skills as oceangoing sailors to piracy and raiding. As early as the year 9 CE, Germanic tribes on the continent had destroyed and butchered more than 15,000 of Rome's finest legionnaires in one of the worst military disasters in Roman history. These tribes eventually helped bring down the Roman Empire.

Although separate Norse groups developed throughout northern Europe, all Norsemen shared the same way of life. It was a harsh culture, in which women and slaves were second-class citizens and unwanted children were exposed to the elements and left to die. This brutal culture had a mythology to match—of fierce war gods, often demanding blood sacrifice. There are stories of a sacred site to the Norse gods in Uppsala, in Sweden, where sacrificed men hung in trees. One account of a Viking king's burial includes the sacrifice of a slave concubine who is strangled and added to the funeral pyre after all of the dead king's companions had sex with her. The Vikings believed that a warrior's death ensured passage to a fighters' paradise called Valhalla. There, in the great Hall of the Slain, the Norsemen thought they would live among the gods, fight by day and feast by night, until the world came to an end in one all-encompassing, apocalyptic Battle of the Gods.

But the fighting didn't wait until after death for the Vikings. Known as "Danes," "Norsemen," or "Northmen," they terrified most of Europe as they conquered or looted parts of France, Germany, Italy, and Spain. The "Northmen" became "Normans" when they established a base in France (Normandy) and then invaded England under William the Conqueror in 1066. The Swedish branch of the Viking family tree settled in eastern Europe and was called the Rus, and Russia was named for them. The name "Viking" probably came later from Vik, in southern Norway. The expression "to go a-viking" meant to head off to fight as a pirate.

But in spite of their well-deserved reputation for ferocity, the great majority of Norsemen were simple farmers who lived in villages. These villages comprised a society that was roughly divided into three social classes—nobles, freemen, and slaves—who had little upward mobility.

The freemen included farmers, merchants, and traders, and the slaves were often those who had been captured in Viking raids and battles. All Vikings spoke a Germanic language with two major dialects that everyone understood. They also had an alphabet system called runes, a strange script that was used primarily by priests for secret ritual purposes. Like the Celts, the Vikings didn't record any of their myths and legends until after they had been Christianized.

Even so, there is a vast body of Norse literature collected in two works called Eddas that were set down during the Christian era from an earlier oral tradition. The Poetic, or Elder, Edda is a collection of poems composed anonymously between 1000 and 1100 CE. Twenty-four of the thirty-eight poems in the Poetic Edda are heroic tales, many of which recount the exploits of the great hero and dragon slayer Sigurd (Siegfried in German; see below). The other fourteen poems include accounts of the creation and the end of the universe in a fiery conflagration known as Ragnarok, in which the gods die.

The second collection is the Prose, or Younger, Edda, written during the 1200s by Snorri Sturluson (1179–1241), an Icelandic poet, historian, and courtier. Sturluson's Prose Edda was designed as a primer, or textbook, for other poets, and consists of a preface and three sections. The first of these sections tells about the Norse deities, while the second and third parts provide techniques for aspiring poets. Besides the Prose Edda, Sturluson also wrote a history of the kings of Norway stretching from early times to his own day. A wealthy and powerful man in Iceland, as well as a medieval Icelandic poet, Snorri Sturluson became involved in Norway's court intrigues and was murdered in 1241, apparently on the orders of the Norwegian king.

A final source of Norse myths are the *Skald* (the Icelandic word for a type of minstrel or bard), a complex form of Icelandic poetry that survives from the period from 900 through the 1200s CE. Most court poets in Scandinavia came from Iceland, and hundreds of these poems—many of which deal with contemporary rather than mythic figures—are preserved in the Icelandic sagas of the 1100s and 1200s. However, the *Skald* were composed after the Scandinavian countries began the conversion to Christianity, and so, as with the Christian-era Celtic literature, many of these myths have been layered over with Christian traditions, symbolism, and interpretation.

How do a giant's armpit and a cow help create the Norse world?

According to the Eddas, two places exist before the creation of life—Muspel ("world's end"), a fiery region in the south, and Niflheim ("dark world"), a northern land of ice and freezing mists. Between them lies Ginnungagap—the "beguiling void"—a great emptiness where the two worlds of heat and ice collide, congeal, and all things are created. Out of the merging of these two places comes the first living thing, a primordial frost giant called Ymir, who is soon joined by a primeval cow named Audhumla, whose four streams of milk keep Ymir alive. In time, Ymir gives birth to three beings, born from the sweat of his armpits and from one of his legs. Meanwhile, a second giant, Buri, is released from the primordial salty ice blocks of Niflheim after the cow Audhumla licks him free. Buri creates a son named Bor, who marries the giantess Bestla, and they have three sons—Odin, Ve, and Vili—who begin the first race of gods.

In a story with echoes of the Greek Creation accounts, Odin grows to manhood, joins with his brothers, and kills Ymir. The incredible flow of the primal giant's blood creates a great flood that kills all of the frost giants except for Bergelmir and his unnamed wife, who escape the deluge in a boat and re-create the race of ice giants. Although the gods defeat the giants in this Creation battle, the giants' descendants plan revenge on their conquerors—an enmity between these two races that permeates all Norse myths. (It is not known if this Norse Flood story predates the Christian era, or is an example of a biblical influence on Norse traditions.)

Having dispatched Ymir, Odin becomes—like Zeus—supreme ruler of the world, and goes on to create the earth from Ymir's body and the sky from his skull. The giant's blood becomes the oceans, his ribs the mountains, and his flesh the earth. The gods then happen upon two logs lying on the beach and turn them into the first two humans, Ask (ash) and Embla (elm or vine).

Supporting the entire creation is a giant ash tree known as Yggdrasil, which has three roots. One root reaches into Niflheim, the world of ice. Another grows to Asgard, the realm of the gods. The third extends to Jotunheim, land of the giants. Three sisters called Norns live around the base of the tree, and control the past, present, and future, determining the fates of men. A giant serpent, Nidoggr, loyal to the defeated race of giants, lives near the root in Niflheim and continually gnaws at the root, attempting to bring down the tree, and the gods of Asgard with it.

After the world is created, Odin and his brothers construct their heavenly home in Asgard. Odin and the other gods of Asgard are called the Aesir (the sky gods), but there is another race of lesser gods called the Vanir (earth gods), most likely fertility gods who existed before the Vikings took control of the region, though little is known of their origins. A bridge called Bifrost—usually described as a rainbow but sometimes associated with the Milky Way—connects Asgard to the earth, or Midgard, where men live. Within the walls of Asgard, the gods build their palaces and halls, including Valhalla, the Hall of the Slain. Here the kings and heroes fallen in battle are brought by the Valkyries—"choosers of the slain"—to spend their time feasting and fighting, but always ready to defend Asgard against attack by the giants. That day will come in the fearsome, apocalyptic battle called Ragnarok, literally the "fate of the gods," which is known in German as Götterdämmerung, or "the twilight of the gods." Fans of Wagnerian opera are familiar with these places, which appear in the *Ring Cycle*.

Ragnarok is somewhat unique in mythology, as it gives a complete account of the end of the world—a great battle fought between the gods and goddesses of Asgard and the giants who wait to avenge the attack of Odin on their ancestors during the Creation. When Ragnarok comes, most of the gods, goddesses, and giants are killed, and the earth is destroyed by fire. After the battle, the god Balder and his wife are reborn and with several sons of dead gods they form a new race of deities. During Ragnarok, a man and a woman also take refuge in the World Tree, Yggdrasil, and sleep through the battle. After the earth again becomes fertile, the couple awakens and begins a new race of humanity.

WHO'S WHO OF THE NORSE GODS

Balder Known as "the good" or "the beautiful," Balder is the favorite son of the supreme god Odin and is famed for his good looks and wisdom. Eloquent and full of grace, he is otherwise an ineffectual god, whose death is the most important feature of his story. When Balder has troubling dreams, his mother, Frigg, sees he is fated to die and asks that every living thing and all other objects swear an oath not to harm her fair son. Knowing he is invulnerable, the other gods amuse themselves by hurling stones and other things at Balder, but he is unharmed. Envious of Balder's invincibility, the Trickster Loki discovers that mistletoe—considered the "all-heal" by the Celtic Druids—has not sworn the oath to Frigg. So, Loki forms a dart from a sprig of the plant and gives it to Balder's blind brother, **Hod**. As Loki guides Hod's aim, the mistletoe dart hits Balder, killing him instantly. As the gods mourn Balder's death, his wife, **Nanna**, instantly dies of grief and is burned with Balder on his funeral pyre. Hel, the goddess of the underworld, agrees to release Balder from death if every person and thing in the world weeps for him. But the malevolent Loki—now in the guise of an old giantess—refuses to cry and Balder remains in the underworld. It is said that when the world is made new after the Battle of Ragnarok, Balder—who fits the dying-and-reborn-god archetype—and Nanna will return to begin another golden age of the gods.

Bragi God of poetry and eloquence, Bragi is called the "braggart" by Loki, and the word "brag" is derived from his name. He is married to Idun.

Freyr (**Frey**) The god of agriculture, fertility, and plenty, Freyr ("lord") and his twin sister and consort, Freyja ("lady"), the goddess of love and fertility, are Vanir—or deities of earth and water rather than sky gods (Aesir). But they are significant enough to have a place among the other gods in Asgard. The twin children of **Njord**, the sea god, and **Skadi**, the goddess of mountains and forests, Frey ensures the success of a harvest, while Freyja blesses marriages.

As fertility gods, Freyr is associated with rites that may have involved orgies, while his sister, Freyja, is linked with sexual freedom in the pre-Christian world of Europe. A sexual free spirit in the mode of Inanna and other Near Eastern fertility and love goddesses, Freyja sleeps with four dwarves on successive nights in return for her prized possession, a "flaming necklace," the symbol of her fertility. In some accounts, Freyja is counted as the leader of the Valkyries, the women sent to choose who will die in battle, and bring them to Valhalla, where they become heavenly cocktail waitresses. Freyja also selects from among the dead warriors who will live with her in her palace at Asgard.

Frigg The mother goddess, Frigg is the principal wife of Odin, father of the gods. Ruler of sky and clouds, Frigg also protects the household and marriage and is the bestower of children. Choosing not to live with Odin, she resides in a modest home of her own, where she and her handmaidens spin golden thread and weave clouds. Clairvoyant, Frigg knows of events in the present and future, but cannot affect them. When she learns that her son Balder is fated to die, she tries to alter his destiny by extracting a promise from all things in creation not to harm him. But she neglects the mistletoe, thinking that it is too young and weak to threaten her son. Loki uses Frigg's omission to bring about Balder's death and his exile to the underworld. In some traditions, Frigg's tears become the berries of the mistletoe. When Frigg learns that Balder will be restored to life, she hangs the mistletoe and promises to kiss all those who walk beneath it—adding to the ancient source of the Christmas tradition of mistletoe, which, in Celtic rites, was a sign of goodwill.

Frigg is related to another earlier Germanic goddess, named **Frea**, and her name is the source of the word "Friday."

Heimdall Known as "world brightener," Heimdall is god of dawn and another of Odin's sons. He is famed for his acute hearing and vision—he can hear grass grow and see for hundreds of miles, day or night—and serves as the lookout on Bifrost, the Rainbow Bridge that leads to Asgard, ready to blow a horn signaling that the battle Ragnarok is to begin. Heimdall is also credited with creating social order among humans during his visits to Midgard. In one tale, Heimdall sleeps in a poor man's hovel and is given meager food. Nine months later, a woman gives birth to **Thrall**, the first of the race of serfs, or slaves. Next, he sleeps in a farmhouse, where the hardworking people treat him well and he fathers **Karl** (source of the word "churl"), first of the race of free peasants. Finally, he sleeps in a fine hall, where he is well treated and fathers *Jarl* (source of the word "earl"), first of the race of noblemen.

When the Battle of Ragnarok finally comes, Loki steals his sword, but Heimdall manages to kill the trickster before dying of his wounds.

Hel The daughter of Loki and a giantess, Hel is the sinister goddess of death and the underworld, who is said to be half-black and half-white. She is cast into the cold regions by Odin, who decrees that she will rule over those who die of sickness or old age. Hel is also the sister of the monstrous wolf **Fenrir** and the serpent **Jormungand**, the other creatures who lead the final battle against the gods.

Hel rules the land of the dead, which bears her name. According to descriptions of it, the road to the Norse Hel is a freezing river filled with blocks of ice and weapons, its entrance guarded by a mighty dog similar to the Greek Cerberus.

Idun Wife of Bragi, Idun is goddess of immortality and keeps the golden apples of youth that preserve the gods' eternal youth. When Loki is coerced into luring Idun away from Asgard so a giant can steal the apples and weaken the gods, Odin and the other gods begin to wither with age. Using a magical falcon skin and citing the secret words—the runes—Loki becomes a bird and flies to the giant's palace and returns with Idun and the apples, rescuing the gods after having put them in peril in the first place.

Loki The supreme trickster god of uncertain parentage, Loki might be the offspring of the giants—the sworn enemies of the gods. But he is a frequent companion of the gods Odin and Thor. At times destructive and mischievous, Loki is also an appealing god who helps the other deities out of difficulties—usually the very ones he has created.

In the Eddas of Snorri Sturluson, Loki is described as "pleasant and handsome in appearance, wicked in character and very changeable in his ways. He had much more than others that kind of intelligence that is called cunning and stratagems for every eventuality. He was always placing the Aesir into the most difficult situations; and often extracts them by his wiles."

In a typical story, Loki taunts Odin, who wants to sleep with Freyja after she has bedded the four hideous dwarves in exchange for the marvelous "necklace of the Brisings." Turning into a fly, Loki finds Freyja asleep, so he next turns into a flea and bites her breast. When the goddess rolls over, Loki undoes the clasp on the necklace and takes it to Odin, who agrees to return it to Freyja only if she will stir up a war among men.

As time goes by, Loki becomes so bitter at the gods' dislike and mistreatment of him that he triggers Ragnarok. This great battle is set in motion after Loki brings about Balder's death by learning how to harm him with mistletoe. Loki's punishment for this crime is to be secured to three rocks with the intestines of his own son, which harden like metal to bind him. A snake is then placed above Loki's head and drips poison on his face until the day Loki breaks free and leads the giants in the apocalyptic Battle of Ragnarok. During the fight, Loki's offspring, the monstrous wolf Fenrir, swallows the sun and bites the moon while another of Loki's children, the venomous serpent Jormungand, which swims in the great ocean surrounding the world, stirs up the ocean depths and fights with Thor.

Odin Also known as Woden or Wotan, Odin is derived from an earlier Germanic god, and is chief among the Norse pantheon. The father of Thor, Balder, and other gods, Odin lives and rules in Asgard, the home of the gods.

In order to learn the secret knowledge of the runes, Odin pierces himself with his own spear and then hangs from the World Tree, Yggdrasil. After nine days and nights of pain and self-sacrifice, Odin learns all the hidden knowledge and casts off death. An old myth recorded after Christian times, this story would clearly connect Odin with the figure of Jesus, who also is pierced by a spear and dies hanging on a wooden cross often referred to as a tree before being resurrected.

Associated with battle, magic, poetic inspiration, and known as the protector of kings and heroes, Odin is a one-eyed god who had traded his other eye for a drink from the spring that provides clairvoyance. When the time comes for a warrior to die, Odin dispatches the Valkyries to the battlefield to select who will be brought to Valhalla, the Hall of the Slain in Asgard. His most devoted warriors are known as "Berserkers," which probably means "bear shirt," to describe those who wore bear or wolf pelts into battle. The Berserkers were renowned and feared for their ecstatic state of battle frenzy, possibly brought on by hallucinogenic mushrooms. The battle ecstasy also connects Odin to the state of inspiration that was believed to grip poets in their artistic frenzy. At the climactic Battle of Ragnarok, Odin is killed and swallowed by the monstrous wolf Fenrir.

His Germanic name, Woden, is the source of the word "Wednesday."

Thor Second in importance in the Norse pantheon after Odin, Thor is ruler of the sky, the god of lightning and thunder. He causes thunder with his great ax-hammer, Mjolnir, "the destroyer," a weapon of such devastating power that it can destroy giants and mountains with a single blow. When Thor throws the hammer, it magically returns to his hand like a boomerang. He is probably derived from an older Germanic god named Donar, and has also been associated with the Celtic thunder god Taranis. (See above.)

Immense in stature, with a great red beard, Thor has flaming eyes and a huge appetite. He is the most popular Viking god, because his life reflects the values of Viking warriors. A generous and gentle giant, he flies into a great rage when provoked.

In one popular tale, a giant steals Thor's hammer and will only return it in exchange for Freyja, the goddess, as a bride. Thor and Loki go to the giant disguised as Freyja and a handmaiden with the massive Thor hidden beneath a bridal veil. At the wedding feast, Thor almost reveals his identity when he eats and drinks in his usual insatiable way—he was capable of drinking an ocean—but Loki cleverly explains that "she" hasn't eaten in eight nights in anticipation of the wedding. The "bride to be" asks to see the fabled Hammer of Thor, which he then uses to crush the giant's skull and then shows no mercy on the wedding guests.

During Ragnarok, Thor dies by drowning in the venom that pours out of his victim, the dying World Serpent, Jormungand.

Thursday was named after Thor.

Who is the most important hero in Norse myth?

The quintessential Norse warrior Sigurd was a King Arthur of the Northern world, a figure with a possible historical origin who became the magnet for many stories, as Norse-myth authority Kevin Crossley-Holland describes him. Son of a warrior and grandson of a great king favored by Odin, Sigurd was a handsome, stately dragon-slayer and rescuer of women—a mortal with possible divine ancestors. He ranks as the most important human hero in Norse mythology—more than half of the Poetic Edda are about him—and the stories of his exploits had an impact well beyond the Viking myths. Tolkien clearly borrows much from these heroic tales of a gold ring made by dwarves, which increases the wealth of its owner but comes with a dreadful curse. The stories of Sigurd also became the model for the mythical German hero Siegfried, who appears in the *Nibelungenlied* ("Song of the Nibelungs"), a famous German epic composed around 1200, which, in turn, served as the basis for Wagner's opera cycle *The Ring of the Nibelung* (1869–1876).*

The stories about Sigurd probably originated in Germany around the Rhine in the 400s CE and reached Scandinavia, where they were given poetic treatment in the Elder Edda, the collection of poems composed in Iceland between 1000 and 1100. The prose *Saga of the Volsungs*, written in Iceland during the 1100s or 1200s, tells the stories more fully.

According to these myths, Sigurd is born after his father, Sigmund, is murdered. As he is dying, Sigmund predicts that his unborn son will accomplish great deeds and his name will never be forgotten. Raised by a king, Sigurd is tutored by the dwarf Regin, who gives him a magical

*Composed over a period between 1853 and 1874, the cycle begins with *Das Rheingold* (*The Rhine Gold*), which serves as a prologue to the three main operas: *Die Walküre* (*The Valkyrie*), *Siegfried*, and *Die Götterdämmerung* (*The Twilight of the Gods*). All four works were first performed as a cycle in 1876 for the opening of the Festival Opera House, built by Wagner in Bayreuth, Germany.

horse and forges a wondrous sword, Gram, from the shards of his dead father's sword—a gift from Odin—which he then uses to avenge his father's death.

The central story in Sigurd's adventures is his killing of the dragon Fafnir. At first a powerful, greedy, and violent man with magical powers, Fafnir is the son of a farmer who also had magical skills. Fafnir and his brother Otr are both shape-shifters. One day, while out with the other gods, Loki kills an otter, which is actually Otr (source of the word "otter") in the animal's shape. When Otr's father realizes what has happened, he demands compensation from the gods, and Loki agrees to fill the otter skin with gold. Conveniently, the trickster finds a nearby dwarf with a large treasure, including a gold ring, which Loki commandeers. Stripped of his treasure, the dwarf curses the ring, dooming whoever possesses it. Once given the treasure, the farmer is first to die when Fafnir kills his father, steals the gold, and then turns himself into a dragon, spending the rest of his life hoarding the treasure. (Readers of Tolkien will surely recognize these themes as similar to the tale of the Ring of Power, jealously guarded by whoever possesses it.)

Eager to take the treasure for himself, Fafnir's brother Regin, the dwarf who tutors Sigurd, instructs the young warrior in how to kill Fafnir, planning all the time to kill Sigurd after Fafnir is dead. As a dragon, Fafnir only leaves his lair and the treasure hoard occasionally to drink from a nearby river. Sigurd digs a hole in the path that leads to the river and hides inside. When Fafnir passes over the hole, Sigurd stabs Fafnir in the heart. Having killed Fafnir, Sigurd roasts the dragon's heart, as Regin had instructed, but he accidentally burns his fingers and puts them in his mouth. After tasting the dragon's magical juice, Sigurd is able to understand the language of some nearby birds, who warn him that Regin plans to kill him. Sigurd lops off Regin's head, drinks some of his blood, eats more of the dragon's heart, and then discovers Fafnir's lair and the ring of gold.

Possessed of the ring, Sigurd also falls under its curse. He is loved by a Valkyrie, Brynhild (Brunhilde in the Wagnerian version), whom he promises to marry, but who is imprisoned in a ring of flame for offending Odin—à la Sleeping Beauty. Sigurd rescues her, but does so to give Brynhild to another man. When Brynhild discovers that she has been tricked, she has Sigurd killed before immolating herself in his funeral pyre.

The gold ring and the rest of the treasure are then hidden in the Rhine, where it has been ever since.

MYTHIC VOICES

The two humans who hid themselves deep within Yggdrasill will be called Lif and Lithrasir. . . . Lif ("Life") and Lifthrasir ("thriving remnant") will have children. Their children will bear children. There will be life and new life, life everywhere on earth. That was the end; and this is the beginning.

—*from* The Norse Myths, *Kevin Crossley-Holland*

BRIDGE TO THE EAST

Oh, East is East, and West is West, and never the twain shall meet,
Till Earth and Sky stand presently at God's great Judgment Seat;
But there is neither East nor West, Border, nor Breed, nor Birth,
When two strong men stand face to face,
tho' they come from the ends of the earth!

—RUDYARD KIPLING, "The Ballad of East and West"

The stories, legends, and myths of northern Europe, the Mediterranean world, and the ancient Near East are mostly tales of long-dead religions. True, some of their gods, rituals, concepts, and theories remain alive today, borrowed by later faiths, including Judaism, Christianity, and Islam, and kept alive in traditional celebrations and superstitions. A trendy but powerful New Age "revival" of goddess worship, Wicca, and "neo-druidism" has also attempted to resuscitate ancient mythic beliefs, worship, and other "old ways." But it is fair to say that most of the mythologies of Europe and the ancient Near East are, well, ancient history.

When we come to Asia, however, the story is a very different one—especially in India, China, Japan, and other places where Hinduism, Buddhism, and Shintoism are vigorous, widely practiced religions, deeply rooted in the myths of a very distant past. Nearly as old as Egypt and Mesopotamia, the civilizations of India and China, in particular, retain aspects of mythical systems that were born in the deep mists of prehistory. While understanding these ancient traditions—seemingly so "foreign" to Western experience—has always been intriguing and important, the need to gain a firmer grasp of the beliefs that form the soul of so much of Asia is greater now than ever before.

The reasons why should be fairly obvious. For one thing, our world is changing. Fast. Travel has made the globe smaller. Technology has made it spin more rapidly. When you call your bank or computer maker's "tech support" from New York, the phone may be answered in New Delhi. The decisions made in Beijing and Bombay—more than ever before—affect people in Boise, Buenos Aires, Berlin, and the Bronx. The clashing ideologies of East meeting West have complicated our lives. And sophisticated weaponry has made it all the more dangerous.

Yes, "East is East and West is West." But the twain now meet in cyberspace, on telephone call centers, and, certainly, in superstores, where Western shelves are loaded largely with Eastern-manufactured consumer goods.

Then there are the simple, raw numbers—populations are shifting and exploding. Though currently the world's most populous nation, with more than 1.3 billion people in 2003, China will eventually be surpassed for that dubious distinction by India, which passed the 1 billion mark in 1999. Together, these two countries already account for nearly one-third of the planet's population. And their ranks are swelling rapidly, even as Western birth rates slow or shrink.

Historians are often asked what the most important event at any given moment might be. Although it is impossible to answer definitely, it would be safe to guess that some of the most important things happening in the world in the early years of the twenty-first century won't be happening in the capitals of Europe or America. Chances are they will happen in India, China, or elsewhere in Asia, where booming populations and economies are changing global realities. Viewed not so long ago as developing "third-world" nations, these countries are quickly industrializing and taking the lead in science and engineering. Like the Western superpowers, they possess nuclear arsenals and have ambitions in space. And with a very old tradition as innovators in science and technology, they will gain economic strength and vie for a leadership role in the world.

So where does myth fit into the geopolitical picture? Arguably, front and center. To understand where the world is going, we need a better understanding of where this part of the world has been. How better to gain some insight than to know their myths and see how these myths reveal some part of their collective soul?

CHAPTER SIX

THE RADIANCE OF A THOUSAND SUNS

The Myths of India

If the radiance of a thousand suns were to burst forth at once in the sky, that would be like the splendor of the Mighty One.

—Bhagavad-Gita

For certain is death for the born
And certain is death for the dead;
Therefore over the inevitable
Thou shouldst not grieve.

—Bhagavad-Gita

If I were asked under what sky the human mind has most deeply pondered over the greatest problems of life, and has found solutions to some of them which well deserve the attention of those who have studied Plato and Kant—I should point to India. And if I were to ask myself what literature . . . is most wanted in order to make our inner life more perfect, more comprehensive, more universal, in fact more truly human a life, again I should point to India.

—MAX MÜLLER

How do we know what the ancient Indians believed?

What role did myth play in ancient India?

If it's all an endless cycle of birth and destruction, where does the Hindu Creation begin?

How do you get ten gods in one?

Who's Who of Hindu Gods

What kind of hero doesn't want to fight?

Why would a hero banish his loving wife?

What is Nirvana?

MYTHIC MILESTONES

India

Before the Common Era

c. 4500 Introduction of irrigation techniques in Indus Valley region in northwestern India.

Rice is cultivated south of Ganges River.

Pottery is made with corded decoration.

c. 2500 The emergence of civilization in the Indus Valley lowlands at the early cities of Mohenjo-Daro and Harappa, centered in the Indus River plain between what is now Pakistan and northwestern India; walled towns develop.

Earliest known woven cotton cloth found in Mohenjo-Daro.

2000 Collapse of Indus Valley civilization.

1500 Indo-Aryan nomadic invaders arrive and settle northwestern India.

Composition of the Sanskrit hymns of the Rig-Veda begins (completed c. 900).

1030 Aryans in India expand down the Ganges Valley.

c. 1000 Aryans establish small states in India.

c. 900 Composition of late Vedas, Brahmanas, and Upanishads begins.

c. 800 Rise of urban culture in Ganges Valley.

c. 600 Sixteen Aryan kingdoms are spread across northern India.

Emergence of Hinduism.

563 Birth of the Buddha.

540 Birth of Mahavira, founder of Jain religion.

c. 500 Religious law codes composed.

Caste system introduced in India.

c. 483 Death of Buddha.

c. 400 Composition and compilation of epic poems *Mahabharata* and *Ramayana.*

326 Alexander the Great crosses the Indus River into India; farthest advance of his empire.

321 Chandragupta founds Mauryan Empire.

297 Chandragupta, the first man to unite the Indian subcontinent, abdicates in favor of his son, Bindusara.

273 Reign of Ashoka after he seizes throne.

262 Ashoka converts to Buddhism; renounces violence; Buddhism becomes state religion.

232 Ashoka dies.

c. 100 Composition of seven-hundred-verse Bhagavad-Gita.

When the first atomic bomb was successfully tested in New Mexico's desert in July 1945, Robert Oppenheimer—the brilliant young physicist who directed the Los Alamos laboratory—recalled the moment like this:

> We waited until the blast had passed, walked out of the shelter and then it was extremely solemn. We knew the world would not be the same. . . . I remembered the line from the Hindu scripture, the Bhagavad-Gita: Vishnu is trying to persuade the Prince that he should do his duty and to impress him he takes on his multiarmed form and says, "Now I am become Death the destroyer of worlds." I suppose we all thought that, one way or another.*

Consider that scene. One of the most significant moments in human history has just occurred. and it is marked not by a passage from the Bible, or a Greek philosopher, or Shakespeare, but by an obscure reference from an ancient mythic tradition. To many Westerners accustomed to Judeo-Christian doctrines and the rationalism that began in Greece and flowered in Europe's Enlightenment, India's mythic legacy remains inscrutable. It is a magical mystery tour of the exotic and wondrous. A blue-tinged god with surplus arms. An elephant-headed deity who rides around on the back of a rat. A terrifying goddess adorned with severed body parts. A monkey king who would be at home in *The Wizard of Oz*. An awesome divinity who dances the world into destruction. And a thousand-year-old temple adorned with a host of X-rated figures in bewildering contortions.

It is the unfurling of millions of yoga mats in gyms around the world, turning a three-thousand-year-old path to enlightenment into the latest fitness craze. It is "Instant Karma" and the *Kama Sutra* combined, a picture muddled for many Westerners by saffron-robed groups hustling spare change at the airport as they chant "Hare Krishna."

And what is the story with those "sacred cows"?

**The Making of the Atomic Bomb*, Richard Rhodes, p. 676.

Occupying a triangular peninsula about the size of continental Europe that juts down from Asia's landmass into the Indian Ocean, India is a place of enormous physical contrasts—extraordinary mountains, a great desert, broad plains, winding rivers, tropical lowlands, and lush rain forests watered by life-giving but sometimes destructive monsoons. Part of this diverse country's fortunes in ancient times lay in the fact that it was largely set apart by its physical boundaries—the Arabian Sea and a large desert, the Thar, to the west; the Bay of Bengal to the east; and to the north, the towering, snowcapped Himalayas that separated India from China.

Yet remote, obscure India beckoned to the West for centuries. First for its silks and spices. Then, later, for its approach to contemplating the "Big Questions"—eternity, good, evil, and the meaning of life. With a cosmic view completely at odds with traditional Western thought, India has long been interested in the transcendent and the immortal, the idea that creation and destruction are an endless cycle, that the soul is an essence searching for perfection through reincarnation. These ideas found expression in what mythologist Arthur Cotterell has called "a range of myth and legend which is unrivaled anywhere else in the world."

The roots of those Indian myths are also very old, stretching back more than 4,500 years to the broad plains of the Indus River Valley. Once centered in what is now the border region between northwestern India and southern Pakistan, the ancient Indus Valley civilization flourished for a thousand years. Most likely, it was anchored by a very ancient, fertility-based, goddess worship, as well as the worship of cows deemed sacred for the milk they provided and the dung that helped fertilize their crops. This civilization lasted until a group of warlike nomads swept in around 1500 BCE. Speaking a language called "Sanskrit," which is at the root of all other Indo-European languages, these new arrivals probably originated near the Caucasus Mountains in central Asia. They called themselves arya (meaning "kinsmen" or "noble ones"), and eventually came to be known as Aryans.* Just as the people later

*Due to its unfortunate association with Hitler and Nazism, the word "Aryan" has acquired a taint. Hitler and the Nazis used the term to refer to Germans and other northern Europeans, whom they considered racially superior to all other people.

called "Mycenaeans" had barreled into Greece bringing some of their own gods with them and absorbing some of the local deities they found, the Aryans conquered the remnants of the Indus civilization and imposed their "alpha male" pantheon of gods on the locals. That is, at least, the prevailing view; another school of thinking holds that this was a kinder, gentler Aryan migration.

Once settled, the Aryans spread to the south and east, eventually extending their rule over most of India. Over time, the gods and culture of the Aryans gradually combined with those of the existing local cultures, and what Westerners later called "Hinduism" evolved from this ancient marriage. Although the Aryans never developed a great and voracious imperial government intent upon world conquest—just as no dominant state emerged in ancient Greece—their myths eventually knit together the people of this vast and diverse "subcontinent" as no single state or government bureaucracy ever could. Their beliefs and sacred rituals, the Sanskrit language, the holy temples to the cosmos of gods and goddesses—and the unshatterable "caste" system their beliefs cemented rigidly in place—formed the soul of Indian culture.

Yet to talk about "Hinduism" as a monolithic religion is a mistake. It has no pope or hierarchy. No founder or central prophet. No uniting creed. No Vatican or Mecca or Jerusalem. As it exists today, Hinduism—along with its two most significant offshoots, Buddhism and Jainism—is a complex collection of beliefs with a vast pantheon of gods and differing schools of thought. Its dizzying diversity has led writers such as historian Ninian Smart to comment, "Even to talk of a single something called Hinduism can be misleading because of the great variety of customs, forms of worship, gods, myths, philosophies, types of ritual, movements and styles of art and music contained loosely within the bounds of the religion . . . It is as if many Hinduisms had merged into one. It is now more like the trunk of a single ordinary tree; but its past is a tangle of most divergent roots."

This racist use of the term continues among white supremacist groups, such as the Aryan Nation in the United States. Even the swastika, adapted as the symbol of Nazi power, has its roots in a similar ancient but benign Hindu symbol, which originally meant "let good things happen." Among the other people who referred to themselves as Aryans were the Iranians; the name "Iran" itself comes from the word "Aryan."

From those ancient roots—the stories, legends, and ancient myths—comes a vibrant, pulsing religion with a collective consciousness that has few parallels in other cultures or belief systems, either East or West.

MYTHIC VOICES

Scholars of India are puzzled by why their culture, so ancient, so rich in sculpture and architecture, in works of mythical and romantic literature, should have been so lacking in critical historical writings. Some suggest that the ancient Indian works of history written in Sanskrit may, for still unexplained reasons, have suffered wholesale destruction. A more plausible explanation is that they never existed. . . . The main interest of Hindu Indians in their past was not in the rise and fall of historical empires, but in the rulers of mythical golden age. . . . The lack of a historical record reveals not merely the Hindu preoccupation with the transcendent and the eternal, but also the widespread sense that social life was changeless and repetitive. . . . In a society that did not know change, what was there for historians to write about? When real events were recorded, they were usually transmuted into myth to give them a universal and enduring significance.

—DANIEL BOORSTIN, The Discoverers

If the slayer thinks he slays,
If the slain thinks he is slain,
Both these do not understand;
He slays not, is not slain.

—*Katha Upanishad*

How do we know what the ancient Indians believed?

The Egyptians left us their Book of the Dead, the Mesopotamians their *Gilgamesh*. The Greeks gave us Homer and Hesiod. The Celts left stories that were later preserved by monks. But when it comes to the myths of ancient India, we have a vast collection of mythic and religious writing that dwarfs all others. If anybody deserves the sobriquet "people of the book," it may well be the compilers of India's vast libraries.

When the Aryans arrived in the Indus Valley sometime between 1700 and 1500 BCE, they brought along Sanskrit, the oldest known written language of India. Although Sanskrit died out as a "living language" by about 100 BCE, it was used—like the Latin of medieval Europe—as the "learned language" of poetry, science, philosophy, and religion. Forming the core of Hinduism's beliefs and practices, the collections of Sanskrit hymns, poetry, philosophical dialogues, and legends all exist in an imposing set of texts that include, most significantly, the Vedas and Upanishads, the epic poems *Ramayana* and the *Mahabharata*—which contains an important section called the Bhagavad-Gita—and the Puranas.

The oldest sacred Sanskrit writings, the Vedas were thought to be composed beginning about 1400 BCE over a period of nearly 1,000 years, an era in India's history called the "Vedic period." The Vedas are considered to be older than the sacred writings of any other major existing religion, including the Hebrew Old Testament. Only the ancient Egyptian pyramid texts are older. Like many mythic and religious documents, the Vedas probably first existed in oral form for centuries, and may go back as far as 4000 BCE. Hindu tradition holds that they were composed in 3500 BCE, in the time of Krishna, an earthly incarnation of the Hindu god Vishnu, before they were finally written down by some anonymous scribes.

There are four Vedas, beginning with the oldest and most famous, the Rig-Veda. (The later Vedas include Sama-Veda, Yajur-Veda, and Atharva-Veda.) Written in archaic Sanskrit and first translated for the West by Max Müller in the mid-nineteenth century, the Vedas have been studied not only for their religious significance, but for their connection to the early history of the Indo-European languages, including the Greek, Latin, Germanic, and Slavic language families, which are derived from archaic Sanskrit. Ancient Sanskrit is also the original source for many languages spoken in modern India, including Hindi and Urdu. To many Westerners, Sanskrit is more obscure and indecipherable than Greek or Latin. But many linguists consider ancient Sanskrit a highly polished and systematic language with precise rules of grammar.

The word "Veda" means "knowledge," and sacred knowledge in particular. Roughly equivalent to the Hebrew Psalms of the Old Testament, the Vedas are poetic collections that provided the songbook for the holy rites of the early Vedic religion. The Rig-Veda contains more than one thousand hymns, totaling more than ten thousand verses—an enormous number, compared to the 150 biblical Psalms.

Later additions to the Rig-Veda include two other important texts, which were composed as commentaries on the Vedas—the Brahmanas and the Upanishads.

Brahmanas are long prose essays; they explain the myths and theology behind the sacred rituals that include offerings to gods, chanting, pilgrimages, and acts of charity or self-denial, such as food taboos. According to Devdutt Pattanaik's *Indian Mythology*, "The human custodian of these manuals was known as the brahmana. As keepers of Vedic lore . . . brahmanas served as the link between the material and spiritual realms. They knew the secret of the cosmos. . . . As people, communities and tribes mingled and merged, the Vedic brahmanas tried to retain their superior position and their spiritual purity by not sharing food or their daughters with nonbrahmanas." Organized around this priesthood, the system came to be called Brahminism, led by the Vedic priests who came to be known as Brahmins (also spelled Brahmans), a hereditary priesthood occupying the highest place in society. And just as Christianity's sacred religious language, Latin, was written and read almost exclusively by the priesthood, Sanskrit became the preserve of the Brahmins. Knowledge is, was, and always has been Power.

Upanishads are deeply philosophical works, one hundred and eight of which have been preserved; they appeared between 800 and 600 BCE and formed a basic part of Hinduism as it evolved. "Upanishads" roughly means "sitting near devotedly," or "to sit close to." They were composed, like certain works of Greek philosophy, as dialogues between a teacher and student.

At the core of the Upanishads is the notion of Brahman, the divine universal power that lives in the whole of creation, including the human soul, which is believed to be eternal. Expressing the idea that knowledge brings spiritual uplift, the Upanishads also introduced the notion that one lifetime is not enough to gather all the necessary knowledge. By accumulating knowledge over many rebirths, one can finally be rejoined with Brahman and achieve *moksha*, the ultimate "release" or "salvation" that is the true goal of all human beings.

Another key source of India's myths is the *Mahabharata*, one of the longest literary works in history, more than seven times the combined length of the *Iliad* and the *Odyssey*. One of India's two epic poems, the *Mahabharata* was said to have been dictated to Ganesh, the elephant-headed god of wisdom. In fact, it is a collection of Sanskrit writings by several authors who lived at various times, and parts of it may be more than 2,500 years old.

Mahabharata literally means "Great King Bharata," and the poem recounts a cataclysmic family feud between the descendants of King Bharata—two related families, the Pandavas and the Kauravas, who lived in northern India, perhaps about 1200 BCE. The Pandava brothers lose their kingdom to their Kaurava cousins and engage in a mighty struggle to win it back.

The main narrative of the *Mahabharata* is frequently interrupted by other stories and discussions of religion and philosophy, one of which is the enormously important work called the Bhagavad-Gita. Perhaps the most widely read, beloved, and significant piece of Hindu scripture, the Bhagavad-Gita (Song of the Lord) is presented as a conversation between the warrior hero Arjuna and the god Krishna, who has taken the mortal role of Arjuna's chariot driver. The Gita, as it is known, sets forth Krishna's teachings to Arjuna, who faces a moral crisis as the two armies prepare to do battle.

The second of India's two epic poems is the *Ramayana*, which supposedly describes events that took place 870,000 years ago. The poem contains 24,000 couplets—again, originally written in Sanskrit—and attributed to a sage Valmiki, who wrote it about 500 BCE. Very simply, it is the story of Rama, a prince whose father exiles him for fourteen years because of a dispute over the throne. Like the *Iliad*, it is largely about a war over a woman, as the main plot line is about the conflict between

Prince Rama and a demonic king called Ravana, who kidnaps Rama's beloved wife, Sita. The Hindi translation, written by the poet Tulsidas in the late 1500s CE, remains the most popular version of the *Ramayana* today.

Finally, there is a large collection of Sanskrit texts called Puranas, which were compiled between the early centuries of the Common Era and as recently as the sixteenth century. Mainly written in verse, they present an encyclopedia of Hindu lore, often taking the form of a dialogue—just as the works of Plato do—between a sage and a group of disciples. There are eighteen major and eighteen minor Puranas, and each is a long book that consists of various stories of the gods and goddesses, hymns, cosmology, rules of life, and rituals. Essentially extensive references and guides to religion and culture, the Puranas also describe the Hindu beliefs about Creation and how the world periodically ends and is reborn.

Just as many Christian churches traditionally used a catechism to teach their basic tenets, the Puranas were used to disseminate Hindu religious principles and practices to the majority of illiterate people as well as those prohibited from the older Vedic traditions, including women and the socially inferior people in India's strict caste system. Many of the Puranas are especially important in understanding myth, because they were composed to explain the connection between particular places with mythological events, such as the origin of a sacred site where a deity had manifested itself.

Words, of course, are not all that we have to go on when it comes to India's vast mythology. Along with the thousands of Hindu temples still in active use, there is archaeology. During the late-nineteenth-century era of British colonial rule in Pakistan and India, British scholars were the first Westerners to discover vestiges of whole cities full of ancient artifacts buried in huge earthen mounds in the Indus Valley region. By the 1920s, archaeologists uncovered the remains of a previously unknown civilization, now called "Harappan," in the area's two central cities, Harappa and Mohenjo-Daro. (More recently, other significant discoveries have been made at sites including Kalibangan, Lothal, and Surkotada—in India and Pakistan.) Arranged around a citadel, and built on a grid, these carefully planned cities featured paved streets and underground sewage drains. Excavations have also revealed large baths with connecting rooms, where ancient purification rites may have taken place, along with prominent phallic symbols and large numbers of statues of goddesses, suggesting an early focus on fertility rites. With a uniform system of weights and measures, and covering a larger geographical area than either the civilizations of Egypt or Sumer did, Harappan civilization had broken down by about 1700 BCE. There are few records or historical clues to explain this decline, but the breakup may have been due to changing river patterns that disrupted local agriculture and the Indus Valley economy.

The demise of the Harappan civilization roughly coincides with the arrival of the Aryans. And from the fusion of these two ancient cultures came the eventual rise of Hinduism—a mythology, a religion, and a philosophy that completely shaped India's future and identity.

Mythic Voices

In Vedic religion, people had experienced a holy power in the sacrificial ritual. They had called this sacred power Brahman. The priestly cast . . . were also believed to possess this power. Since the ritual sacrifice was seen as the microcosm of the whole universe, Brahman gradually came to be a power which sustains everything. The whole world was seen as the divine activity welling up from the mysterious being of Brahman, which was the inner meaning of all existence.

Brahman cannot be addressed as "thou"; it is a neutral term, so it is neither he nor she; nor is it experienced as the will of a sovereign deity. Brahman does not speak to mankind. It cannot meet men and women; it transcends all such human activities. Nor does it respond to us in a personal way: sin does not "offend" it, and it cannot be said to "love" us or be angry. Thanking or praising it for creating the world would be entirely inappropriate.

—Karen Armstrong, A History of God

By the Lord all this universe must be enveloped,
Whatever moving thing there is in this moving world.
Renounce this and you may enjoy existence,
Do not covet anyone's wealth.
Even while doing deeds here
One may wish to live a hundred years;
Thus on thee—this is how it is—
The deed adheres, not on the person.

—*from the Upanishads*

What role did myth play in ancient India?

A better question might be, "What role didn't myth play in ancient India?"

Although there is oddly no equivalent word for "myth" in India's numerous languages, few other places were as engulfed and pervaded by their myths as was ancient India. From the vegetarian diet many Indians embraced, to their view of the Ganges River as sacred water, to the rigid social classes into which their people were divided, religious ideas born of myth completely dictated life in ancient India. As Anna Dallapiccola writes in *Hindu Myths*, "Myths permeate the totality of Indian culture, mementoes of mythical events dot the whole country, old myths are told anew and new myths are created . . . Each story is connected to many more, one more exciting than the previous; each merges in an ocean of stories."

The power of myth in ancient India's everyday life grew out of the Vedic traditions, which formed the heart of the country's religious practices for centuries. Stretching back to before 1500 BCE, when the Vedas were written, the Vedic traditions were steeped in an older generation of gods, but were ever-present in the actions of priests who petitioned the gods for favors by chanting and making offerings of flowers, food, and gifts. They also oversaw such rites of passage as marriage, childbirth, and death, and—perhaps most important—made sacrifices at fire altars in the hopes of currying the favor of the gods. Tolerant of local customs and beliefs, the Vedic priests—later the Brahmins—accommodated the local cults that worshipped trees, snakes, mountains, rivers, and other regional deities as they spread across India. Bringing these localized cults into the Vedic fold not only expanded the number of worshippers in India, it also swelled the vast pantheon of gods.

With the introduction of the Upanishads between 800 and 500 BCE, a striking shift in India's mythic mind-set took place. The emphasis was no longer on the simple, ancient belief in sacrificing to individual gods who could provide protection, send a good husband, or bring rain to make the plants grow. The emergence of the Upanishads ushered in a new era of far more abstract belief, in which the many gods of ancient times were reduced to the single concept called Brahman, and the emphasis was placed on escaping an endless cycle of death, rebirth, and reincarnation in order for the human soul to link with Brahman, the Absolute Godhead.

Making that cosmic leap involved another notion introduced with the Upanishads—that of karma, the law of cause and effect which dictates that every action has consequences that influence how the soul will be reborn. Unlike the Egyptian or Christian notion, in which proper behavior might guarantee a pleasant afterlife, this Indian concept—simply put—held that living a good life means the soul will be born into a higher state in its next incarnation. An evil life did not mean eternal damnation but a rebirth of the soul into a lower state, possibly even as an animal. This ongoing cycle of life-death-reincarnation continues until a person ultimately achieves spiritual perfection, at which point the soul enters a new level of existence called *moksha* ("release" or "salvation"), in which it is joined with Brahman, the divine godhead.

As these more abstract religious concepts took hold, the old rituals were not abandoned, but made part of a new order that was contained within a concept called dharma—an all-inclusive sense of moral and spiritual "duty" with implications of truth and righteousness as well. In essence, dharma means the correct way of living. Maintaining dharma is believed to bring rhythm to the natural world and order in society. When dharma is not upheld, the result is uncertainty, natural disaster, and accidents—what *Star Wars* would call "a great disturbance in the Force," or as Lemony Snicket of children's book fame might put it, "A

Series of Unfortunate Events." Essential to maintaining dharma was careful adherence to sacred religious observances and the social order. Every man was supposed to do his duty as defined by his station. For women, as Devdutt Pattanaik notes, "There was only one dharma: obeying the father when unmarried, the husband when married, and the son when widowed." Not exactly a modern feminist's idea of Nirvana, but certainly in line with the notions of most other male-dominated ancient societies.

The core of Brahmanism's order was the Brahmin social structure, which evolved into the Hindu caste system. A highly rigid division of social classes, the caste system may have existed in some form before the Aryan invaders—or immigrants—arrived in the Indus Valley. But as the Aryans and their descendants gradually gained control of most of India, the caste system was used, at first, to limit contact between themselves and the aboriginal people known as Dravidians. The Sanskrit word for caste means "color," and it is widely thought that the tall, fairer-skinned, and possibly blue-eyed Aryans imposed this system on the darker-skinned aboriginal Dravidians.

The three original divisions later became four principal castes—gradually divided into many layers of subcastes—each with its own rules of behavior, particularly regarding marriage. Marrying outside of one's caste—like an English aristocrat marrying a "commoner"—just wasn't done. It was not dharma.

On top of the caste system were the Brahmins, the priests and scholars concerned with spiritual matters; next came Kshatriyas, the rulers and warriors who administered the society; beneath them were the Vaisyas, the merchants and professionals who managed the society's economy; and then the Sudras, the laborers who serviced the society. For centuries, one large group has ranked below even the lowest, Sudra caste. Known as Dalits ("broken" or "ground down"), they were the "untouchables" who performed the most menial tasks and existed outside the four castes—giving us the English word "outcast."*

*Dalits, or untouchables, have traditionally held such occupations as tanning, street sweeping, and other menial jobs forbidden to members of the four castes. In 1950, untouchability was constitutionally outlawed, but discrimination against the Dalits is deeply ingrained and a form of caste-based "apartheid" still exists in

Just as priests ruled the European medieval world, and the imams and ayatollahs dictate to modern Islamic governments in places like Iran, the Brahmin caste of priests, philosophers, and scholars held the high ground in ancient Indian society. Elite and powerful, they attained and held their status through religious principle. In *Guns, Germs, and Steel*, his groundbreaking view of human history, Jared Diamond coined the word "kleptocracies" to describe powerful ruling classes and the ways in which they were able to transfer wealth—and power—from commoners to themselves. Far from limited to India's Brahmins, Diamond's fairly cynical view of these systems neatly sums up the underpinnings of the caste system: "[One] way for kleptocrats to gain public support is to construct an ideology or religion justifying kleptocracy. Bands and tribes already had supernatural beliefs, just as do modern established religions. But the supernatural beliefs of bands and tribes did not serve to justify central authority, justify transfer of wealth, or maintain peace between unrelated individuals. When supernatural beliefs gained those functions and became institutionalized, they were thereby transformed into what we term a religion."

Whenever myth morphs into religion, elaborate rituals usually emerge.* This was certainly apparent in India, where the sacred Ganges, a river originating high in the Himalayas and revered as the physical manifestation of the goddess Ganga, had been associated with purification since ancient times. Bathing in the waters of the Ganges is still a lifelong ambition for Hindu worshippers and, each year, thousands visit such holy cities as Varanasi (Benares) and Allahabad in pilgrimages to do just that. Temples line the banks of the Ganges and ghats (stairways) lead down to the river, where the pilgrims come to bathe and carry home some of its water. While some come only to cleanse and

India, where, according to Human Rights Watch, Dalits are often the victims of violence. "Pariah," a Tamil word used for people with no caste, has also come to mean a social outcast.

*There is a school of mythology called ritualism, which suggests that rituals precede myths—merely stories created to justify the ritual. A "which came first, the chicken or the egg?" debate, the ritualist concept does not alter the fact that rituals and myths combine as powerful forms of belief and social order.

purify themselves, the sick and crippled come—just as thousands of Christian pilgrims flock to such "miraculous" sites as Lourdes—hoping that the touch of the water will cure their ailments. Others come to die in the river, because the Hindus believe that those who die in the Ganges will have their sins removed.*

Another later symbol of the order permeating Indian society was the construction of Hindu temples, which began to be built around 300 CE, during the period of the Gupta Empire (c. 320–550 CE), a period known as India's Golden Age for its accomplishments in literature, science and mathematics, the arts, and architecture. Constructed to venerate a particular deity, these temples, now located across India, housed the god, whose devotees came to the temple for a glimpse of the divine in order to absorb the god's power and carry that power with them in their daily lives. When they came to the temple, worshippers expressed adoration, made offerings, and sought blessings. Often adorned with erotic sculptures celebrating the Hindu pantheon, these temples represented another step in India's evolving society. As Devdutt Pattanaik points out, "Not satisfied with approaching the divine through trees, animals, rivers, and natural rock formations, the kings sponsored the making of idols of Gods and Goddesses in metal and stone that were enshrined in temples. Between 800 and 1300, vast temple complexes came into being. They were controlled and managed by brahmins, who once again came to dominate society. . . . Caste hierarchy manifested in the temple tradition too, with caste based on occupation determining whether one was allowed to enter the temple or not. With rituals came the idea of pollution. Those at the bottom of the caste hierarchy—sweepers, cobblers, and other menial laborers—were the most polluted."

*Unfortunately, the Ganges has also become an industrial chemical dump, an open municipal sewer for the millions who live along its great length, and a depository for animal carcasses and human remains. It may indeed be a divine river, but it is a river seriously soiled by human hands.

MYTHIC VOICES

In the beginning a lotus bloomed. Within sat Brahma. He opened his eyes and realized he was all alone. Afraid, he sought the origin of the lotus he sat on. It emerged from the navel of Vishnu, who slept in the coils of the serpent Ananta-Sesha on the surface of a boundless ocean of milk. Having been formed by Vishnu, Brahma set about creating living beings.

—VISHNU PURANA

There was neither being nor nonbeing then, neither atmosphere nor the sky above. What stirred? Where? Under whose protection?

There was neither death nor immortality then. Day was not separate from night. Only the One breathed, without an alien breath, of Himself—and there was nothing other than He.

Was there below? Was there above?

Who really knows? Who will here proclaim it? Whence was it produced? Whence is this creation? The gods came afterwards with the creation of the universe. Who knows then whence it has arisen?

Whence has this creation arisen—perhaps it formed itself or perhaps it did not? He whose eye watched over it from the summit of heaven, He alone knows. Or perhaps even He doesn't know.

—*Rig-Veda 10:129*

If it's all an endless cycle of birth and destruction, where does the Hindu Creation begin?

Maybe the "One" knows. Maybe the "One" doesn't know. It all depends.

If these kinds of cosmic conundrums hurt your hair, welcome to the world of Eastern thinking. In Hindu tradition, as in other civilizations, explanations can run the gamut from the sublime to the profound to the profoundly enigmatic, and everything in between. Step up to the buffet table of Hindu Creation stories.

For starters, try the "cosmic egg" variety of Creation tales, of which there are several popular variations. In one ancient folkloric version, a supreme goddess lays three eggs in a lotus, and from them emerge three worlds and three gods—Brahma, Vishnu, and Shiva. When the first two of these gods refuse to make love to their "mother," she reduces them to ashes with her scorching gaze. But Shiva agrees to do the deed in exchange for the goddess's fiery third eye. Once he has received it, Shiva shows no mercy—he uses the third eye to incinerate his mother and revive his two sibling gods. Deciding to populate the world, the godly trio realizes that they need wives. So, they divide the remains of the cremated goddess into three ash heaps and, using the power of the third eye, create three goddesses. Together, these three gods and three goddesses populate the cosmos.

In another cosmic-egg story, a golden egg floats in the primordial waters. The golden egg is broken in half by the god Brahma in his role as the creator. The two halves of the egg shell then form heaven and earth. The mountains, clouds, and mists originate from the egg's membranes, the rivers from its veins, and the ocean from the egg's fluid.

There are at least two other cosmic-egg accounts of Creation. In one of these, all of Creation is simply contained within the unbroken egg. In the other, which is included in one of the Puranas, the Creation begins when the god Shiva appears in an androgynous form, and deposits his fiery seed in his female half. A cosmic egg is born of this union.

Unscrambling all of these eggs is tricky. So, set aside the divine hatchery and move to another popular vision of Creation, drawn from the *Mahabharata*, and frequently depicted in Indian art. In this tale, the god Vishnu lies resting on a many-hooded serpent—often a mythic symbol of regeneration, since it sheds its skin—whose numerous coils symbolize the endless cycles of time. When Vishnu assumes the form of an all-consuming fire that destroys the universe, rain clouds appear and extinguish the flames, leaving behind a great sea. Lying on the serpent floating in this immense sea, Vishnu falls into a deep sleep. A lotus sprouts from his navel, and within the lotus is Brahma, the creative force that sets in motion the process of regeneration once more.

Finally, in one Hindu version of Creation, man appears. Manu is the first man, son of Brahma and Sarasvati, and his story has clear parallels to that of Noah, Deucalion, and the other Mesopotamian flood survivors. When the world is threatened by a flood, Brahma takes the form of a fish and tells Manu to build a large boat and store on it all the seeds of living things on earth. As the floodwaters rise, everything is submerged, but Manu's boat lands on the highest peak in the Himalayas. Eventually the floodwaters recede, and Manu makes an offering to the gods, which produces a beautiful woman named Parsu. She and Manu become parents of the human race.

So, then, one might ask: where did the Hindu Creation begin?

One may never know. Even if the "One" knows.

How do you get ten gods in one?

Simple. Count their "avatars."

In the breadth of Indian myth, gods often appear in many physical forms called avatars. Based on a Sanskrit word meaning "descent of a deity from heaven" (*American Heritage Dictionary*), an avatar isn't simply a disguise that a god slips on and off—like Zeus becoming a thunderbolt or a swan and then turning back into Zeus again. Nor is it a simple manifestation, such as the goddess Ganga appearing as the Ganges River. An avatar is an entirely separate entity. In Hindu myth and theology, an avatar can be human or animal and have its own name, personality, physical characteristics, and purpose in life.

That means a goddess like Devi could be a benevolent mother—but her avatar Durga could be dark and destructive. A god's avatar could also take the form of a fish or a boar. Those are just two of the avatars of Vishnu, the central god most associated with these incarnations. Vishnu comes in at least ten different varieties, ranging from tortoise and dwarf priest to king and warrior hero. As Vishnu and the other divinities in this "Who's Who" amply demonstrate, each avatar provides the Hindu gods epic opportunities for adventures and miraculous doings.

WHO'S WHO OF HINDU GODS

Although there are virtually thousands of gods in the Hindu pantheon, these lists include some of the chief deities worshipped in India throughout its long history. Part I includes the earliest gods in the Vedic pantheon. These gods are prominent in the Rig-Veda, and are part of the oral tradition that dates to the Aryan arrival in 1500 BCE. Part II comprises the gods and their manifestations who took a dominant role in the period after the establishment of the Hindu pantheon from about 600 BCE on. The earlier gods were not replaced but usually demoted to lesser rank and power.

Part I: "The Old Gods"

Agni The god of fire, Agni is one of the three chief deities of the ancient Rig-Veda. Although he appears in many guises, he is usually depicted with seven arms and a goat's head, or as a red man with many arms and legs, riding a ram, belching, and emitting light. Agni is more than simply a bringer of fire—he is the vital spark in nature that sometimes consumes in order to create. Manifested both as lightning and the spark of human imagination, Agni symbolizes the Hindu idea of rebirth through destruction. In his role as a "guardian" deity, he is believed to have made the sun and filled the night sky with stars.

When Hindus burn the bodies of their dead, they believe Agni dispatches their souls to heaven in the form of smoke and grants immortality. Among his symbols are a phallic stick used to start a fire by rubbing it in a wooden hole—a metaphor for the heat of the sexual act.

Indra In the early Vedic hymns, Indra is the king of the gods, chief god of sky, storms, and thunderbolts—much like the Greek Zeus. Possibly based on a historical Aryan warrior, he is a great fighter, a lusty drinker of soma—the nectar of the gods—and is often shown with a bloated belly full of the intoxicating beverage (see below). Already tall and powerfully built, Indra grows to enormous size and fills the heavens and earth when he drinks soma.

Indra's position and power result from his defeat of Vritra, the lord of chaos, a serpent-dragon who swallows the world's water and causes a drought. During their intense battle, the serpent swallows Indra and retains the upper hand until the other gods join the battle and gag him. When Indra jumps out of Vritra's mouth, he kills him using his thunderbolt, and then unleashes the monsoon, India's life-sustaining rains. Killing the dragon allows Indra to separate the waters from the land and causes the sun to rise every morning. (Once again, the story of a powerful god's victory over a sea monster or dragon of chaos is a very ancient and widely shared myth: Marduk-Tiamat, Seth-Apep, Zeus-Typhon, and Yahweh-Leviathan are all examples.)

Over time, as the myths of India evolved, Indra was reduced in rank, and many of his functions and powers were then taken over by Vishnu (see below). In a brief story symbolic of this transfer of power, Vishnu lifts an entire mountain with a single finger and uses it like a parasol to protect the people from Indra's torrential rains. Clearly bested by Vishnu's power, Indra assumes his lower station as a rain god.

Soma A certain American beverage calls itself the "king of beers." Soma might be the "god of beers." A most unusual god, Soma is the name of both a deity—the Vedic moon god—and a sacred beverage. As a deity, Soma is said to be the creator and father of the other Vedic gods—a sign of the importance of soma, the beverage of the gods.

In its liquid form, soma was evidently an ecstasy-inducing potion. Judging from its frequent mention in the Rig-Veda, soma was obviously a significant, if not indispensable, element of the ancient Vedic ritual. Either highly intoxicating or hallucinogenic, the "active ingredient" in soma has been the subject of considerable conjecture. In *The Encyclopedia of Psychoactive Substances*, author Richard Rudgley catalogues many of the possible candidates for the source of this powerful party punch—including cannabis, ginseng, opium, some sort of "magic mushroom," and a plant called Syrian rue. Most of these, according to Rudgley, have been rejected, and the truth behind soma remains a mystery.

If "soma" sounds vaguely familiar, you may recognize it for two reasons. It is the brand name of a modern muscle relaxant, and it is also the narcotic widely used in the novel *Brave New World* by Aldous Huxley, the English writer who was both a student of Hinduism and a noted experimenter with hallucinogenic drugs.

Surya Another of the oldest among Indian gods, Surya is the sun god, a dark red man with three eyes and four arms, who rides in a chariot. In the ancient hymns of the Rig-Veda, Surya is a god of almost unbearable intensity who causes the great heat of India's dry season. When Surya's intensity becomes too much for his wife, **Sanjina** ("conscience"), to bear, she transforms herself into a mare and goes to live in the forest. Surya follows her, transforms himself into a horse, and they mate, giving birth to the warrior Revanta and twin sons who are the ever-young and handsome messengers of dawn.

When Sanjina's father later comes on to the scene, he cuts away some of his son-in-law's brightness and these blazing fragments of the sun god fall to earth. These "sun-drops" are transformed into the weapons of the other gods—the discus of Vishnu and the trident of Shiva.

Together with Surya, Sanjina also produces the underworld god Yama.*

Yama Originally thought of as the first man in Vedic lore, Yama has a twin sister, **Yami**, who desires her brother. For resisting the incestuous desire, Yama is immortalized and comes to judge those who enter the underworld. As the god of the dead, he represents judgment, bringing happiness to the virtuous and righteous but bestowing suffering on sinners.

*__Mitra__ was a minor Hindu Vedic sun god worth mention because he eventually travels far beyond India. His twin brother was **Varuna**, the guardian of the cosmic order, and both were thought to be young, handsome, shining deities. Mitra ruled the day while Varuna ruled the night. The god of friendship and contracts, Mitra was good-natured and seen as a mediator between the gods and man. While Mitra occupied a more significant place in pre-Vedic times, his prominence faded with the coming of the Indo-Aryans. But in Persia, he had a longer run as Mitra, a Persian god who was later adoped by the Romans as Mithra. (See page 169, "*What are three Persian magicians doing in Bethlehem on Christmas?*")

Part II: The Second Generation/Later Gods

Brahman Set aside notions of God as a white-bearded man on a throne. Or any of the many other tangible forms that gods take in myths. Prepare for a separate reality. The most absolute, abstract form of God, Brahman is a concept—the soul of the universe, the essence of life, and the divine force that sustains the entire cosmos. Glorified in the Upanishads over all other forms of God, Brahman ("One that is multiple") is the absolute godhead—infinite, changeless, and impersonal.

But in Hindu mythology, Brahman becomes a real, living entity that gets involved in the affairs of the world by manifesting through a trinity of gods called the Trimurti. They are Brahma, creator of the universe; Vishnu, its preserver; and Shiva, its destroyer.

Confused by that? Consider for a moment the Christian notion of God—all-powerful, omniscient, creator of everything. But orthodox Christianity also teaches that this God exists in "three persons"—Father, Son, and Holy Spirit.

Brahma, the Creator One of the three manifestations of Brahman, Brahma is called "lord and father of all creatures" and is regarded as the greatest of all sages as well as the first god. Born in one Creation account from a golden egg that floated in the primeval waters, Brahma is said in another version of the Creation to be the welling up of the Brahman's primeval essence. In yet another Creation account, Brahma emerges from a lotus that grows from a seed in the navel of the god Vishnu. The image of the lotus, a beautiful flower that floats above swampy waters, represents the Hindu ideal of living in the world without being corrupted by it. Brahma is said to have thought up the world while meditating, and is the father of both gods and men.

When Brahma is born, he has only one head, but he grows five faces so that he is always able to gaze on the beautiful Sarasvati. (In a later legend, one of these five faces is destroyed, still leaving him with the four he is usually depicted as having.) An ancient agricultural fertility goddess, Sarasvati—which is also the name of another of India's most sacred rivers—is born from Brahma's side and is also goddess of the creative arts, poetry, music, science, and language. Not only does she get credit for inventing Sanskrit, she gives birth to the first man, Manu, who is sired by Brahma.

In one legend, Brahma and fellow god Vishnu argue about which of them created the universe. As they debate, a great lingam—the word for "phallus" in Hindu terminology—appears, rising out of the ocean, crowned with flame. Staring into its vastness, Brahma and Vishnu see a cave deep within this creative phallus in which the god Shiva resides. Awed by his sight, they concede that Shiva is the ultimate creator.

Finally, a word about Brahma, the cosmic clock-keeper. If you think that time spent waiting at the doctor's office or in a supermarket checkout line is long, consider the awesome mystery of what might be called "Brahma-time." In the incredibly complex mathematics of the Hindu universe, a day in the life of Brahma—called a *kalpa*—lasts the equivalent of 4,320 million earth years. A "night of Brahma" is the same length. Divided into constant, smaller cycles, each of these *kalpas* ultimately ends as the world is consumed by fire and the universe is destroyed and recreated. According to Hindu thought, the current age is called the Kali-Yuga, the final act of a *kalpa* begun eons ago, a dark age that is approaching its end, after which the world will be destroyed once more and prepared for another cycle of creation.

In *Midnight's Children*, his prizewinning mythical novel of modern India, Salman Rushdie captures a sense of the vastness of this Indian concept of time and its impact on people:

> Think of this: history, in my version, entered a new phase on August 15, 1947—but in another version, that inescapable date is no more than a fleeting instant in the Age of Darkness, Kali-Yuga, in which the cow of mortality has been reduced to standing, teeteringly, on a single leg! Kali-Yuga—the losing throw in our national dice-game; the worst of everything; the age when property gives a man rank, when wealth is equated with virtue, when passion becomes the sole bond between men and women, when falsehood brings success (is it any wonder, in such a time, that I too have been confused about good and evil?) . . . Already feeling

somewhat dwarfed, I should add nevertheless that the Age of Darkness is only the fourth phase of the present Maha-Yuga cycle, which is, in total, ten times as long; and when you consider it takes a thousand Maha-Yugas to make just one day of Brahma, you'll see what I mean about proportion.

As Indian myth evolves, Brahma gradually recedes from the picture, and is overshadowed by two more active gods in the Hindu trinity—Shiva and Vishnu—along with a powerful mother goddess, Devi.

Devi The great mother goddess of Hindu myth, Devi is thought to be derived from the original Mother Earth goddess probably worshipped in the Indus Valley before the Aryans arrived. Devi, or Mahadevi ("the great goddess"), is the creative force, but also demands sacrifices. Like the male deities, she has many avatars, some of whom became wives and consorts—or Shakti—of the three gods. Many of the countless goddesses of Hinduism are considered aspects of this great goddess.

Durga The goddess Durga, which means "inaccessible" or "unapproachable," is a dark avatar of the mother goddess Devi. Emerging from the flames shot from the mouths of the male gods Vishnu and Shiva when they are battling a powerful buffalo-demon named Mahisha, Durga is fierce and physically imposing, with yellow-tinged skin and vampirelike teeth. Riding a lion while carrying a club, a noose, a sword, and a trident in her four hands, she seduces the buffalo-demon, captures it with a noose, and beheads it.

As Shiva's consort, Durga combats evil, rids the world of demons, and destroys ignorance. But in spite of her fearsome, violent, and combative origin and nature, Durga is also a goddess of sleep and creativity, and in that spirit is credited with introducing yoga to mankind.

Durga may not be aware of what she has wrought. A quick Internet search under "yoga" produces about 20 million results! There are probably few health clubs left that don't offer some form of yoga exercise, making this ancient Hindu form of discipline one of the most widely shared aspects of Hindu tradition in the world today. Essentially an Indian secret until the eighteenth century, yoga may predate the Aryan arrival in the Indus Valley, according to archaeological evidence. In essence, all forms of yoga are disciplines designed to link the physical body and mind with the unconscious soul, stilling the mind to allow a glimpse of enlightenment. The ancient Sanskrit root of "yoga" is the same for the English word "yoke," as in animals yoked together to work as one. While there are several types of yoga, the one most familiar to Westerners is hatha yoga, the series of breathing techniques and stretching exercises developed as a way to liberate the spirit by channeling energy through the spinal column to the rest of the body. It was originally intended as a preparation for the intensive meditation that is part of raja yoga ("royal yoga"), one of the four main forms of traditional yoga. Other popular derivations of yoga techniques are Transcendental Meditation, a form of yoga using the constant repetition of a divine name (mantra), and bhakti yoga, which involves the dedication of all actions and thoughts to a chosen god. Perhaps the best-known practitioners are members of the Krishna Consciousness movement, who constantly chant the name of Lord Krishna to achieve an ecstatic state.

Ganesha (**Ganesh**) A short, potbellied man who rides around on the back of a rat (or a mouse, in some traditions) and removes obstacles to success, the elephant-headed Ganesha is the god of wisdom, literature, and good fortune. The child of Shiva and Parvati, Ganesha is told by his mother to guard the door as she bathes. When Ganesha refuses to allow his father, Shiva, to enter the bath, Shiva angrily decapitates Ganesha. To calm his angry wife, Shiva then replaces his son's head with the first one he finds—that of an elephant.

By invoking Ganesha's name at the beginning of any activity, a devotee opens the door to material success and spiritual growth.

Ganga Ganga is the water goddess of purification, divinely manifested in the Ganges River. Married to the ocean, Ganga must be careful not to descend to earth too swiftly from the sky—an obvious allusion to the threat of flooding—or she will wash away the earth. In legend, Shiva protects the earth from this threat when his matted hair breaks Ganga's fall to earth. Shiva then divides Ganga into seven rivers (the Ganges and its tributaries).

Kali When Indiana Jones has to confront the bad guys in *The Temple of Doom*, he is up against a very evil deity who demands the heart of sacrificial victims. This bloodthirsty Kali is not the product of the fertile imaginations of Hollywood's Spielberg and Lucas. The goddess Kali, who is known as "the black one," is the offspring of Durga and another dark avatar of the great goddess Devi.

Kali may be the most horrific of all goddesses—not just in India, but in all world mythology. Born from the forehead of Durga while she is fighting another demon, Raktavija, Kali springs forth to win the battle, destroy the demon, and then drink all of his blood so that it doesn't fall to earth and produce more demons. (In another version of Kali's birth, she is said to be the result of Shiva's teasing his wife Parvati about her dark complexion. In contemplation, Parvati sheds the dark skin, which becomes Kali.)

A goddess of destruction, usually portrayed with a fearsome and grotesque collection of accessories—a necklace of skulls and a belt of severed arms or snakes—Kali is connected to human sacrifice and is often depicted as dancing on Shiva's sexually aroused corpse. But even in this image of pleasure and pain, there is regeneration. And in the Hindu vision, destruction and creation are two sides of the same coin. When she dances on Shiva's corpse, Kali actually reanimates him.

Kali also is responsible for a colorful word in English—thug. For centuries, bands of professional assassins in India were known as Thugs—the term derives from the Hindustani word "thag," meaning a thief—a criminal society in India, whose members committed murder and robbery in honor of Kali.

Krishna An avatar of Vishnu, the dark-blue-skinned Krishna is also worshipped as a god in his own right, and is one of the most popular Hindu gods, a Hindu Heracles. Krishna is often shown with a flute in his hand and his consort, the milkmaid **Radha**—a manifestation of Vishnu's wife, **Lakshmi**—standing at his side.

Krishna is born, in one account, from a single black hair that

Vishnu plucks from his head and places in a woman's womb. Created to rid the world of evil, Krishna battles with a bull-demon, a horse-demon, and Kansa, his uncle, the evil king, who has been told by an oracle that he will be murdered by one of his sister's children. Kansa decides to kill the children before they harm him, but through an incredibly complex series of events, Krishna and all of the children are saved by the other gods. Still trying to do away with Krishna, King Kansa sends the demon Putana to nurse the newborn with poison milk, but even as a baby, Krishna is unusual. He kills the demon Putana by sucking all the life from her body as he nurses.

As a young man, Krishna is known for his irresistible good looks and virility. In a story that is often depicted in Indian art, he steals the clothes of a group of milkmaids as they are bathing. The women come before him, naked, and bow, and Krishna returns their clothes to them.

The legend of Krishna's death echoes that of the Greek hero Achilles. As Krishna sits in meditation, a hunter pursuing an antelope sees the soles of the god's feet and thinks they are an animal's ears. When the hunter shoots an arrow into Krishna's foot, he hits a vulnerable spot and kills him—so an Achilles' heel could also be called a Krishna's sole.

Lakshmi Wife of Vishnu, Lakshmi is the goddess of good fortune and bestower of wealth. Goddess of perfect beauty, she is born fully formed from the froth of the ocean—just as Aphrodite, the Greek's ideal beauty, was. Symbolized by the lotus flower—which represents the female principle: the womb, fertility, and life-giving waters—Lakshmi is the personification of maternal benevolence. In very ancient traditions, Indian rulers underwent a symbolic marriage to Lakshmi, just as Mesopotamian kings married Inanna in a rite to ensure the bounty of the earth.

Parvati Another avatar of the divine mother Devi, Parvati ("mountain") is the reincarnation of Shiva's first wife, Sati, and becomes his second wife. Daughter of the sacred Himalayas, Parvati is also the affectionate mother of Ganesha, and a recipient of Shiva's tremendous spiritual and sexual energy, which she releases to the world. In one myth, Shiva initially rejects Parvati because of her dark skin, but he changes his mind when she makes her body glow. This suggests that Parvati may have originated as a pre-Aryan aboriginal goddess who was absorbed into the Hindu pantheon.

Sati The daughter of an ancient god called **Daksha**, Sati marries Shiva over her father's objections. This leads to an argument over inviting in-laws to dinner—talk about archetypes!—and a bloody feud when Daksha summons all of the other gods to a special sacrifice, but snubs his son-in-law. Enraged, Sati throws herself onto the sacrificial fire. Learning of this tragedy, Shiva kills many of the guests at the feast and decapitates Daksha, replacing his head with that of a goat. Daksha repents and becomes a loyal attendant to Shiva, who performs a dance of destruction after which Sati is reincarnated as Shiva's second wife, Parvati.

Are you dizzy yet?

The legend of the dutiful, loyal Sati lives on in an unfortunate reality. In the traditional Indian practice known as suttee, Indian widows throw themselves on their dead husband's funeral pyre in suicidal self-immolation. Known in other ancient cultures, the practice of suttee may have been introduced into India as late as the first century of the Common Era, and became a fairly widespread practice after that. Banned by British colonial authorities in the nineteenth century, the tradition continued sporadically. A recent criminal case in India involved the prosecution of eleven people who were accused—but later acquitted by a special court—of encouraging a widow to commit suttee in 1988.

Shiva, the Destroyer The all-knowing punisher of the wicked, Shiva is the four-armed god of great power known as the Destroyer because he periodically destroys the world so it can be re-created. Shiva possesses a "third eye," from which comes the fire that destroys the Creation.

Often depicted dancing, Shiva haunts graveyards and lives with demons and other supernatural beings. But Shiva is beyond simple distinctions of good and evil, and his followers consider him a merciful god, despite his fearsome characteristics. In Hindu philosophy,

Shiva avoids taking an active part in human affairs, and Hindu art often shows him in solitary meditation on a mountain.

Vishnu, the Preserver One of the main gods of Hinduism, Vishnu has a kindly nature, and is called the Preserver by worshippers who believe that he tries to ensure the welfare of humanity.

In the complexity of Hindu mythology, Vishnu creates, preserves, and destroys the world over and over in a pattern of yugas, which are ages of time. The current period is called the Kali-Yuga, a dark age characterized by dissension, war, and strife, in which materialism rules desires, virtue is nonexistent, and the only pleasure is found in sex. Vishnu sometimes descends from heaven to the earth as one of his avatars when the universe faces a catastrophe or when humanity needs comfort and guidance. In several myths, he must battle some sort of *asura* (demon) who is threatening either the gods or the universal order. While Vishnu has countless avatars, or physical incarnations, these ten are considered of principal importance:

1. **Matsya** is the fish avatar who plays a role in the story of Manu, the first man, by warning him of the flood that is coming.
2. **Kurma** is the tortoise avatar who supports a sacred mountain on his back during a battle with demons.
3. **Varaha**, the boar avatar, uses his tusks to lift the earth, in the form of a beautiful woman, out of the ocean after she falls in. In another version, a demon who has stolen the Vedas pushes the earth into the sea, and the boar rescues the earth and the sacred scriptures with its tusks.
4. **Narasimha**, the half-man-half-lion avatar, kills the invulnerable demon who brings terror to the world.
5. **Vamana**, the dwarf-priest avatar, tricks an *asura* by requesting the amount of land he could cover in three steps. The demon, named Bali, agrees, and Vishnu assumes his full size, covers the whole earth in two steps, and crushes Bali with the third step.
6. **Parashurama** is a brave human of the Brahmin caste who carries a great battle-ax given to him by Shiva to punish all those in the warrior caste (Kshatriyas) who have become arrogant and are sup-

pressing the Brahmins. In winning twenty-one battles, Parashurama proves the supremacy of Brahmins.

7. **Rama**, who is usually depicted as a king carrying a bow and arrow, is a popular mortal hero in Hinduism and the central figure in the *Ramayana* (see below).
8. **Krishna** is Vishnu's other divine avatar, and the central character in the *Mahabharata* and the Bhagavad-Gita, in which he assumes the role of Arjuna's charioteer and they engage in lengthy philosophical discourse (see below).
9. **Buddha** is the only avatar who can be connected to an actual historical person—the great religious teacher who founded Buddhism (see below). Scholars suggest that Buddha was made an avatar in order to bring his worshippers back into the Hindu fold.
10. **Kalki** is the coming avatar who will end the current evil age (Kali-Yuga). In an apocalyptic vision, Kalki will ride a white horse and carry a great sword to punish all evildoers in this world, and usher in a new Golden Age.

Mythic Voices

Facing us in the field of battle are teachers, fathers and sons; grandsons, grandfathers, wives' brothers; mothers' brothers and fathers of wives.

These I do not wish to slay, even if I myself am slain. Not even for the kingdom of three worlds: how much less for a kingdom of the earth!

—*Bhagavad-Gita* 1: 34–35

The author of the Mahabharata has not established the necessity of physical warfare; on the contrary, he has proved its futility. He has made the victors shed tears of sorrow and repentance, and has left them nothing but a legacy of miseries.

—Mohandas Gandhi

What kind of hero doesn't want to fight?

A war epic might seem like an unlikely favorite of one of the twentieth century's most notable apostles of nonviolence. But Mahatma Gandhi (1869–1948), the leader of the peaceful resistance movement that secured India's independence from England in 1947, was said to be profoundly influenced by the Indian poem of war and peace, the *Mahabharata*.

Presumably based on a much older oral tradition, *Mahabharata* was first recorded between 500 and 400 BCE and was continually refined and edited until as late as 500 CE. At least four times the length of the Bible, it recounts an epic feud between two related families—the Pandavas and the Kauravas—who are the descendants of King Bharata. Over centuries, the word "Bharata" has become synonymous with India, so the epic is considered the story of India itself. Although some of the poem's heroes are taken from history, the dating of the war it is said to be based on—once placed at 3102 BCE—has now been discredited.

One relatively small but enormously important piece of the *Mahabharata* is the beloved Hindu scripture, Bhagavad-Gita ("song of the lord or blessed one"). The hero of the Bhagavad-Gita is the warrior hero Arjuna, the "Achilles" of the Pandavas—the semidivine son of the ancient god Indra and a mortal woman. As Arjuna prepares to do battle in the ongoing war between the Pandavas and Kauravas, he has an extended conversation with the god Krishna, who has taken on the role of Arjuna's friend and chariot driver. Arjuna is caught in a moral dilemma which he voices in the opening of Bhagavad-Gita. As a member of the warrior caste, Arjuna knows he must defend his brother, the king. However, arrayed on the opposing side are his cousins, other relatives, and teachers, and he is frozen by the thought of killing these acquaintances and relatives for the reward of a kingdom.

As Arjuna wonders what to do, Krishna teaches him—in eighteen-verse chapters as the battle awaits—that people can achieve freedom by following their prescribed duty without attachment to the results of their action. Summing up the Gita, religious historian Peter Occhiogrosso wrote, "Its chief moral argument is that bodies can be killed, but not souls. Since warfare is Arjuna's dharma, or class duty, it's all in a day's work." As Lord Krishna tells Arjuna, "It is better to die engaged in one's own duty, however badly, than to do another's well."

Ultimately, Krishna reveals himself to Arjuna in his universal form: all-devouring time. Recognizing his duty, Arjuna rejoins the battle. Fought over eighteen days, the battle claims the lives of many heroes on both sides. Finally, largely due to some devious tactics suggested by Krishna, the Pandavas emerge victorious.

Why would a hero banish his loving wife?

How pure and perfect must a devoted wife be to please her husband—and the neighbors? That question is central to the *Ramayana*, the second of India's two great epic poems. At about one-quarter the length of the *Mahabharata*, it is also more accessible and has been popular for centuries. Set in Adodhya, in northern India, and featuring Rama, the seventh avatar of Vishnu and the oldest son and heir of an Indian king, the poem, like other popular hero legends and folktales, is the story of a dispossessed prince, victim of an evil stepmother.

The trials of Rama begin when his stepmother demands that her own son, Bharata, Rama's half-brother, rule as king. Rama's father has promised his wife a wish, and must concede. Rama dutifully accepts his role—his dharma—and is forced into exile, living in the forest for the next fourteen years with his devoted and beautiful wife, Sita—bound by her dharma to remain with her husband—and his loyal brother, Lakshmana. (Bharata, meanwhile, recognizes Rama's right to rule. He places his half-brother's sandals on the throne and agrees to rule from a small village until the day Rama returns.)

While in the forest, Sita is abducted by the demon king Ravana, who takes her to his island kingdom of Lanka (identified as what is now Sri Lanka, once known as Ceylon). Rama goes to war against Ravana and his armies, enlisting the aid of monkey troops led by the shape-shifting monkey general Hanuman. One of the most popular of Indian gods, Hanuman is a gifted healer with supernatural powers who understands the curative qualities of herbs. Rama defeats the forces of Ravana, kills Ravana with an arrow, and rescues Sita. But Rama is initially skeptical of

Sita's faithfulness to him. After she undergoes a trial by fire and proves her innocence, Rama takes her back and they return to Adodhya, where Rama is consecrated as a king.

But even then, in the last book of the *Ramayana*, there is gossip about Sita's "infidelity." Knowing the rumors are unfounded, but feeling duty-bound as ruler to respect the people's wishes, Rama banishes Sita, who is pregnant. Having suffered so much, Sita asks Mother Earth to recall her, the ground opens beneath her, and she vanishes forever. Dividing his kingdom between his two sons, Rama enters a river and yields his life, merging his human existence back with the divine Vishnu.

What is Nirvana?

No, not Kurt Cobain's band. The concept of being peaceful and blessed that describes one's state of mind. A state without desire. Of perfection. In Buddhism. But wait. That's getting ahead of the story.

By the 500s BCE, the Brahminism, which had evolved out of Indian myth, was undergoing the usual growing pains that occur when faithful followers take what they have learned and make it their own. Or decide that there might be another version of Truth. Two of the most profound reactions to the theology and social order of the Brahmins emerged almost simultaneously, and both would have lasting influence. The first—Buddhism—developed during the sixth century BCE out of the teachings and beliefs of a religious and social reformer, Siddhartha Gautama, who became known as the Buddha ("the enlightened one"). The second—Jainism—was developed sometime after 580 BCE by Mahavira, whose name means "the great hero."

Known to millions from those rotund little statues that show him sitting with his legs crossed, in the lotus position, his eyelids serenely closed, the palms of his hands turned up, Buddha is a universally recognizable character. He was born Siddhartha Gautama around 563 BCE on the Nepal-India border, about 145 miles (233 kilometers) southwest of Katmandu, according to archaeological excavations completed in 1995. Beyond those meager details, however, there is little concrete information about his life. Buddhist legend suggests that the Buddha's mother, Maya, dreamed of her son coming into her womb in the form of a white elephant. According to folklore, earthquakes attended the Buddha's birth. And Buddha himself claimed that he was an incarnation of the ancient Hindu god Indra.

And then there is the well-known "biography" that starts with Buddha's decadent youth in the palace of his warrior-caste father, King Suddhodhana. When Suddhodhana receives a prophecy that his son will not become a great ruler if he sees the pain of the world, the father tries to shelter his son, even prohibiting the use of the words "death" and "grief" in Siddhartha's presence. Each time his son leaves the palace, Suddhodhana orders the servants to go before him, sweeping the streets and decorating them with flowers. Another legend says that Siddhartha is given three palaces and between 10,000 and 40,000 dancing girls to keep him occupied.

But reality catches up with Siddhartha. After he marries the princess Yasodhara and has a newborn son, the twentysomethingish Siddhartha has a series of visions—or actual encounters. In the first vision, he sees an old man. In the second, he sees a sick man, and in the third, a corpse. In the fourth vision, he meets a wandering holy man. The first three visions convince Siddhartha that life involves aging, sickness, and death—that "everything must decay." The vision of the holy man convinces him that he should leave his family and seek spiritual enlightenment.

Following these insights, Siddhartha renounces his family and wealth, and becomes a wandering monk practicing extreme forms of self-denial and self-torture for the next six years. Living in filth and eating only a single grain of rice some days, he pulls hairs from his beard, one by one, to inflict pain. But Siddhartha eventually realizes that extreme self-denial and self-torture can never lead to enlightenment, and abandons the practices.

One day, Siddhartha wanders into a village and sits under a shady fig tree, known as the bo, or bodhi, tree ("tree of wisdom"), determined to meditate until he gains enlightenment and completes his quest for the secret of release from suffering. As he sits in meditation, Siddhartha is tempted by the evil demon Mara, much as the biblical gospels tell of the temptations of Jesus in the wilderness. First, the demon sends his beautiful daughters to seduce Siddhartha. But Siddhartha resists. Then the demon threatens the young man with devils. But Siddhartha stands firm. In a final act, the devil throws a fiery discus at Siddhartha's head, but it is transformed into a canopy of flowers.

After sitting for five weeks and enduring a world-shattering storm, Siddhartha finally achieves enlightenment. The roots of suffering are desires, he discovers, and one only has to reach a state without desire to overcome suffering. Released from all suffering and from the cycle of reincarnation, Siddhartha becomes Buddha and decides to show other people the way, preaching a doctrine of compassion and moderation.

In a religious coming-out ceremony near the holy city of Varansi, Buddha preaches his first sermon to five holy men. This sermon, which includes the "saving truth" of Buddha's message, is one of the most sacred events in Buddhism.

As Buddha continues preaching throughout northern India, he attracts disciples and his fame increases. Soon stories begin to spread among his followers, describing his religious insight and compassion—along with tales of his magical powers. His followers believe that Buddha has lived many lives before being born as Siddhartha Gautama, and the stories describing the events of these lives, called *jatakas*, become the popular means of understanding Buddha's message, which includes the concept of Nirvana. No, not Kurt Cobain's band.

According to Buddhist belief, the perfect peace and blessedness is a state called Nirvana. Attaining Nirvana enables a person to escape from the continuous cycle of death and rebirth caused by an individual's worldly desires, such as craving for fame, immortality, and wealth. In Buddhism, people attain Nirvana only when such desires are completely eliminated.

Buddha preached that Nirvana can be attained by following a Middle Way between the extremes of ascetic self-denial and sensuality, yet living in the world with compassion and by practicing the Noble Eightfold Path, which consists of:

1. Perfect understanding, or Knowledge of the truth
2. Perfect aspiration, the intention to resist evil
3. Perfect speech, or saying nothing to hurt others

4. Perfect conduct through respecting life, morality, and property
5. Perfect means of livelihood, or holding a job that does not injure others
6. Perfect endeavor, striving to free the mind of evil
7. Perfect mindfulness through controlling one's feelings and thoughts
8. Perfect contemplation through the practice of proper forms of concentration

"Through observation and effort," summarizes Jonathan Forty, author of *Mythology: A Visual Encyclopedia*, "a person can break out of the laws of karma. . . . The aim of Buddhists is to step outside this wheel of karmic rebirth and attain nirvana, or release from it and reunification with the One."

At about the age of eighty, Buddha became ill and died. His disciples gave him an elaborate funeral, burned his body, and distributed his bones as sacred relics.

In Indian history, Buddhism reached a high mark of sorts when an Indian emperor named Ashoka converted in 262 BCE, renounced violence, and named Buddhism the state religion (Ashoka died in 232 BCE). In Buddhist tradition, Ashoka had become horrified at the cost of empire-building and embraced Buddhism. Today, Buddhism is one of the major religions of the world and it has been a dominant religious and social force in most of Asia for more than two thousand years. There are an estimated 364 million followers today.

Reacting to the growing popularity of Buddhism, the Brahmins later tried to absorb it by depicting the Buddha as the ninth avatar of Vishnu, which would be a little like traditional Judaism finding a way to absorb Jesus or Mohammed into its list of prophets. It was successful in many respects, because Buddhism gradually faded as a dynamic influence in India. When leaders of the Gupta Dynasty reunited northern India around 320 CE, they brought about a revival of Hindu religious thought, caste lines were reinforced, and Buddhism eventually disappeared as a force in India.

Emerging in about the same era as Buddhism did, the second major offshoot of Hinduism is Jainism. Like Buddhism, Jainism is traced to a man who is believed to be an actual historical individual. Mahavira is said to have been born to aristocratic parents in 540 BCE and was a contemporary of Buddha, though they may have never met. Nonetheless, as with Buddha, certain myths developed about Mahavira. At his birth, the gods were said to have descended from heaven and showered flowers, nectar, and fruit on his father's palace. There are many legends about his extraordinary childhood, but as an adult, he is said to have lived an ordinary life until his parents died. Then, at the age of thirty-two, he gave away his possessions, left his wife and child, and became a wandering monk. The sky glowed like a lake covered in lotus flowers when this happened.

Mahavira's teachings form the basis for Jainism, which is centered on the belief that every living thing consists of an eternal soul called the *jiva* and a temporary physical body. Attaining release from the world of sorrows can be achieved by renouncing sin and violence, engaging instead in strict penance and extreme, disciplined, nonviolent conduct. In Jainism, sadhus (holy men) and sadhvis (holy women) try to separate themselves from the everyday world through a vow of poverty and may not own any property except a broom, simple robes, bowls for food, and walking sticks. They may not live in buildings except for brief periods and must beg for all their food. They are not allowed to kill any living creature, and Jain monks wear a veil or mask over their mouths, so they don't accidentally swallow any insects.

Small in number, with some 4 million adherents worldwide, Jainism has been influential, nonetheless. Laypeople, or followers who are not priests or holy men and women, observe a less rigorous code of conduct, and support the priesthood. Many of them are businesspeople who have flourished, in no small part, because Jainists enjoy a reputation for scrupulous honesty in commercial activity that does not directly involve killing any living thing.

CHAPTER SEVEN

EVERYWHERE UNDER HEAVEN

The Myths of China and Japan

The mountain rests on the earth: the image of splitting apart. Thus those above can insure their position by giving generously to those below.

—I Ching (Book of Changes), twelfth century BCE

To be able to practice five things everywhere under heaven constitutes perfect virtue . . . gravity, generosity of soul, sincerity, earnestness, and kindness.

—Confucius, 551–479 BCE

The Way of Heaven has no favorites.
It is always with the good man.

—Lao-tzu, c. 520 BCE

Before heaven and earth had taken form all was vague and amorphous. Therefore it was called the Great Beginning. The Great Beginning produced emptiness and emptiness produced the universe. . . . The combined essences became the yin and yang, the concentrated essences became the four seasons, and the scattered essences of the four seasons became the myriad creatures of the world.

—Huai-nan Tzu, second century BCE

What are oracle bones?

How did the ancient Chinese think the world began?

What role do "family values" play in Chinese myth?

Who's Who of Chinese Gods

What do fortune cookies have to do with Chinese religion?

What religion shunned the Confucian approach?

Who was Japan's first divine emperor?

How did Shinto become an "Asian fusion" religion?

Who's Who of Japanese Gods

MYTHICAL MILESTONES

China

Just as Egyptian history (see chapter 2) is traditionally divided by the dynastic ages, ancient Chinese history is also characterized by long periods of ruling dynasties following the prehistoric and so-called legendary periods.

Before the Common Era (BCE)
Prehistoric and Legendary Eras

c. 8500 Earliest Chinese pottery is created.

c. 7000 First farming villages formed in the Yellow River basin.

c. 3500 Wet rice farming begins near east coast.

First planned villages appear in northern China, with distinct residential and burial areas.

c. 2700 Silk weaving practiced.

c. 2500 First walled cities are built.

2205–2197 Reign of Yu, legendary emperor of first Xia Dynasty.

c. 1900 Boldly painted burial urns are first used in western China.

Shang (Yin) Dynasty (1523–1027)

1300 Oracle bones made from deer bones and tortoise shell, with written inscriptions, are used in divination.

Chinese script is created.

Royal burials with human sacrifice are practiced.

Zhou Dynasty (1027–221)

c. 1000 Chinese bronze casting is at advanced level unrivaled elsewhere in the world at this time.

Origin of the Yi Jing (I Ching, Book of Changes).

841 Beginning of accurately dated history in China.

State-sponsored exploration and early mapping of China's geography.

c. 650 Silk painting, lacquer work, and ceramics become highly skilled.

c 563–483 Life of Buddha.

c. 551–479 Life of Confucius, most influential philosopher in Chinese history.

c. 520 Speculative birth of Lao-tzu (Laozi), philosopher and traditional founder of Taoism.

513 First mention of iron; casting techniques allow for production of huge quantities of tools and weapons.

371–289 Life of philosopher Mencius, who continues Confucian teachings.

c. 360 Widespread use of crossbow in warfare.

325 Prince of Qin takes the title of *wang* (king), and claims to rule all of China.

c. 300 Cavalry introduced.

Qin Dynasty (221–207)

214 Work on Great Wall begins.

206 Great Wall is completed.

The first emperor of all China, Shi Huangdi, unites China. At his death, he will be buried in a vast, man-made mountain. This tomb, discovered in 1974, was guarded by a now-famous army of seven thousand painted terra-cotta warriors.

Han (Western) Dynasty (206 BCE–9 CE)

206 Liu Bang proclaims himself emperor of new Han Dynasty; capital established at Chang'an.

165 First official examinations for selection of Chinese civil servants.

141 Han power expands into western China under Emperor Wudi.

136 Confucianism becomes state religion.

111 China conquers and incorporates northern Vietnam.

c. 110 Opening of Silk Road across Central Asia. It links China with southwest Asia and eventually Europe.

108 China takes control of Korea.

98 State establishes a monopoly on alcohol.

Common Era
Han (Eastern) Dynasty (25–220)

2 First census of Chinese population of more than 57 million, mainly concentrated in the North.

57 Ambassador from king of Nu (Japan) is recognized by the Han emperor.

65 First evidence of Buddhism in China.

106 Invention of paper by Cai Lun, a eunuch serving in the imperial court. With this invention, the Chinese could discard expensive bamboo blocks and silk and adopt a cheap, easily transported writing medium.

168 Following death of Emperor Huandi, Han Empire begins a period of rapid decline, similar in some respects to decline of Roman Empire. It collapses in 220.

220–265 Three kingdoms (Wei, Shu, and Wu): period of disunity.

c. 250 First known use of lodestone (magnetic) compass.

265–589 Period of division between Wei (northern) and Qi (southern) Empires.

c. 350 Invention of rigid, metal stirrup in China; vitally important innovation in mounted warfare.

399 Chinese monk and pilgrim Fa Xian journeys to India to study Buddhism.

444 Taoism is made the official religion of Wei Empire after the conversion of the emperor.

446 Rebellion in Buddhist monastery against Taoist reforms. Wei emperor orders the execution of every monk in the empire; but many escape.

477 Buddhism becomes Chinese state religion. In 489 huge cave temple complex is built in in northern province of Yungang.

Sui Dynasty (590–618)

589 Reunification of China begins.

c. 600 Beginning of book printing.

Tang Dynasty (618–906)

618 Under Tang Dynasty control, China becomes a vast empire of some 60 million people.

626 Tang court adopts Buddhism.

Rise of scholar officials.

Expansion into Korea, Manchuria, Central Asia.

630 First Japanese ambassador welcomed at Tang court.

907–960 Five Dynasties Period.

Song Dynasty (960–1279)

1215 Mongols seize most of North China; Genghis (Chinggis) Khan rules an empire from the Pacific Ocean to the Caspian Sea in the west before his death in 1227.

1260–94 Rule of Kublai Khan.

1275–1295 Marco Polo in Mongol-ruled China.

Japan

Before the Common Era (BCE)

c. 10,000 Earliest known pottery vessels made in Honshu.

c. 660 Jimmu-tenno ("divine warrior emperor") is the legendary first human emperor of Japan.

c. 500 Rice cultivation spreads to Japan from China.

Common Era

57 Ambassador from king of Nu is recognized by China's Han emperor.

247 Civil war between rival kingdoms.

260 Temple of Amaterasu founded in Ise, the most sacred and revered shrine of Shinto religion.

c. 300 Emergence of Yamato state in Japan.

478 First Shinto shrine appears.

538 Buddhism reaches Japan via China and Korea.

592 Conflict between clans over Buddhism and local deities leads to execution of the emperor.

630 First Japanese ambassador at China's Tang court.

685 Buddhism becomes state religion of Japan; in 741, Buddhist temples are established throughout the land by government decree.

January 1, 1946, the Japanese emperor Hirohito (d. 1989) denied his own divinity. In 1947, the Japanese Constitution ended official state Shinto. Modern Japan is a parliamentary democracy, in which the emperor is the head of state, but the elected prime minister is the head of government.

In 1793, King George III sent an emissary to the court of Chinese emperor Qian Long. Arriving in what was then called Peking, the British ambassador displayed a lavish array of gifts for the Chinese ruler, including six hundred cases of scientific instruments. The emperor, a member of the Qing Dynasty, who had been on the throne for nearly sixty years, was polite but unimpressed. "There is nothing we lack," he told Lord McCartney, the British envoy. "We have never set much store on strange or indigenous objects, nor do we need any more of your country's manufactures."

It was seemingly true. For much of its nearly 4,000-year-long history, China had thrived in splendid isolation, a mysterious and unwelcoming empire that neither needed nor desired contact with outsiders. Cut off from most of its neighbors by natural boundaries—the Himalayan mountains to the west, the Gobi desert and forbidding Mongolian territory to the north, and the Pacific Ocean to the east—China had expended vast sums and countless lives building walls.* Behind these formidable barriers of earth and stone, sheltered from the gaze of potential invaders and eager Christian missionaries, China's successive dynasties had developed a civilization that was in many ways far more advanced than any of its contemporaries. Not only did the Chinese create a vast network of rural villages held together by a remarkable bureaucracy and a single written language, they invented paper, printing, gunpowder, fireworks, the seismograph, noodles, the compass, and ships capable of sailing the world long before Westerners did. China was, as historian Daniel Boorstin once called it, "an empire without wants."

Yet, in spite of its early history of writing, China did not leave the world's richest written mythic legacy. Unlike the Egyptians, who stored

*The existing Great Wall of China dates from the Ming Dynasty (1368–1644 CE). But records of wall-building by the Chinese go back as far as 600 BCE, and the idea to construct a large Great Wall began during the Qin Dynasty (221–206 BCE).

thousands of funeral texts in their grand tombs, the Chinese were seemingly far less concerned with elaborate burial rites, and left no detailed road maps to the afterlife, although they built expansive tombs. Though nearly every great ancient civilization composed epic poems of love and war, there is no ancient Chinese *Gilgamesh*, *Iliad*, or *Ramayana*. China certainly had its cornucopia of Creation and Flood accounts—as many as six separate Creation stories and four Flood narratives, each featuring different characters. But, intriguingly, these tales don't emphasize heavenly retribution for sinful behavior. And while the Chinese acknowledged a wide range of nature gods, mythical semidivine rulers, and prophetic priest-kings in their fourth-century treatise *Questions of Heaven* and their third-century encyclopedia of gods, *The Classic of Mountains and Seas*, it was human ingenuity—not divine intervention—that was seen as the solution to most problems.

No surprise, then, that myth never formed in China the deep cultural identity that is typically associated with Greece or India. Or became the monolithic state religion, or powerful priesthood associated with Egypt. Quite to the contrary, China's vast size and regional differences checked the development of a single "national" mythology that could unite the country. Even as generations of Chinese students immersed themselves in the myth-tinged works of Chinese literature called the Five Classics, their goal was not to become priests. They were preparing for the ancient Chinese SATs, or "civil service exams," required to climb the imperial bureaucratic ladder or advance in the army. (The Chinese even had an "examination god" named Kui Xing, who was called upon by scholars for divine assistance at test time.)

But where myth failed to unify China, philosophy succeeded. Far more important than China's poets and storytellers were its sages and wise men. Think China, and you think Confucius, not a poet like Homer.

The two great strands of native Chinese philosophy, Confucianism and Taoism—both introduced around 500 BCE—clearly shaped China's history, government, and culture more than any myth or religious belief did. Emphasizing social order, loyalty to family and king, and ancestor worship, Confucianism is a moral code of proper behavior designed to achieve an ideally gentle world in which every individual has a place within the family and every family has a place within the society. Confucianism places the virtue of a disciplined communal order above the need to appease the gods, while Taoism, the second major school of Chinese thinking, stresses the importance of individuals living simply and close to nature. By far the more influential of the two, Confucianism was made the state religion in 136 BCE, during the powerful Han Dynasty—a 400-year period in Chinese history often equated with the Roman Empire in terms of its size and prestige. Just as Confucianism was being institutionalized, Buddhism was imported from India, adding a spicy new accent to China's philosophical potpourri and creating in China a picture markedly different from other great civilizations, where myth was often all-pervasive.

One other significant but very modern factor has diminished China's mythic legacy. The study of Chinese myths and mythic sources was severely stunted under Communist rule. The all-powerful official Chinese Communist Party that has governed China since 1949 largely suppressed all religions, which were regarded as mere superstition. Classical Confucianism was opposed by the Maoist "powers that be"—which created a mythology of its own to lionize Chairman Mao—because it emphasized the past and, in the Party view, justified social inequality.

The Five Classics, studied by aspiring bureaucrats and functionaries for two thousand years, were set aside in favor of Karl Marx and Chairman Mao's Little Red Book. During decades of strict Communist authoritarian rule, the Party turned Buddhist and Taoist temples into museums, schools, and meeting halls, and the study of mythology and other ancient Chinese traditions suffered. An entire generation of academics, scholars, and researchers was largely eliminated in the violent upheaval of the Great Proletarian Cultural Revolution of the 1960s, when universities were shut down for years, and foreign embassies closed. Some 7 million students, teachers, and others in the professional classes were sent to be "reeducated" on rural collective farms, where many did not survive the purges and repression of the Red Guard era.*

*With the reforms that have come to China during nearly thirty years since China was "reopened" in the post-Mao era, government attitudes toward religion have softened considerably. Now recognizing the value of the Confucian emphasis on correct moral behavior, the Party has returned some temples to the control of religious groups. But religion in China is still very much in official hands. The

The diplomatic "opening" of China that followed President Richard Nixon's historic 1972 visit has also unlocked China in other ways that relate to its mythic past. Ancient Chinese healing arts, such as acupuncture, "energy healing," and herbal medicines, are now a growing part of the Western medical arsenal. Many Westerners now decorate their homes and offices with careful attention to feng shui (*fuhng shway*), the ancient Chinese art of placing objects with the goal of creating a sense of balance and harmony. According to feng shui, the life-force energy called chi flows from every living and inanimate object, and can be promoted with the careful placement of furniture and the proper use of colors.

In the arts, a band of Chinese filmmakers known as the Fifth Generation has introduced American audiences to Chinese history and folk traditions in such films as *Raise the Red Lantern*, *Yellow Earth*, and *Farewell My Concubine*. Chinese-American filmmaker Ang Lee wowed the world with his legendary folktale *Crouching Tiger, Hidden Dragon*. Acclaimed Chinese-American writers such as Amy Tan, author of *The Joy Luck Club* and *The Kitchen God's Wife*, have reached wide audiences as they explored Chinese mythic and family traditions and their impact on a generation of Chinese-Americans. Even Disney got on the Chinese dragon-wagon with its animated *Mulan*, a 1998 "girl power" version of a Chinese folktale, whose heroine takes her father's place in battle and is the same Fa Mu Lan whom Maxine Hong Kingston wrote about in her award-winning memoir *Woman Warrior*.

During the past thirty years, along with this vanguard of a new "cultural revolution," a generation of archaeologists and scholars has also been allowed to peek over the "Great Wall" that surrounds China's ancient history. As they do, they have begun to reveal the rich, imaginative, and colorful myths born in "the empire without wants."

government initially approved of a movement that combined Buddhist and Taoist philosophies with deep breathing and martial arts exercises. However, Falun Gong, as it is known, has come under official attack since 1999, when 10,000 of its followers demonstrated in front of Party headquarters.

MYTHIC VOICES

Thunder comes resounding out of the earth:
The image of Enthusiasm.
Thus the ancient kings made music
In order to honor merit,
And offered it with splendor
To the Supreme Deity,
Inviting their ancestors to be present.

—from I Ching,
Richard Wilhelm and Cary F. Banes, translators

What are oracle bones?

Near the end of the nineteenth century, a large number of so-called "dragon bones" began showing up in apothecary shops throughout China. Ground into powders to be used in folk medicines and aphrodisiacs, these bones were thought to possess magical powers, because they had strange markings on them. Scholars became aware of the "dragon bones" in the early twentieth century, and by the 1920s, when extensive excavations were made near some of the oldest human settlements along China's Yellow River, more than 100,000 of these bones had been unearthed in what proved to be an archaeological gold mine.*

Made from the bones of deer and oxen, and tortoise shells, these artifacts were later identified as "oracle bones" used by royal priests—and even by early Chinese kings themselves—in making prophecies, communicated through dead ancestors. Representing the earliest form of Chinese religion, the bones were marked with Chinese characters from one of the earliest known forms of written Chinese. In ancient divining

*Early human beings apeared more than a million years ago in what is now China. By about 10,000 BCE, a number of New Stone Age cultures had developed in the Yellow River area and, from them, a distinctly Chinese civilization gradually emerged. One of these was the Longshan culture, which spread over much of what is now the eastern third of the country. China's first dynasty, the Shang Dynasty (1523–1027 BCE), arose from the Longshan culture during the 1700s BCE.

sessions, the bones were marked with a shallow cut—which evolved into written "questions"—and then heated until the bone or shell cracked. The resulting fissures were then "read" by a priest, who made predictions based on the configuration of the cracks. Usually a simple "fortune-telling" question about the weather, the success of a hunt or battle, or the sex of an expected child would be asked. Interestingly, a "reading" of "yes" meant boy, while "no" meant girl—an indication of a very old Chinese preference for male children, still a concern in China today, where modern sex-screening methods are used to abort female fetuses.

Although they probably represent traditions that go back much further in Chinese history, many of the oracle bones are dated to 1300 BCE, during the Shang Dynasty (1523–1027 BCE), one of the first kingdoms in Chinese history for which there is significant archaeological evidence. Based in the Huang He Valley, the Shang was organized as a city-state with a king who probably served as high priest, similar to the organization in ancient Mesopotamia. Other finds from the Shang Dynasty include sophisticated bronze drinking vessels that show a high degree of metalworking skill. On the grimmer side of the archaeological ledger are Shang tombs in which kings and nobles were buried with treasures that included war chariots—often complete with horses and charioteers. Clearly, human sacrifices were made during the Shang period, and the remains of sacrificial victims, ceremonially beheaded in groups of ten, have been found in these tombs. Patricia Ebrey, a scholar of Chinese family and kinship, writes that one tomb of a Shang king, who ruled around 1200 BCE, "yielded the remains of ninety followers who accompanied him in death, seventy-four human sacrifices, twelve horses, and eleven dogs. . . . Some followers were provided with coffins and bronze ritual vessels or weapons of their own, some (generally female) with no coffins but with personal ornaments; others were provided with no furnishings and were beheaded, cut in two, or put to death in other mutilating ways." (Human sacrifice was apparently abolished during the next dynastic period, that of the Zhou Dynasty, 1027–221 BCE.)

During this very early period in Chinese history under the Shang, "religion" was largely based on the idea that each person had two souls—one a "physical" soul, and one an "eternal" soul—that could be kept alive through sacrifices performed by a male family member. With proper sacrifice, the eternal soul became a deity of power and influence that could respond to divination requests or perform other heavenly favors. But if a deceased ancestor's soul was neglected or treated poorly, that soul could become a demon and haunt the living. The vast majority of China's people throughout history have been rural peasants, and this farmer class had little to do with these lofty beliefs, which were reserved for the landed wealthy. Instead, the religion of the peasant farmers centered on the worship of local deities of soil and water and shamanistic cults featuring spirit mediums—practices largely dismissed by the upper classes of China. Just as there was human sacrifice at the royal level, there was a grim side to these local rites. In many farming villages, there were river festivals in which beautiful girls were selected as the "bride of the river." Set afloat in a boat, they were ultimately drowned as an offering to the river god.

Chinese religion developed without a powerful priestly class, and the sacrificial cult services were performed by the head of the family, or by state officials. The sacrifices included a variety of domesticated animals, or wine poured as a libation. The concept of proper sacrifice was so important, according to historians W. Scott Morton and Charlton Lewis, that the downfall of a kingdom would be attributed to times when "the sacrifices were interrupted."

The methods of divination that produced the ancient oracle bones became more sophisticated over time, and were formalized in an important Chinese classic called the I Ching (also commonly called Yi Jing, among other various Romanized spellings), or Book of Changes. Counted among the earliest and most influential of the ancient Chinese texts called the Five Classics, the I Ching probably originated about 1122 BCE, early in the Zhou Dynasty, which ruled China for more than 800 years, including the period in which Confucius lived. Grouped together with the I Ching were The Classic of History, material about early kings of questionable authenticity; The Classic of Poetry, a collection of folk and ceremonial songs; The Collection of Rituals (or Rites); and The Spring and Autumn Annals, attributed to Confucius. This family of books constituted the basis of study for the imperial examination that had to be mastered by anyone wanting to advance in the Chinese imperial bureaucracy right up to the early twentieth century.

Like oracle bones, the I Ching was first used to predict the future. A person with a question followed a specific ritual that involved tossing special sticks or coins and then referencing the appropriate commentary in the I Ching. Over time, with the growing influence of Confucianism, the function of the I Ching evolved, and by the 500s BCE, the I Ching was viewed as a book of philosophy.

Traditionally, it was believed that the principles of the I Ching originated with Fu Hsi (Fu Xi), a creator god said to be one of China's legendary early rulers. (See page 375, *Who's Who of Chinese Gods*.) It was also long accepted that Confucius himself had either written or edited the I Ching. During the past fifty years, however, discoveries in archaeology and linguistics have reshaped theories of the book's history. Scholars have been helped immensely by the discovery in the 1970s of intact Han Dynasty–era tombs in Hunan Province. (The Han ruled China for roughly 400 years from about 200 BCE to 200 CE.) One of these tombs contained more or less complete second-century BCE texts of the I Ching that are centuries older than the previously discovered texts. Mostly similar to the well-known I Ching, these tomb texts include additional commentaries on the I Ching, previously unknown and apparently written as if they were meant to be attributed to Confucius. The bottom line is that, after considerable investigation, many modern scholars doubt the actual existence of the mythical ruler Fu Hsi (Fu Xi), and think that Confucius had nothing to do with the Book of Changes.

The oracle bones and divination texts of the Shang period contained another important Chinese mythical religious concept that dictated Chinese history for two thousand years. The Shang Dynasty had ruled because of the belief that they had "family connections." In the view of the ancient Chinese, the founders of China had been deities, and the Shang ancestors had joined these divine rulers in heaven. To the Shang, heaven was very active in earthly matters, and they ruled with the intercession of a supreme god they called Shang Di—the Lord on High.

The idea that heavenly connections guided an earthly king's reign evolved into a Chinese concept called "the mandate of heaven." In essence, the mandate was a sign of divine approval. If a king ruled well, he continued in power; if he ruled unwisely, heaven would be displeased and would give the mandate to someone else—sort of like a divine board of directors canning the CEO. The first people to exercise the mandate were the Zhou Dynasty (1027–221) from western China, when they overthrew the Shang Dynasty. The Zhou made it clear, in explaining the mandate to the defeated people of the Shang Dynasty, that if their king had not been so evil, his mandate would not have been withdrawn. The same logic was later used to overthrow the Zhou.

One significant consequence of the idea of the mandate of heaven was that it was not necessary for a person to be of noble birth to lead a revolt and become a legitimate emperor. In fact, a number of dynasties were started by commoners, including the mighty Han, whose first emperor was a rebel army officer who seized power during a civil war. If the emperor ruled unwisely or failed to perform the proper rituals, he was out the door—whether a noble or commoner—and, most likely, without a generous "severance package."

On the other hand, the mandate of heaven also promoted the "might is right" idea, since any dynastic founder possessed the mandate by virtue of his success, and any failed ruler was considered to have lost it, no matter how great his personal virtue. The mandate also encouraged both Chinese unity and a disdainful attitude toward the outside world, since there was only one mandate, and so only one true ruler of humankind, the emperor of China.

When the Shang Dynasty was overthrown by the Zhou, the essential continuity of Chinese civilization continued. It was during the Zhou Dynasty that the major philosophers of Chinese history, Confucius and Lao-tzu, both lived and formulated the two schools of thought that would shape Chinese civilization—Confucianism and Taoism.

Mythic Voices

People say that when Heaven and earth opened and unfolded, humankind did not yet exist. Nü Kua kneaded yellow earth and fashioned human beings. Though she worked feverishly, she did not have enough strength to finish her task, so she drew her cord in a furrow through the mud and lifted it out to make human beings. That is why rich aristocrats are the human beings made from yellow earth, while ordinary poor commoners are the human beings made from the cord's furrow.

—*from* Chinese Mythology, *Anne Birrell*

How did the ancient Chinese think the world began?

Eggs and mud. Yin and yang. A giant, a gourd, and children in peril. These ancient elements all figure prominently in China's Creation stories, which run the gamut from the profoundly primordial and primitive to the folkloric and fanciful, as they attempt to explain how the world got started.

Like other civilizations, China has several Creation stories that emerge from its many regions and long history. These stories come down through a variety of sources, including the Five Classics, *Questions of Heaven* (an ancient text from the fourth century BCE), and an anonymous compilation called *The Classic of Mountains and Seas*. The last of these works, collected between the third century BCE and second century CE, is the closest thing there is to a Chinese "encyclopedia" of myths, including more than two hundred mythical figures.

The most influential of China's Creation stories describes the universe simply coming into being from a cloud of vapor that is suspended in darkness. Out of this primordial chaos come the two essential forces, yin and yang. Dual opposites—stuck together like cosmic peanut butter and jelly—these forces profoundly affected Chinese culture and society, especially in the philosophic system that later emerged, called Taoism.

Often represented by a circle with dark and light areas, yin and yang exist in a delicate balance and underlie the entire Chinese universe. While yin is associated with the qualities of the "feminine"—cold, heaviness, darkness, and earth, yang is linked with the "masculine" qualities—warmth, light, brightness, heaven, and the sun. The interaction of these opposites is believed to have created a major portion of the universe, the seasons, and the natural world. Yin gave birth to water and the moon; yang gave birth to fire and the sun.

As historian Alasdair Clayre writes in *The Heart of the Dragon*, "Thinking in yinyang terms means analyzing the universe into pairs of fluidly netting opposites, such as shadowed and bright, decaying and growing, moonlit and sunlit, cold and hot, earthly and heavenly or female and male. . . . Men and women are not seen as exclusively yang or yin: each has only a predominance of the one aspect or the other. . . . The relation of the two elements of a yinyang pair is not a static one, but is thought of as a continuous cycle in which each tends to become dominant and responsive in turn."

The Creation is described in several other stories that were well known to the ancient Chinese. Two of most popular involve a pair of China's most important gods, Panku (Pan Gu, P'an Ku), a gigantic primeval deity described as the child of yin and yang, and Nü Gua (also Nü Kua, Nu Wa), a popular deity known as "gourd woman" or "woman Gua." The latter name refers to snail-like creatures that lose their shells and symbolize regeneration.

In the Creation myth of Panku, whose story became the widely accepted Chinese Creation myth by the third century Common Era, the world is an enormous egg filled with chaos, in which the giant Panku has been sleeping for 18,000 years. When Panku—whose name is translated as "coiled antiquity"—grows large enough to crack the egg, its clear, translucent fluid (the ethereal yang matter) oozes out and floats up and becomes the heavens. The yolk and heavy (yin) parts drip down to become the earth. Afraid that sky and earth might converge, Panku pushes the sky up with his head and the earth down with his feet. Like the Greek Atlas, he remains that way for another 18,000 years, until he realizes that the sky is high enough and won't fall. Exhausted by his efforts, Panku lies down to rest and dies in his sleep. As he is dying, his breath becomes the wind and the clouds, his voice the thunder. One of his eyes becomes the sun and the other the moon. His limbs become mountains and his veins turn into the roads. No part of his giant body goes unused in the creation of the world. In later versions of the story, even the flies, fleas, and other parasites on his body are transformed into the ancestors of man.

Prefer a Creation story that is a little more "dirty"? The chaste Chinese don't have a very sexy Creation tale, but there is one that involves playing in the mud.

Try a very old tale featuring Nü Gua, the fertility deity and mother of Creation, who is lonely after the world has come into being. Scooping up some wet clay from the bank of the Yellow River, Nü Gua presses it into tiny figures and impregnates each one with the force of yin or yang, so the figures come to life. Those who receive yang become men, and those who receive yin become women. When Nü Gua tires of molding the figures one by one, she spins clay off the end of a rope, or vine, that she has dragged in furrows across the muddy ground. The misshapen figures that come from the gobs of falling mud become humans born into poverty, while the handsome figures, molded by the goddess's hand from the clay of the Yellow River, become the Chinese nobility.

Nü Gua also appears in one of several Chinese Flood stories, appropriate for a region prone to violent flooding. (The Yellow River has been called "China's sorrow" for the ferocity of its devastating floods.) Unlike Flood tales in other civilizations, in which men such as Noah and Deucalion play the dominant roles, the Chinese version features Nü Gua, along with her brother, Fu Hsi, and their father. In this legend, a thunder god who is a fishlike deity with a green face, scales, and fins is angry with Nü Gua's family. Fearing the god, Nü Gua's father builds an iron cage outside his house and waits with a pitchfork in case of an attack by the fish god. In the midst of a great storm, the thunder god arrives and threatens Nü Gua's father. But the clever man is able to trap the god in the cage and plans to cook the fish god. With the thunder deity contained, father goes off to buy spices, so that the thunder god will taste delicious once he is stewed up, but first warns his children not to give the god anything to drink. When the thunder god whimpers that he is thirsty, Nü Gua takes pity on him and gives him a drink. Swallowing the water helps the thunder god regain all of his power, enabling him to break out of the cage. Before he escapes, he gives one of his teeth to the children. They plant it, and a tree soon grows, bearing an enormous gourd.

When the father returns, he sees that the god is gone and a tree is growing. Fearing that the thunder god will take revenge on him, he builds an iron boat and gets into it while the children climb inside the great gourd. When the incessant rains come, both the boat and the gourd float up toward heaven. The father bangs on the door of the king of heaven, who is so surprised by his unexpected visitors that he stops the rain. Instantly, the boat and the gourd fall one thousand miles back to earth. While the father dies, his two children in their gourd are spared, and Nü Gua and her brother Fu Hsi then repopulate the world.

In her book *Chinese Mythology*, Anne Birrell presents a slightly different version of this myth, in which Nü Gua and her brother Fu Hsi—

whose name means "prostrate or sacrificial victim"—create humanity but are ashamed of their incest:

> Long ago when the world first began there were two people, Nü Gua and her older brother. They lived on Mount K'un-lun. And there were not yet any ordinary people in the world. They talked about becoming man and wife, but they felt shamed. So the brother at once went with his sister up Mount K'un-lun and made this prayer:
>
> *Oh Heaven, if Thou wouldst send us two forth as man*
> *and wife,*
> *then make all the misty vapor gather.*
> *If not, then make all the misty vapor disperse,*
>
> At this, the misty vapor immediately gathered. When the sister became intimate with her brother, they plaited (wove) some grass to screen their faces. Even today, when a man takes a wife, they hold a fan, which is a symbol of what happened long ago.

By the time of the Han Dynasty (206 BCE–220 CE), these two gods were often depicted as serpent figures with human heads and interlocking tails. And in Chinese tradition, Nü Gua is also the goddess of matchmaking and a go-between, who helps arrange marriages.

Another popular Flood story involves a god with the body of a serpent and the head of a human, named Gong Gong (Kung Kung or "common work"), who stirs the waters of earth so violently that they threaten the world with chaos. Gong Gong next tries but fails to overthrow his father, Zhu Rong, the benevolent lord of the cosmos. When Gong Gong angrily butts against one of the mountains of heaven that prop up the sky, it causes the cosmos to tip. This myth explains why the rivers on earth flow in a southwestern direction. As the protective creator goddess, Nü Gua restores order by filling the hole in the sky and then propping up the sky with the legs of a giant tortoise.

What role do "family values" play in Chinese myth?

Compared to the sex-obsessed, whoring, cheating, philandering, and otherwise sexually rapacious gods of Greece, Rome, Egypt, Mesopotamia, India, and the Celts, the Chinese gods seem like models of decorum. Sure, Chinese myths have their share of bad guys, evildoers, and greedy siblings. But there is no serial adulterer like Zeus in the Chinese pantheon. Nor is there a Cuchulainn on the lookout for any opportunity to deflower a maiden. As Anne Birrell comments, "Chinese heroic myth differs from other mythologies in its early emphasis on the moral virtue of the warrior hero." Chinese myth is rather G-rated and squeaky-clean.

Even when there is a suggestion of incest in Chinese mythology, as in the Flood story featuring Nü Gua and her brother, who repopulate the world, the two siblings feel shame over their behavior and "screen their faces." Sleeping around is not sanctioned in Chinese myth; "love children" have no place; and little tolerance is shown for feuding in families, which are meant to be honored.

Instead, the Chinese gods are usually hardworking, creative types. Nü Gua's brother, the god Fu Hsi, for instance, invents nets and teaches the people how to fish. The engineer god Yu figures out how to stop the dangerous flooding of the river and is rewarded with immortality. Huang Di ("Great God Yellow") invents clothing and coins, while Shen Nong reveals the medicinal value of plants and even dies in the attempt to make new medicines. Author Anne Birrell suggests that the clean-living nature of this pantheon may lie with those who later compiled China's myths—a post-Confucian set who placed virtue above racy storytelling. "The theme of love is rare," she writes, "and is narrated in a sexually non-explicit manner, which may suggest early prudish editing."

WHO'S WHO OF CHINESE GODS

In the Chinese pantheon, there are literally hundreds of gods, both major and minor, who were worshipped locally as well as nationally. *The Classic of Mountains and Seas* specifies more than two hundred different gods. Some of these deities clearly emerge from distant Chinese history and may have been actual early rulers whose accomplishments entitled them to be elevated to the status of a god. Some of these mythical emperors/deities were even assigned dates of their reigns. Three of these were called the "three sovereigns" and three were called the "sage kings."

Four Ao The four Ao are water gods who take the form of dragons and are under the command of the Jade Emperor. In control of the rain and sea, each was given an area of land and sea to control.

Fu Hsi (**Fu Xi**) First of the three sovereigns and the brother of Nü Gua, Fu Hsi (translated as "great brilliance" in some traditions and "sacrificial victim" in others) is a god who, from the fourth century BCE on, is deemed an important creator and protector of the human race, especially in floods and other calamities. Fu Hsi is believed to be responsible for the invention of writing, hunting, and, most important, with the process of divination through the oracle bones, which later became the Book of Changes (I Ching). When Fu Hsi observes the markings on all the birds and beasts, he contemplates the divine order of things, and creates the first written markings from which humans can make prophecies.

In one charming tale, Fu Hsi watches a spider spinning a web and is inspired to devise nets from knotted cords, which he uses to teach humans how to hunt and fish.

By the time of the Han Dynasty, Fu Hsi was declared to have been the first emperor, who ruled from 2852–2737 BCE.

Guan Di (**Kuan Yu**, **Kuan Kung**) A figure out of Chinese Confucian folklore, Guan Di is a war god who may have once been an actual army general, executed as a prisoner of war during a time of division in China. While perhaps once a man, the god was supposedly nine feet tall with enormous strength, and Guan Di is usually depicted with a red face and a forked beard. But unlike the fierce war gods of other mythologies, Guan Di is known for courtesy, faithfulness, and being most contented when peace prevails. In 1594, Guan Di was recognized as a god by the Chinese emperor, who offered sacrifices to him.

Hou T'ou (Ti, She) Known as a "prince of the earth," Hou T'ou is the agricultural and fertility deity who manifests itself as the whole planet. Each year in ancient China, the emperor and village officials all around the country turned over the first spadeful of earth at planting time as part of a fertility ceremony, reflecting China's preoccupation with feeding its many people.

Huang Di (Huang-Ti) The third of the three sovereigns and a mythical leader whose name means "great god yellow," Huang Di is also called the "yellow emperor."* Credited with bringing civilization to China, Huang Di is the supposed inventor of upper and lower garments, weapons, the compass, coins, and government. A peace-loving warrior who has four faces so he can see everything, Huang Di fights four battles. In one battle, he uses water to defeat his brother, the fire god **Yan Di**, the "great god flame," and gain the sovereignty of the world. In another battle, Huang Di uses drought to defeat the war god "Jest Much," who has the weapon of rain.

In the Taoist tradition, Huang Di becomes the supreme god and dreams of a paradise where people live in harmony with nature.

Jade Emperor A deity sometimes known as Yu Huang or Huang Shang-Ti, the Jade Emperor becomes the divine ruler of heaven during the Song Dynasty (960–1279 CE). He lives in a heavenly palace similar to those on earth, and governs through a civil service just like that of China. His consort is **Xi Wang Mu**, "the queen mother of the West," who is more like the Wicked Witch of the West. A powerful tyrant, Xi Wang Mu sends plagues and punishment down to earth, keeps the elixir of immortality, and presides over paradise.

Lung Although not really a god, Lung is the benevolent dragon associated in Chinese folklore with clouds, mist, rain, and rivers. Less like the demonic creature done in by St. George and more like the

*Yellow was considered a regal color in recognition of the rich, yellowish silt, or loess, that is deposited by the Huang He, or Yellow River. The yellow loess was the life-sustaining source of good crops in China, just as the Nile's black earth was for Egypt. When the imperial Forbidden City was later built, its roof tiles were yellow.

benevolent "Puff, the Magic Dragon" of song fame, Lung is such an appealing creature that sometimes the gods take the form of dragons, which eventually will become the symbol of Chinese royalty. The Chinese dragon probably evolved from the serpent, an early royal symbol deemed immortal, since it was able to renew itself when it shed its skin.

Nü Gua (**Nü Kua, Nu Wa**) The great creator goddess, Nü Gua is a very ancient fertility deity who has remained popular in myth and legends throughout China's long history. As the divinity who created humans and saved the universe from catastrophe when Gong Gong (Kung Kung) threatened, she is a powerful protector. In the later Han Dynasty times (202 BCE–220 CE), she is viewed as both Fu Hsi's sister and his wife. In the latter role, she is credited with teaching people how to procreate and raise children.

Panku (**Pan Gu**) The primal creator god who is a child of yin and yang, Panku is born from a cosmic egg in one of China's most important Creation myths. With his death, his body parts become the various bits of the universe and earth, and the insects that come from his body become "the black-haired people" (the Chinese). Many scholars think that Panku may have originated elsewhere in Central Asia and arrived in China in the second or third century CE.

Shen Nong Depicted as a divine being with a bird's head, Shen Nong ("fiery emperor") is the second of the three sovereigns, a legendary emperor who is the inventor of the cart and who teaches people how to farm. Shen Nong is also the ancient god of the pharmacy, who reveals the healing properties of plants to humanity. In myth, he has a see-through stomach, which enables him to view the effects of his experiments with medicinal herbs. Unfortunately, he tests a kind of grass that causes his intestines to burst.

Shun (**Yu Di Shun**) One of the three sage rulers of antiquity, Shun is another virtuous ruler-god to whom heaven sends birds to weed his crops and pull his plow.

Tsao Chun (**Zao Jun**) The very ancient "kitchen god" of Chinese myth, Tsao Chun is the most important domestic deity in China and lives in the niche near the cooking stove in Chinese homes. Portrayed as a kindly old man surrounded by children, he supplies the chi, or energy, that aids nourishment. Every New Year, he is said to visit heaven and give an accounting of each household. Before he goes, each household tries to "bribe" him and smears the mouth of Tsao Chun's idol with sweet paste or honey, to help him speak "sweet words" and avoid saying anything bad when he arrives in heaven.

In her novel *The Kitchen God's Wife*, Amy Tan has an American character ask if the kitchen god is like Santa Claus. An elderly Chinese woman replies in a huff, "He is not Santa Claus. More like a spy—FBI agent, CIA, Mafia, worse than IRS, that kind of person. And he does not give you gifts, you must give him things. All year long you have to show him respect—give him tea and oranges. When Chinese New Year's time comes, you must give him even better things—maybe whiskey to drink, cigarettes to smoke, candy to eat, that kind of thing. You are hoping all the time his tongue will be sweet, his head a little drunk, so when he has his meeting with the big boss, maybe he reports good things about you." Then the Chinese mother adds, "His wife was the good one, not him."

Yao (**Tang Di Yao**) Another of the three sage rulers of antiquity, Yao is a mythical emperor who is elevated to the status of a god. Yao lives frugally and always cares for his people. But because his son is not worthy to ascend to the throne, Yao chooses his son-in-law as his successor, and Confucius singles him out for praise as a model ruler.

Yi (**Hou I**, **Hou Yi**) Perhaps the greatest of the Chinese hero-gods, Yi is the great archer who figures in a myth dating from the sixth century BCE. In this tale, there are ten suns, each one the son of the ruler of heaven. When they all appear at the same time, their intense heat withers the crops and the lord of heaven sends the archer Yi to restore order. But instead of commanding the suns to go home, Yi shoots nine of them with his arrows. Even though the farmers are happy, Yi is banished by the lord of heaven to live as a mortal on earth with his wife, **Chang E**. Upset at losing her immortality, Chang E acquires a special elixir from the Queen Mother of the West and consumes it all, even though half is meant for her husband. For her disobedience, Chang E is sent to the moon and becomes the moon goddess. Yi accepts his mortality, but in some accounts, goes back to heaven after being forgiven.

Yu (**Da Yu**) Another of the three sage rulers of antiquity, Yu is a god and an engineer who appears in a foundation myth. When the Emperor Shun (above) asks Yu to work on containing the waters of the great flood, he leaves his wife and children to do the job. Instead of building a boat to escape the deluges, Yu spends thirteen years creating canals to control the floodwaters that periodically threaten parts of China. Yu is awarded the throne for his work. He is said to have founded the legendary first Chinese dynasty, the Xia, between 2205 and 2197 BCE, but there are no confirmed historical accounts of any such dynasty.

Mythic Voices

The Master said, At fifteen I set my heart upon learning. At thirty, I had planted my feet firm upon the ground. At forty, I no longer suffered from perplexities. At fifty, I knew what were the biddings of Heaven. At sixty, I heard them with docile ear. At seventy, I could follow the dictates of my own heart, for what I desired no longer overstepped the boundaries of right.

—*from* The Analects of Confucius

> **He [Confucius] has had a greater influence on China than any other human being. Yet almost nothing is known about him as a man. . . . The central teaching of Confucius was that nothing is more important to man than man. He himself refused to have anything to do with four kinds of thing: what was violent, what was disorderly, what was strange and what had to do with the supernatural. "One should revere the ghosts and gods," he once said, "but still keep them distant."**
>
> —*from Alasdair Clayre,* The Heart of the Dragon

What do fortune cookies have to do with Chinese religion?

"Confucius say . . ."

For years, those words have combined the wisdom of fortune cookies with the humor of old Charlie Chan movies, effectively reducing Confucius and his philosophy to a series of witless jokes. Too bad. Because in Chinese history, the legendary philosopher Confucius is one of the most significant people who ever lived, responsible for both the ethical practices and political philosophy that governed Chinese history for 2,000 years.

As with the life of Jesus or Buddha, some of Confucius's biography must be taken on faith. According to tradition, Confucius was born in 551 BCE, in Lu, in the northeastern Sandong Province. His name was Kong Fu Zi ("great master Kong") and the name "Confucius" is the Latinized form first used by Jesuits who came to China in the seventeenth century. While Confucius is said to have practiced archery and music—activities of the Chinese nobility—he seems to have been born into relatively humble circumstances. According to one tradition, his parents died when he was a child, but in his works, there is little reference to father, mother, or a wife, though Confucius is believed to have had a son who died, as well as a daughter. When Confucius himself died, he was largely unknown in China.

Although there is no evidence that Confucius ever wrote anything himself, he was long thought to have edited the collection of ancient wisdom books called the Five Classics, including the I Ching, the ancient divination guide to which he supposedly added his own commentaries. (This is now in dispute.) His conversations and sayings were also included in a book of his thoughts and anecdotes called the Analects which was compiled by his disciples. These disciples included the early Confucian philosopher Mencius (371–289 BCE), who believed that people were born good and simply needed to preserve "the natural compassion of the heart" that makes them human; and Xun Zi (mid-200s BCE), who believed people could live together peacefully only if their minds were shaped by education and clear rules of conduct.

But Confucianism itself is the centerpiece of the philosopher's contribution. Begun as a code of conduct that only later evolved into what might be called a "religion," Confucianism has no organization or clergy. Nor does it teach a belief in a deity or in the existence of life after death. Instead, Confucianism stresses moral and political ideas, putting an emphasis on respect for ancestors and government authority while insisting that women belong in the home. These ideas were not new or radical by any means, but Confucius placed them in a new framework by suggesting that the individual has a proper place in the political, societal, and family hierarchies, and that within these hierarchies one must venerate those above and care for those below.

Confucius further argued that tradition and order have to be respected to maintain the equilibrium of the universe. That meant practicing piety, ethical norms, and human benevolence—or *jen*, a concept that encompasses love, goodness, integrity, and loyalty—which apply to every aspect of life. Adhering to *jen* depends on following the "middle way"—or moderation. Central to this idea was the Confucian version of the Golden Rule: "What you do not want done to yourself, do not do to others."

By about 200 BCE, the first large, unified Chinese empire had begun under the Han Dynasty. The Han rulers approved of Confucianism's emphasis on public service and respect for authority. In 124 BCE, the government established the Imperial University to educate future government officials in the Confucian ideals found in the Five Classics. Candidates applying for government jobs had to pass rigorous examinations based on the Five Classics, and a second set of texts called the Four Books.* Mastery of these classics was also proof of moral fitness and the chief sign of a Chinese gentleman, even one not born into nobility. Under the Han Dynasty, the idea that the emperor's authority came from heaven was also given greater clout, and Confucianism increasingly became the state "religion" of China from the 100s BCE until the mid-twentieth century. When the Chinese Communists gained control of China in 1949, they opposed Confucianism, because it encouraged people to look to the past rather than to the future. It was among the "four olds"—old ideas, habits, customs, and culture—rejected by the Party in the 1950s. Official opposition to Confucianism ended in 1977. Since then, the Communist government has relaxed some of its policies against religion, and so Confucianism has enjoyed a revival on the mainland.

MYTHIC VOICES

Truthful words are not beautiful.
Beautiful words are not truthful.
Good men do not argue.
Those who argue are not good.
Those who know are not learned.
The learned do not know.

The sage never tries to store things up.
The more he does for others, the more he has,
The more he gives to others, the greater his abundance.
The Tao of heaven is pointed but does no harm.
The Tao of the sage is work without effort.

—Lao-tzu, Tao Te Ching 81
(translated by Gia-Fu Feng and Jane English)

*All post-Confucian, the Four Books are Great Learning; the Mean, on moderation; the Analects, a collection of Confucian sayings; and the Mencius, the collected wisdom of Confucius's successor. Their study remains influential today.

What religion shunned the Confucian approach?

It has been turned into a way to raise cats or children, invest, paint, understand physics, heal yourself, and even reinterpret *Winnie-the-Pooh*. Stick the word "tao" in a book title, and it conveys an image of some secret inner knowing. Not bad for a philosophy that was conceived in mystery and myth. Taoism is a philosophy with obscure, legendary beginnings in China during the 300s BCE—although many practitioners claim its oral roots go back thousands of years—and that it acquired the qualities of a religion by the 100s BCE.

While Confucianism stressed that a good life only comes from living in a well-disciplined society that emphasizes ceremony, duty, morality, and public service, the Taoist ideal rejected conventional social obligations and urged the individual to lead a simple, spontaneous, and meditative life close to nature, and to see change as the way of the universe. The word "tao" (also spelled "dao" and pronounced *dow*) originally meant "path" or "way." Tao was all about getting in rhythm with the great cycles in nature, and learning to live in harmony with the changing seasons.

The beliefs of Taoism as a philosophy are showcased in the Tao Te Ching ("the classic of the way and the virtue"). Tao Te Ching is a collection from several sources, but its authors and editors are unknown. Unreliable accounts say that a man named Laozi lived during the 500s BCE and wrote these works. A legend tells how Laozi, supposedly a keeper of imperial archives some six centuries before the Christian era, could foresee the imminent decay of society. He was preparing to leave China for the fabled land of the West. A guard at the frontier asked this master for an account of his ideas, and Laozi responded with Tao Te Ching. However, the Tao Te Ching, made up of eighty-one brief sections, was probably compiled and revised during the 200s and 100s BCE—well after Laozi had died. (A legendary meeting between Laozi and Confucius is also most likely just that—a legend.) Chuang-tzu, his disciple, lived around 329–286 BCE and expanded on the tao with a second book, called Chuang-tzu.

Composed largely in verse, the Tao Te Ching describes the unity of nature—the tao, or "way"—that makes each thing in the universe what it is, and determines its behavior. Enigmatic and elusive, this unity can be understood only by mystical intuition. Because, in Taoism, "yielding eventually overcomes force," the book teaches that a wise man desires nothing. He never interferes with what happens naturally in the world or in himself. One passage in the Tao Te Ching says: "The highest good is like water. Water excels in giving benefit to all creatures, but never competes. It abides in places that most men despise, and so comes closest to the Tao." The Tao Te Ching also teaches that simplicity and moving with the flow of events are the keys to wise government.

Over time, Taoists began to add more mystical practices in the hope of helping adherents reach a transcendent state. As Taoism evolved into a form of worship, it took on aspects of traditional folk religion, including the growth of a hereditary priesthood that used rituals to submit the people's prayers to various folk gods. Working in trance, the chief priest prayed to other divinities, who were aspects of the Tao, for favors for the people. Taoist groups also sought to attain immortality through magic, meditation, special diets, breath control, or the recitation of scriptures. Besides looking at these "alternative" avenues, many believers pursued their search for knowledge in various pseudosciences, such as alchemy and astrology.

Who was Japan's first divine emperor?

It is the land of shoguns and samurai to most Westerners, a string of four main islands and thousands of smaller ones, which roughly equals the size of the state of California. In the mid-nineteenth century, Japan emerged from hundreds of years of near-isolation and became one of the great empires of modern times. After it fell in the fiery destruction of World War II, it then rose, like a phoenix, from the ashes to become a modern financial and trading empire.

According to Japanese legend, the first emperor of this island nation was Jimmu-tenno, or "divine warrior emperor," who is traditionally dated from 660–585 BCE. Believed to be the great-great-great-grandson of the divine sun goddess Amaterasu, Jimmu and his elder brother supposedly marched eastward from a region of Kyushu Island, intent on consolidating their power. After his brother is killed in battle, Jimmu presses on, guided by a heavenly crow. His army continues its march until he reaches Yamato, traditional home of the Japanese emperors. The consensus today is that Jimmu-tenno did not exist, that there were no emperors at that time, and that more than a dozen of Japan's earliest reputed emperors were inventions. Historians today assert that the imperial line actually began in the fifth or sixth century of the Common Era.

When the Yamato emperors were actually established, in a public-relations move designed to establish their authority, they proclaimed Amaterasu as the ancestress of their clan. Stories connecting the gods and the emperor provided the core of the state religion that became known as Shinto ("the way of the gods").

Japan's highly militaristic traditions—begun with the legend of Jimmu and other warrior emperors—continued for centuries, carrying over into the two iconic military institutions of the samurai and shogun. Both inspired legends, but neither had a place in true Japanese mythology. The samurai—immortalized in the films of Akira Kurosawa—were members of a hereditary warrior class, more like the knights of medieval Europe. The early samurai defended the estates of aristocrats, and around 1000 CE, they began to develop a code of strict values and self-discipline, prizing horsemanship, archery skills, and bravery. Above all, they valued total obedience and loyalty to their lords, and personal honor. Dishonor brought an obligation to commit ritual suicide.

The samurai began to grow more powerful in 1192, when the emperor gave the title shogun ("great general") to the military leader Yoritomo of the Minamoto family. Yoritomo established the first shogunate, or warrior government. These militaristic governments then largely ruled Japan from the late 1100s to the mid-1800s. In 1867, as Japan struggled toward modernity, the shogunate was overthrown and powers were restored to the emperor. This scenario became the background for the Tom Cruise film *The Last Samurai*, a highly romanticized view of the traditional samurai attempting to stave off modern times.

Writer Stefan Lovgren burst that Hollywood "myth" when he wrote in *National Geographic*, "Mythology colors all history. Sometimes, legend and lore merely embellish the past. Other times, mythology may actually devour history. Such is the case with the samurai, the military aristocracy of feudal Japan. The samurai are known as strong and courageous warriors, schooled with swords. In reality, they were an elitist and (for two centuries) idle class that spent more time drinking and gambling than cutting down enemies on the battlefield."

How did Shinto become an "Asian fusion" religion?

China's influence on ancient Japan was so profound that it is difficult to separate Japanese ideas from those that arrived on the islands from China over the centuries. While an early form of the Japanese belief system of Shinto probably existed before the arrival of Buddhism and Confucian teachings from China, Shinto can rightfully be thought of as an "Asian fusion" religion, because it only becomes a unified religion with a complete mythology after the Chinese influence is felt. There are, for instance, many similarities between Japanese and Chinese Creation accounts, including the idea of a cosmic egg, and a god whose eyes form the moon and sun.

No written records of the origin of Shinto exist, and no one knows when or how Shinto began. A mixture of different beliefs, Shinto means the "way of the gods." It seems to have combined the ancient practices of the Ainu, Japan's earliest inhabitants, now reduced to a small number living in Hokkaido, Japan's northernmost island, with those of the prehistoric people who migrated to Japan from other parts of Asia, including Mongolian people from Siberia. What resulted is a religion centered on nature—mountains, rivers, rocks, and trees. Shinto also acknowledges the force of gods, known as *kami*, in such processes as creativity, disease, growth, and healing. Emphasizing rituals over philosophy, Shinto pays little mind to life after death.

Beginning about the 500s CE, the Chinese philosophies of Buddhism and Confucianism began to influence Shinto, which absorbed Buddhist deities into its fold and also identified them as *kami*. Shinto shrines adopted Buddhist images, and Buddhist ceremonies were used for funerals and memorial services throughout Japan. Under the influence of Confucianism, Shinto also emphasized rigorous moral standards of honesty, kindness, and respect for one's elders and superiors.

Shinto myths appear in the *Nihongi* ("chronicles of Japan") and the *Kojiki* ("the record of ancient matters"), both of which were written in the 700s CE. These myths tell how the *kami* created the world and established customs and laws. According to Shinto mythology, the sun goddess Amaterasu was the ancestor of Japan's imperial family. In the late 1800s, the Japanese government invented state Shinto, which stressed patriotism and the divine origins of the Japanese emperor. After Japan's defeat in World War II in 1945, the emperor denied that he was divine, and the government abolished state Shinto.

WHO'S WHO OF JAPANESE GODS

Amaterasu The most significant deity in the Japanese pantheon, Amaterasu is the sun goddess who is also known as "the august person who makes the heavens shine." Born from the left eye of the primal Creator god Izanagi as he bathes in a stream, Amaterasu is assigned to rule the realm of the heavens while one of her brothers, **Tsuki-Yomi**, the moon god, is entrusted with the realms of the night, and another brother, **Susano**, god of storms, is made ruler of oceans.

In a classic family-feud myth with incestuous overtones, Amaterasu and her brother Susano get into an epic fight. In one version of this core Japanese myth, Susano becomes angry, because he has received what he considers a lesser realm, but in another version, Amaterasu and Susano have a fight to see which of them is greater. Amaterasu chews Susano's sword and exhales, creating three goddesses. In response, Susano eats some of his sister's jewels and exhales five gods. As the fight escalates, Susano creates in the heavens a sort of "manic panic"—he uproots rice fields and ruins temples by smearing his excrement on the walls. When he throws the carcass of a horse into the weaving room where Amaterasu and her attendants make divine clothes for the other gods, she is terrified and flees to the cave of heaven, closing the entrance with a great stone and plunging the world into darkness.

As the darkness descends, evil spirits emerge and worsen the destruction of the world. To save the Creation, the other gods attempt to lure Amaterasu out of the cave by getting a young fertility goddess named **Uzume** to dance at its entrance. Gyrating in ecstasy, Uzume—also the goddess of laughter—throws off her clothes, whirling frantically, and the other gods roar their approval at this celestial striptease.

Hearing the merriment from inside the cave, Amaterasu cannot resist peering out. The other gods hold up a mirror and string jewels in the trees outside the cave to entice the sun goddess out of hiding. Once she emerges, the world is once again bathed in light, and the evil forces disappear.

Amaterasu is thought to be the ancestor of Jimmu, the legendary first emperor of Japan. Through an unbroken line of descent, all of Japan's emperors claim to be descended from her. The mirror, string of jewels, and a sword used to draw Amaterasu out of the cave are the traditional symbols of the Japanese royal family.

Benten The deity of luck and wealth, Benten is a goddess associated with music and eloquence. Painfully shy, she marries a dragon prince from the dragon people who surround Japan. The dragon is revolting but, because of her sense of duty—a Japanese concept called *giri*—she reluctantly fulfills her marriage vows. Afterwards, peace comes to the kingdom.

In later times, following the introduction of Buddhism to Japan, Benten became a popular Buddhist deity—goddess of music, eloquence, wealth, love, beauty, and geishas. She also prevents earthquakes by mating with the white snakes that live beneath islands of Japan.

Hachiman Especially popular with the military, Hachiman is the Shinto war god, the protector of the nation and a guardian of children. Nearly one-third of Shinto shrines throughout Japan are dedicated to this deity who is identified with the emperor Ojin (died c. 394 CE), a renowned military leader who was later deified.

Inari The rice god and patron of farmers. Almost every Japanese village has a shrine dedicated to Inari. Depicted as a bearded older man sitting on a sack of rice, and often flanked by two foxes who are his messengers, Inari is regarded as a generous god who oversees wealth and friendship and is revered by merchants, since he brings well-being. His wife, **Uke-mochi**, is the food goddess.

Izanagi (August Male) and **Izanami** (August Female) Descended from a god born from the "boiling ocean of chaos" at the time of Creation, Izanagi is the creator of people. He is helped in this effort by Izanami, his sister, whose first child is a monster and whose second offspring is an island. These curious births occur because Izanami speaks before her brother does, and in Japanese custom the male must go first—which might give you a hint of the traditional role of women in Japan. After they realize their error, all goes well, and the two gods produce people, the islands of Japan, and other gods.

According to the myth, Izanami dies in childbirth when she gives birth to a fire god. However, even in death, she is a powerful creator, whose vomit, urine, and excrement become other gods. Distraught over his consort-sister's death, Izanagi follows her to the underworld, or "land of gloom." In a story with echoes of the Greek Orpheus descending into Hades, he is warned not to look at her, because she has eaten the food of the underworld and is already decomposing. But he does as he pleases. Furious she has been seen covered with maggots, Izanami sends a horde of she-demons after him and promises to kill 1,000 people on earth every day—the mythical reason for death. Able to escape, Izanagi rolls a huge stone over the entrance to the underworld and declares himself divorced—one of the few cases of divine divorce in mythology. This story also reflects a Shinto attitude of horror at death, decay, and dissolution.

As he is bathing after this close call, Izanagi washes the dirt off himself, and it forms harmful spirits. But he makes some good gods as well. Amaterasu comes from his left eye, and Tsuki-Yomi, the moon god, from his right eye. (These stories seem to reflect the influence of the Chinese myth of Panku, whose eyes also become the sun and moon.) The infamous storm god Susano comes from Izanagi's nose and immediately starts to cause trouble.

O-kuni-nushi The god of medicine and sorcery, whose name means "great land master," O-kuni-nushi is credited with inventing healing. He is often accompanied by **Sukuna-Biko**, a dwarf god skilled in both agriculture and medicine, who knows almost everything that is going on in the world.

O-kuni-nushi also figures in an intriguing myth. When O-kuni-nushi stops to help a wounded rabbit that his seventy brothers have passed by, the good deed earns him the right to marry the daughter of the god Susano. This is because the rabbit is actually another god in disguise. Angry that they missed such an opportunity, O-kuni-nushi's brothers kill him, but he is able to regenerate himself. Displeased that his daughter is to marry, Susano subjects his future son-in-law to a series of tests. First, O-kuni-nushi is placed in a room full of snakes, but his bride gives him a magical scarf that protects him. Next, he sleeps in a room filled with poisonous insects, but again, he is saved by his bride's magical scarf. Finally, he is trapped in a great grass fire, but is led to safety in an underground chamber by a friendly mouse.

In return for his father-in-law's tests, O-kuni-nushi ties Susano's hair to the roof beams and makes off with Susano's magic bow and harp. The storm god gains new respect for his son-in-law and allows him to rule over a province in central Japan.

O-wata-tsumi The chief god of the sea, O-wata-tsumi is a god created when Izanagi purifies himself after his descent into the underworld. (In other accounts, O-wata-tsumi is descended from O-kuni-nushi.) O-wata-tsumi is significant in Japanese mythical history because he is considered another divine ancestor of the first emperor, Jimmu.

Susano (Susanowo) The god of storms and the divine embodiment of the forces of disorder, Susano is known as the "valiant, swift, impetuous deity." He is born when the divine father Izanagi clears his nose as he bathes in a stream. When the universe is divided up, and Susano's sister, Amaterasu, the sun goddess, is given the heavens, Susano thinks that he has gotten shortchanged. Banished by his father for his defiance, Susano begins his long struggle to overthrow

Amaterasu and nearly brings catastrophe to the world in what is called "the divine crisis." Terrified by her brother, Amaterasu withdraws into a cave, depriving the world of sunlight.

After the crisis, Susano is expelled from heaven and later wins some measure of respect by defeating the eight-headed dragon, **Yamato-no-orichi**—who had eaten seven of eight daughters of the local king and who sounds like the inspiration for Japan's favorite monster, Godzilla. Susano accomplishes the feat by filling eight bowls with rice wine and luring the monstrous serpent to drink. Once the serpent monster becomes drowsy, Susano cuts open the creature's stomach and finds a magical sword hidden inside. As a reward for his feat, he is given the kingdom he has saved, as well as a princess, **Kusanada-hime**, also called Rice Paddy Princess. Their daughter, who marries the medicine god O-kuni-nushi, is thought to be an ancestor of the Japanese emperors.

Since 1946, when the Japanese emperor Hirohito denied his divinity, after which the Japanese Constitution ended "state Shinto," Japan has been a parliamentary democracy, in which the emperor is the head of state and the prime minster is the elected head of government. But old ideas still die hard. A recent controversy flared in Japan over new regulations requiring teachers to stand in classrooms and face the Japanese flag while singing the national anthem. Banned for three years during the postwar American occupation of Japan, the country's "rising sun" flag is a vestige of the old connection between Japan—or Nipon, which means "rising sun"—and the sun goddess.

But many Japanese feel that the rising-sun flag is a symbol of Japan's militaristic and imperialist past, when troops stormed mercilessly through Asia in the period before and during World War II. The public resistance to the requirement got support from an unexpected source. According to the *New York Times*, Emperor Akihito himself publicly voiced his opposition to the flag law.

ANCIENT PEOPLE, NEW WORLDS

O, wonder!
How many goodly creatures are there here!
How beauteous mankind is! O brave new world
That has such people in't!

—WILLIAM SHAKESPEARE,
The Tempest (act V, scene 1)

For centuries, Africa, the Americas, Australia, and the Pacific islands existed in mysterious solitude, lands completely set apart from the "known" world by vast oceans, jungles, deserts, and wide expanses. Africa's existence had been acknowledged since ancient times, but it was largely impenetrable due to its forbidding geography. The Americas, which occupy 28 percent of the world's landmass, spread from the frozen north of one hemisphere to the "bottom of the earth" in the other. Australia and many thousands of islands in the Pacific were beyond the imaginings of the Western world. Yet all of these places were home to ancient peoples, with long-standing societies, myths, religions, and traditions well insulated from foreign influences.

That all changed forever after the fifteenth century. In the European "Age of Discovery," Portuguese sailors opened up Africa as they made their way to Asia by sea. Christopher Columbus soon followed the Portuguese lead, spurred on by a desire to find still faster routes to the gold, jade, silks, and spicy taste sensations desired by a European palate weary of salted venison. Sailing under the Spanish flag, Columbus set out in 1492 on the first of four voyages that unlocked territories undreamed of and gave Spain the lead in penetrating and then plundering the "New World," first in the Caribbean, then in both South and North America. Spain's dominance in the Americas was soon challenged by the English, French, Dutch, and other Europeans—each staking a claim in the names of their kings and their God to large chunks of land already occupied by tens of millions of people. In the eighteenth and nineteenth centuries, the "discovery" extended to the Pacific islands, where the "Aborigines" of Australia and other natives of the Pacific would meet a similar fate. Millions of these people would be collectively enslaved, converted, displaced, and almost entirely wiped out, along with much of their mythic legacy.

So the story of these "new worlds" is a story of both beginnings and endings. For the Europeans, it was an extraordinary period of empire-building, colonization, and subjugation. But for the people they "discovered," it was the end of cherished traditions. Africa was teeming with cultures, religions, and gods when the Portuguese arrived eager to baptize the "heathens" they found on Africa's West Coast and along the Congo River. But the zealous Portuguese quickly learned that Arabs had already "discovered" much of Africa and begun to import Islam. In time, Africa would become a battleground in the centuries-old conflict between Christianity and Islam, with native myth and belief caught in the deadly crossfire.

The situation was similar in the Americas, where, prior to the European arrival, a stunning array of cultures and civilizations had flourished, from the "Halls of Montezuma" and Mayan pyramids of Central America to the lofty Andean cities of the Incas. Presumably the descendants of the people who wandered from Siberia to the Americas during the waning of the last great Ice Age, the inhabitants of the Americas ranged from the natives of the Arctic region to the tribes of the American Northeast, to the settled farmers of the Southeast, and down to the monumental civilizations of Mexico, Central America, and Peru. But the gods and legends of the Americas, like those of Africa, would soon come crashing headlong before Europe's Christian soldiers, with devastating results for the natives of America. The same scenario would play out in Australia. Home to hundreds of thousands of Aborigines, Australia became a British experiment in exporting its crime problem by converting a whole continent into a prison colony—until gold was discovered there. As miners swept in, missionaries were never far behind.

Apart from this shared destiny of destruction and decimation, however, there are other fascinating parallels between the people of these "new worlds."

First, most of their myths reflect a nonliterate or oral tradition that was not recorded until fairly recently, in most cases. The very survival of these myths is a testament to the deep human desire and ability to hold on to what is sacred. When these mythic accounts were recorded, it was after the introduction of Christianity—as was true in the Celtic and Norse worlds. That does not mean we can't "know" these myths, but we must take into account the prejudices that may have been involved in preserving them, as well as a native desire to conceal and protect their most sacred stories and rituals.*

A second feature often found in many African, Native American, and Pacific Creation stories is a deity who gives shape to the cosmos and then retreats to the background. The African, American, and Pacific-island stories also share a fascination with mischievous animal "tricksters," and animals often play a larger role in these myths than in many other traditions. All of these cultures have many stories involving twins. And, in worlds filled with spirits, the shaman or "medicine man" is often highly revered as the most significant person in the society.

But finally, we come back to the most important parallel of all. Running through the history of all these cultures is the common theme of destruction. The "discovery" of Africa, the Americas, and the world of the Pacific is pervaded by an overwhelming central tragedy—the concerted effort to replace ancient ideas and languages with the conquerors' version of god, truth, and civilization. That effort largely—although not completely—succeeded.

*When the famed photographer Edward S. Curtis filmed a sacred Hopi dance in the American Southwest in the early twentieth century, the performance, never before seen by a white person, turned out to be a complete fabrication for the benefit of the camera.

In spite of that dark history, the myths of these places and people are not lost, dead stories. As elsewhere, ancient folkways and faiths die hard. And there are vivid reminders of these mythic traditions alive today. One example can be seen in the religions that grew up in the Americas. Both voodoo and Santeria, for instance, remain powerful vestiges of the arrival of ancient African myths and deities in the Caribbean and the Americas, brought by millions of Africans who carried their gods, if nothing else, when they were forced into the holds of slave ships. In Latin America, the sacred remains of ancient myths and beliefs poke their heads through "official" Christianity, like some relentless rain-forest flower breaking through the concrete of a modern street.

Museums around the world are also helping to keep ancient myths alive, increasingly recognizing the rich artistic traditions of all these places and people, as well as their impact on art during the past century. Among others, Picasso and Mexican painter Frida Kahlo were profoundly influenced by the imagery of ancient myths. Hollywood, which is often content to ignore such traditions, has also opened eyes with the infrequent big-budget success, such as Kevin Costner's paean to the Sioux, *Dances With Wolves*, while smaller, independent foreign filmmakers have contributed *The Gods Must Be Crazy*, set among the San of the Kalahari, and *The Whale Rider*, which lyrically captured a sense of the disappearing traditions of the Maori. A generation of scholars in the United States, Mexico, Guatemala, and Brazil, among many countries, has also expanded an ambitious effort to recognize and revitalize the study of Native American, African, and other indigenous traditions, expressed in such forms as the increasingly popular celebration of the African harvest festival known by its Swahili name "Kwanzaa," which means "first fruits."

The fact is that myths—like the human soul they often reflect—can be enduring, tenacious, and transcendent. Myths never die. That basic truth is nowhere clearer than in the very ancient places called the "new worlds."

CHAPTER EIGHT

OUT OF AFRICA

The Myths of Sub-Saharan Africa

In the time when Dendid created all things,
He created the sun,
And the sun is born, and dies, and comes again.
He created man,
And man is born, and dies, and does not come again.

—old African song

You who dive down as if under water to steal,
Though no earthly king may have seen you,
The King of Heaven sees.

—traditional proverb of the Yoruba (Nigeria)

Caller-forth of the branching trees:
You bring forth the shoots
That they stand erect.
You have filled the land with mankind,
The dust rises on high, O Lord!
Wonderful One, you live
In the midst of the sheltering rocks.
You give rain to mankind.

—from a traditional prayer
of the Shona (Zimbabwe)

We come upon a curious fact. The pre-colonial history of African societies—and I refer to both Euro-Christian and Arab-Islamic colonization—indicates very clearly that African societies never at any time of their existence went to war with another over the issue of their religion. That is, at no time did the black race attempt to subjugate or forcibly convert others with any holier-than-thou evangelizing zeal. Economic and political motives, yes. But not religion.

—WOLE SOYINKA, Nobel Prize acceptance speech
(December 1986)

Is there an "African" mythology?

What role did myth play in African villages?

Is there an African Creation myth?

Who's Who of African Deities

How did a suicidal king become a god and end up in the Supreme Court?

MYTHICAL MILESTONES

Africa

2.5 million years before present The first known stone tools are used by early ancestors of modern man, *Homo habilis*.

1.7 million years before present Hominids begin to move out of Africa, adapting to a range of environments in Asia, Europe, and the Middle East.

150,000 years before present Migration of early modern humans begins from East Africa.

100,000 years before present Anatomically modern humans with superior "tool kit" emerge in southern Africa.

70,000 years before present Evidence of human burials in southern Africa.

42,000 years before present Ocher, a kind of earth which is ground to a fine powder and used as a pigment, is mined and possibly used for body decoration.

26,000 years before present Evidence of earliest African rock art.

20,000 years before present Evidence of terra-cotta figurines in Algeria (northern Africa).

12,000–10,000 years before present End of the last Ice Age.

Before the Common Era (BCE)

c. 8500 Saharan rock art depicts wide array of elephants, giraffes, rhinoceroses, and other animals long since extinct in this region.

Finely crafted stone arrowheads and other tools are used in the Sahara region.

c. 7500 "Wavy-line pottery," made by dragging fish bones through wet clay, produced in Sahara and its southern fringes.

c. 6500 Domestication of cattle in the Sahara region.

c. 6000 Agriculture begins along the Nile River.

c. 5000 Desertification of Sahara region begins; populations expand south and east.

c. 4100 Sorghum and rice are cultivated in the Sudan and West Africa.

c. 3100 Beginnings of united Egypt (see Mythical Milestones, chapter 2).

c. 1965 Nubia conquered by Egypt.

c. 900 Nubian kingdom of Kush (also spelled Cush) rises along the Nile River in what is now northeastern Sudan. Its founding date is not known, but it existed as early as 2000 BCE. Egypt conquers Kush in the 1500s BCE, and the Kushites adopt elements of Egyptian art, language, and religion.

814 Carthage founded by Phoenicians in northern Africa.

747 Kushites invade and rule Egypt.

c. 600 Capital of Kush moved to Meroë. Kush probably fell about 350 CE after armies from the African kingdom of Axum destroyed Meröe.

c. 500 Daamat, first kingdom in Ethiopian highlands, is founded.

Nok culture begins in northern Nigeria; first known iron working in the sub-Saharan region.

332 Alexander conquers Egypt.

30 Egypt becomes a Roman province.

Common Era

c. 150 Nigerian Nok culture reaches its height.

c. 200 Ghana gains wealth and power through its trade with Berbers of northern Africa.

350 Meroë, capital of Kush kingdom, is destroyed by Ethiopian forces.

c. 451 Ethiopian kingdom of Axum reaches its height.

c. 540–570 Spread of Christianity in Nubia and Ethiopia.

c. 600 Kingdom of Ghana founded.

c. 625 Beginning of Islamic expansion into Africa.

641 Arabs invade Egypt.

c. 700 Kingdom of Ghana grows more powerful and controls trans-Saharan trade routes.

c. 800 Emergence of trading towns on East African coast; trade grows with Arabs and Persians.

c. 850 The construction of the citadel of Great Zimbabwe, in southern Africa, is begun.

c. 1000 Spread of Islam into sub-Saharan Africa, driven by overland trade.

c. 1076 King of Ghana converts to Islam.

c. 1100 Empire in Zimbabwe rises to power in southern Africa, centered in the massive stone-built city of Great Zimbabwe.

c. 1140 Igbo culture flourishes on Niger River.

1150 Yoruba culture flourishes in West Africa, based in capital city of Ilfe.

c. 1240 Rise of empires of Mali in West Africa and Benin.

1350 Mali becomes an Islamic state.

1415 Portuguese capture Ceuta (Morocco), which marks the beginning of Portugal's overseas empire and involvement in Africa.

1431 Chinese admiral Zheng He travels to East Africa.

1441 First shipment of African slaves sent to Portugal.

1485 Portuguese explorer Bartholomeu Dias reaches the Cape of Good Hope.

Four Portuguese Catholic missionaries arrive in Congo.

1498 Portuguese explorer Vasco da Gama rounds the Cape of Good Hope en route to India.

1502 First African slaves are taken to the New World by the Spanish.

Still in the dark about the Dark Continent?

Say "Africa," and the immediate association might be "jungle" or "safari." Or cartoonish images of missionaries in large stew pots. Or a man in a pith helmet, asking, "Dr. Livingston, I presume?" If you grew up in a certain era, your views of Africa were probably shaped by Tarzan movies starring Johnny Weissmuller surrounded by dutiful natives in loincloths saying things like "bwana." Or old issues of *National Geographic* once eagerly perused for pictures of African women with bare breasts. A younger group might identify with the lovable Lion King immortalized in the idyllic animated Disney movie and Broadway musical.

Let's face it. The generally woeful state of American knowledge about the rest of the world is at its nadir when sub-Saharan Africa* is the subject. And the media doesn't help matters. In recent years, Africa has only shown up on the American radar when some catastrophe strikes—an embassy bombing or a Black Hawk down. In the late 1960s, it took a civil war and starving Biafran refugees to make us aware of Nigeria. A rock-and-roll "feel-good" moment like "We Are the World" in 1985 briefly raised consciousness about the troubles confronting Africa. And, of course, when Nelson Mandela was released from prison in 1990, we were very much aware of the peaceful revolution he launched, which removed South Africa's apartheid government.

But the typical and widespread American attitude toward Africa—even during the recent horrific episodes of butchery and genocide—is more like "out of sight, out of mind."

This is historically misguided, because Africa is the place where humanity was born, as well as the fountainhead of a vast and rich tradition of myth, magic, and music. The second largest and second most-populous continent after Asia, Africa is where most evidence of the earliest human ancestors has been found, leaving little doubt that we are all

*"Sub-Saharan" means that part of the African continent that lies south of the Sahara Desert. Ancient Egypt, discussed in chapter 2, and much of northern Africa developed largely separated from the sub-Saharan areas, as the world's largest desert created a mostly uncrossable barrier until the widespread use of the camel around 750 CE.

"out of Africa."* From the many discoveries of bones, stones, and fossils at sites in eastern Africa, there is wide agreement that the earliest human beings lived more than 2 million years ago in eastern Africa, in an area spanning modern Ethiopia, Kenya, and Tanzania. Evidence of emerging "modern" humans from the past 100,000 years, including improved stone tools, artwork on rocks, signs of body decoration, and burials, also appear first at various sites in Africa.

But a clear picture of what happened between the time of those fossilized remains from millions of years ago and last week's headlines remains sketchy at best. By all accounts from the worlds of archaeology, anthropology, and history, a series of migrations took place over hundreds of thousands of years, eventually leading to pockets of very different people dotting the map of Africa. By the time of the Common Era, Africa was not the home of a few scattered tribes living in primitive isolation from the world and each other. Instead, it was a place of many people, hundreds of tribal groups (many of them nomadic, others in thousands of villages), sophisticated cities, and small kingdoms, all with different languages, beliefs, and rituals. These many people included the early Christian kingdoms of Kush and Axum, neighboring Egypt, which claimed to possess the biblical Ark of the Covenant that held the Ten Commandments; the great center of Islamic learning at Timbuktu in Mali; the diminutive Pygmies of the equatorial rain forests; the towering Masai herdsmen of Kenya and Tanzania; the San of the Kalahari Desert;† the cattle-herding Khoi of southern Africa; and the proud Zulu,

*Many questions about human evolution and origins remain unresolved, but the idea that the human species began in Africa is widely accepted. The question of the origin of *modern Homo sapiens* is still open, and recent discoveries have led to two main schools of thought. One argues that modern humans evolved more or less simultaneously from "archaic" humans in several areas. The second holds out for an African origin of all modern humans. According to the Smithsonian Institute, the oldest known evidence for anatomically modern humans comes from about 130,000 years ago, from sites in eastern Africa.

†The term *San* is now preferred to the more commonly used "Bushmen," the tribe prominently featured in the popular 1984 film *The Gods Must Be Crazy*. The film is a kind of modern myth in which a "gift of the gods"—a Coke bottle dropped from a passing plane—becomes a most unwanted gift, and the tribe decides it must be returned to the gods, requiring one man's quest and a perilous encounter with "civilization."

who challenged the might of the British Empire in nineteenth-century South Africa. This variety of people clearly underscores the fact that Africa is not one monolithic "dark continent," but an extraordinary, rainbow-colored cloth woven through with the threads of many beliefs.

Africa's diversity was both transformed and diminished by powerful outsiders—Islamic Arabs, starting in the seventh century, and European Christians in the fifteenth century. In the wake of their arrival, Africa's rich array of native myths and beliefs was nearly eradicated by missionary zeal and then given short shrift by generations of academics and historians. When the African mythic legacy was finally recognized in the twentieth century,* it was brought to life in a panoramic picture of all-seeing deities; mischievous tricksters; tales of death and mortality; powerful ancestors and spirits; the importance of family, friends, and community; and the dominating presence of the African healers, priests, and shamans, once derided as mere "witch doctors."

Along with the revived interest in the role of traditional healers and shamans came the rediscovery of the rich oral history preserved by people like the griot—the musician-storytellers of western Africa who gained notoriety as the inspiration for Alex Haley's *Roots*. Like the village shamans, the griot did not practice their art in a Parthenon, palace, or pyramid. Their sacred stories were expressed as a sort of performance art in song, drumming, and dance—a communal experience still alive today in African village life. Just as the songs of Homer and Hesiod were once sung in Greek villages, the musical tales of the griot captivated African villagers. Encompassing the themes of rain and drought, love and sex, morality and mortality—the same themes that course through all myths and legends—their tales were powerful accompaniments to the belief that all nature was sacred and that spirits inhabited every living thing.

Last but not least, ancient Africa was a preliterate place that produced few texts by which their myths can be studied. There is no ancient *Odyssey* or *Ramayana* written in African tongues. Neither is

*African art and myth had a powerful impact on a generation of modern Western artists, including, among others, Constantin Brancusi, Amedeo Modigliani, and Pablo Picasso, whose 1907 painting *Les Demoiselles d'Avignon* includes figures wearing African tribal masks. New York's Museum of Modern Art mounted one of the first American shows of African sculpture as art in 1935.

there a guide to the afterlife or a native encyclopedia of the gods to help us grasp what the ancient Africans thought.

Fortunately, an extraordinary oral tradition has been maintained throughout Africa to this day. And recent scholarship and a dedication to restoring some of the "lost" African past has cast a bright new light on the dazzling mythology of what was once considered the "Dark Continent."

MYTHIC VOICES

The sun shines and sends its burning rays down upon us,
The moon rises in its glory.
Rain will come and again the sun will shine,
And over it all passes the eye of God.
Nothing is hidden from Him.
Whether you be in your home, whether you be on the water,
Whether you rest in the shade of a tree in the open,
Here is your Master.

—*from a traditional Yoruba song*

The night is black, the sky is blotted out,
We have left the village of our fathers,
The Maker is angry with us . . .
The light becomes dark,
the night and again night,
The day with hunger tomorrow—
The Maker is angry with us.
The Old Ones have passed away,
Their homes are far off, below,
Their spirits are wandering—
Where are their spirits wandering?
Perhaps the passing wind knows
Their bones are far off below.

—*a song of the Pygmies of Gabon*

Is there an "African" mythology?

Good question! But when you stop to ponder it, it's a bit like asking, "Is there a 'European' mythology?" A Greek and an Irishman are both called "Europeans," but share little in the way of ancient myth or national history—or appearance. Similarly, Africa is filled with people who fall under the collective nametag "African," but who look very different and also have an expansive range of traditions and myths.

The wide variety of mythologies that developed among the people living south of the Sahara was a result of constant movement by nomadic populations across enormous geographic barriers in a vast and varied landscape. Occupying a fifth of the world's land—an area three times the size of the continental United States—Africa is a staggering 11,657,000 square miles of territory, divided by deserts, mountains, rain forests, winding rivers, and a massive savannah. Sheer size alone kept myth-mingling to a minimum. As mythologist Arthur Cotterel notes, "Mythologies abound in Africa. Tribes possess their own traditions, and even where they share a language with their neighbors . . . it is the diversity of local belief that surprises rather than the evidence of a common heritage." While Islam and Christianity are widespread today among the more than 850 million Africans, more than 100 million people still practice forms of traditional ethnic religions, according to the *Encyclopedia Britannica Book of the Year* (2004); other estimates of the number of traditional believers in Africa are twice that.

This rich range of ancient beliefs makes it difficult to draw simple conclusions, but some broad parallels can be found in Africa's many traditions. "Central to these," author Chris Romann notes about African religions in *A World of Ideas*, "is a strong sense of the oneness of creation, in which the interconnection between the natural and the supernatural, the physical and the spiritual, the visible and the invisible, the living and the dead are far more important than the differences between them." Traditionally, the majority of African people believe that gods exist everywhere in nature, and that such natural presences as mountains, rivers, and the sun contain a deity or spirit. African religions also tend to be more "here and now," focusing on earthly life instead of an afterlife.

So, put the "oneness of creation" and nature worship at the top of a list of similarities in Africa's many mythic traditions. And bear in mind these other important common characteristics:

- **A Supreme God.** The existence of a supreme being who is omniscient and omnipresent but often disappears from the scene out of annoyance with mankind is a very common theme. For instance, Wulbari, the creator god of the Krachi of Togo in West Africa, gets tired of people asking him for favors and is annoyed by the cook-smoke constantly getting in his eyes, so he leaves the village of people and sets up a heavenly court composed entirely of animals. Another god, We, is irritated because an old woman cuts a piece of him each day to make a good soup. Nyame of the Ashanti (or Asante) of modern Ghana is constantly disturbed by a woman pounding yams who keeps banging on the overhanging floor of heaven. In a move that any New York apartment-dweller with a noisy neighbor can appreciate, Nyame retreats to the sky, and when the people try to build a ladder of calabash gourds to reach him, they tumble down—another common African narrative which mirrors the Tower of Babel story.

- **A Pantheon of Gods.** In many African traditions, the supreme creator may withdraw, but there is still a pantheon of more active and available gods who can be called upon through prayer, sacrifice, or the offering of gifts to gain favors. One of the best examples of this pantheon is the 1,700 divinities—known as orishas—of modern Nigeria's Yoruba, who are part of one of the world's oldest practiced religions, sometimes called Orisha after its gods. Orisha is headed by the supreme god, Olorun (or Olodumare), who couples with Olokun, goddess of the sea, and has two sons, Obatala and Oduduwa. Olorun sends his sons, along with a great palm tree, to create the world, but one of them makes wine from the palm sap, gets drunk, and falls asleep. (The biblical Noah, first man after the Flood, did exactly the same thing.) The other son, Oduduwa, creates earth and separates the land from the seas by having his hen scratch the ground. Oduduwa calls the place he creates Ile-Ife ("wide house"). In time it becomes a great city of the Yoruba, and it is still a major university center in modern Nigeria.*

Also in the Yoruban pantheon is the god of storms and thunder, Shango. According to legend, Shango is a ruler on earth who flees to the forest to escape his enemies but winds up committing suicide and then later being deified. Shango and another orisha, the trickster Eshu, are among the most important gods carried to the Americas by Africans taken as slaves, and are prominent in the African-based fusion religions, such as Santeria, which emerged in the Caribbean, and the voodoo of Brazil, Haiti, and Cuba.

- **A Guardian Spirit**. Many Africans—like the Chinese and other ancestor-worshipping cultures—believe that the souls of their deceased forebears serve as guardians and sources of wisdom for the living. Some believe that ancestors are reborn in living things or in objects. The Zulus, for example, traditionally refuse to kill certain kinds of snakes, because they believe the souls of their ancestors live in these reptiles. In the modern Kwanzaa celebration, ancestor worship plays a part in a ritual that includes pouring a glass of clear water and lighting a candle while praying to the departed ancestor for help and guidance.

- **The Trickster**. One of the most widespread and popular African mythic characters, the trickster often appears in many stories as an animal. The clever, clownish trickster is both a troublemaking hero and a schemer who shows little concern for the consequences of his mischief and fantastic adventures. A typical trickster in African traditions is Turé, the spiderman of the Pygmies, whose loincloth catches on fire from a spark at a blacksmith's shop. Madly dashing through the forest, Turé asks the fire to enter the trees. This explains how humans got fire and why rubbing wooden sticks together produces it. Another famed animal trickster is Anansi the Spider, who was once a Creator god but who now lives by his wits, fooling other animals and mankind. And then there is Hare, whose ability to outwit other animals—and humans—made him the model for Br'er Rabbit, the mis-

*Steeped in the myths and traditions of the Yoruba are the writings of Wole Soyinka (b. 1934), a Nigerian playwright, essayist, and poet who became the first African writer to win the Nobel Prize for Literature, in 1986. He has been imprisoned on several occasions for his political views.

chievous rabbit who constantly outwits Br'er Bear and Br'er Fox in the stories told by plantation slaves in the American South and recorded by Joel Chandler Harris in his Uncle Remus stories (see below).

Eshu, another important African trickster, is not an animal but a god—like the Norse Loki, in many respects—who brings chaos. In one story, Eshu steals vegetables from the Creator god and covers up his theft by making footprints in the garden with the god's own sandals. When the Creator realizes what has happened, he is so angry that he withdraws from earth.

- **Explanations for Death**. In many African traditions, a "mixed message" brings death into the world, usually when an animal courier fails to deliver some important information from the gods to mankind. This is the case in the story of a bird sent by the Creator to tell people that when they get old they should just peel off their skins. On the way to deliver this message, the bird sees a snake eating a dead animal. In return for some of the meat, the bird tells the snake that it can obtain a new life by shedding its skin. While snakes gain the secret of immortality, the message is never delivered to humans—which is why people are mortal. For its failure, the bird is afflicted with a terrible disease. That's why its painful cry is often heard in the tops of trees.

  In a Zulu tale of the origin of death, a lizard carrying the news of death outraces a chameleon who has the message of eternal life. The chameleon arrives only to find that people have accepted the lizard's words as the truth.

- **People with Special Spiritual Abilities**. Magic played a major role in many traditional African religions in which the only way an average person could approach the divine was thought to be through priests or medicine men. As the historian of religion Huston Smith writes, "We can think of shamans as spiritual savants . . . exceptional to the point of belonging to a different order of magnitude. Subject to severe physical and emotional trauma in their early years, shamans are able to heal themselves and reintegrate their lives in ways that place psychic if not cosmic powers at their disposal. Those powers enable them to engage with spirits, both good and evil."

These healers and shamans were usually elders and other individuals singled out for some remarkable ability, and typically were responsible for healing, divination, exorcisms, and escorting the dead to the underworld. As true of shamans in many traditions, the tribal priests of Africa usually performed in an ecstatic trance induced by dancing, drumming, chanting, or with the use of a drug or alcohol. Since there was no established church or clergy in ancient Africa, the appointment of these priests was often hereditary. Among the Masai of eastern Africa, for example, the medicine men all came from one clan.

According to Paul Devereux, an authority on ancient mysteries, "The shaman* was the person who acted as the intermediary between the tribe and the otherworlds of spirits. A shaman would heal the sick tribal members by locating their lost souls, perhaps entering the spirit world to reclaim them, or by deflecting bad spirits and invisible influences. There were also a variety of other reasons for entering the spirit realms, such as accompanying the souls of dying people or seeking information from the spirits or ancestors."

- **Fetishes**. Bones, carved statues, or unusual stones were thought to be inhabited by spirits and contain magical powers, but they were more closely associated with dead ancestors and seen as an integral aspect of ancestor worship in many African traditions. The word "fetish" was coined by Portuguese sailors, some of the first Europeans to encounter these figures among the Yoruba and the Dogon of western Africa. But the use of fetishes was widespread throughout Africa, and in the Congo, for instance, included elaborate, nail-studded statues called *nkisi nkondi* ("power figures").

  Don't snicker if you happen to be one of those people with a "lucky coin" or a rabbit's foot in your pocket. They are "fetishes," too.

*In strict terms, the term "shaman" refers specifically to such people in the Siberian and Central Asian tribes, where the word originates. But it is now used more widely and generally to describe many healers and spiritual leaders, including the Celtic Druids, and especially those who use the ecstatic trance as part of their practice.

What role did myth play in African villages?

It may take "a village to raise a child,"* but what does it take to hold a village together? Just as the temple complexes of Egypt and Mesopotamia, and the agoras of Greece, were the pulsing centers of those cultures, the African village was—and in many places, remains—the heartbeat of Africa. And the myths and stories of African tradition are the connective tissue that hold the village together.

In societies like ancient Africa, where there were no written records, myth played an important role in maintaining a sense of history and cohesion. Just as the bard provided the collective memory of the Greeks, Celts, and Norse, the African storyteller always helped unite the village with sacred stories. These storytellers weren't mere entertainment, trotted out for a once-a-week religious service. Their performances, combining story and song, drumming and dancing, were an integral part of daily village life and helped to convey important messages about the value of family ties, the feats of famous ancestors, the heroes of the past, and the individual's place in society. As folklorist Roger Abrahams explains in *African Folktales*, "In the village the question of the individual in the family and community arises constantly, as does the issue of initiative in a world that must stress the subordination of the individual will to the good of the group."

The good of the group was often tied to the question of food. In a landscape where growing conditions were always challenging, the constant possibility of drought, crop failures, and food shortages was a persistent fear, and social cooperation and collective farming were crucial to survival. African myths and stories were both preoccupied with this theme, as Roger Abrahams notes. "Nothing strains the web of culture so much as the threat of starvation. . . . We see [that] through these tales. Bonds are repeatedly strained because someone steals food, or because children are neglected when crops fail. Therefore no theme is more

*The oft-quoted saying "It takes a village to raise a child" is usually cited as an African proverb. That assertion has never been documented as coming from any specific African source, although there are similar sentiments expressed in many other African proverbs. Now something of a political hot potato since Hillary Clinton used the phrase as a book title, the communal concept behind the proverb can be found in a great many other cultures as well.

important or receives more attention than the building of families and friendship ties to provide that strength which, even in the face of natural disaster or perilous human responses to it, ensures a community's survival. . . . One realizes how great the achievement of family and community is, and how constantly that achievement must be recreated again."

The importance of communal action is clear in a story about the trickster Hare told by the Ewe (of Ghana and Togo) that also appears in other versions in many African traditions. While just an amusing tale on the surface, it underscores the fundamental need for cooperation. When a drought dries up the earth, the animals assemble in a council and all agree to cut off a piece of their ears and extract the fat, which they will sell to buy a hoe to dig a well. All do as they promise except for Hare, the trickster, who reneges. The other animals are surprised, but still manage to buy the hoe and dig until they hit water. Along comes Hare, who first draws some water and then takes a bath, muddying the well. When the other animals realize that Hare has ruined their water, they hatch a plan that involves covering a small statue with "bird lime." Hare comes along and speaks to this "dummy," which, of course, does not respond. Angrily, Hare hits the statue and gets one paw stuck to it, then the other. Next, he kicks at the sticky statue, but only succeeds in getting both feet stuck as well. The other animals, watching from hiding, come out and give Hare a beating, before letting him go. From that day on, Hare never leaves the safety of the grass.*

Given the central importance of communal cooperation, something as seemingly simple as a "work song" takes on a vital role in African mythology. Sung in unison by laborers harvesting crops or hoeing fields, the work song was not just a pleasant diversion from otherwise dreary labor, but a fundamental, cohesive force in tribal life. These songs, still very much alive in Africa today, made their way from thousands of African villages to America for centuries, aboard slave ships, and found voice in the work songs of the plantation slaves as well as in the rhythmic songs of chain-gangs (brilliantly displayed in the opening scene of the

*This story is told in many other versions in which the statue is sometimes called Gum Girl. It is also familiar as the origin of the figure of the "tar baby" in the Uncle Remus stories.

film *O Brother, Where Art Thou?*). These, in turn, powerfully influenced American gospel, rhythm and blues, jazz, and, eventually, rock and roll and Motown. That's just one reason this mythic tradition deserves some R-E-S-P-E-C-T.

Mythic Voices

In the beginning, in the dark, there was nothing but water. And Bumba was alone.

One day Bumba was in terrible pain. He retched and strained and vomited up the sun. After that light spread over everything, The heat of the sun dried up the water until the black edges of the world began to show. Black sandbanks and reefs could be seen, but there were no living things.

Bumba vomited up the moon and then the stars, and after that the night had its own light also.

Still Bumba was in pain. He strained again and nine living creatures came forth: the leopard . . . and the crested eagle, the crocodile . . . and one little fish named Yo; next . . . the tortoise and Tsetse, the lightning, swift, deadly, beautiful like the leopard, then the white heron . . . also one beetle, and the goat.

Last of all came forth men. There were many men, but only one was white like Bumba . . .

. . . When at last the work of creation was finished, Bumba walked through the peaceful village and said to the people, "Behold these wonders. They belong to you." Thus from Bumba, the Creator, the first Ancestor, came forth all the wonders that we see and hold and use, and all the brotherhood of beasts and man.

—a Bantu Creation tale, "The Beginning," Maria Leach

> **The world was created by one god, who is at the same time both male and female . . . named Nana-Buluku. In time, Nana-Buluku gave birth to twins, who were named Mawu and Liza, and to whom eventually dominion over the realm thus created was ceded. To Mawu, the woman, was given command of the night; to Liza, the man, was given command of the day. Mawu, therefore, is the moon and inhabits the west, while Liza, who is the sun, inhabits the east. At the time their respective domains were assigned to them, no children had as yet been born to this pair, though at night the man was in the habit of giving a "rendezvous" to the woman, and eventually she bore him offspring. This is why, when there is an eclipse of the moon, it is said that the celestial couple are engaged in love-making. . . .**
>
> —*Creation tale of the Fon of Dahomey*
> *From* Dahomey, *Melville J. Herskovits*

Is there an African Creation myth?

The Garden of Eden was in Mesopotamia. The Egyptian Creation emerged from the Nile's waters. The Chinese believed that people came from the clay of China's Yellow River. But surely, they all had it wrong. After all, humanity was born in Africa. So it would make complete sense that Africa's Creation stories would be particularly significant.

But myth in a preliterate society can be tricky. Though very old, Africa's myths were not collected or written down until the late nineteenth and early twentieth centuries. Even then, they were recorded by missionaries or colonial administrators, who often had their own agendas. Perhaps this is why the Creator god Bumba is described as white. And why parallels with the Old Testament pop up in African Creation myths. These agendas acknowledged, there are still hundreds of different African stories about how the universe began and humans were created. While few complete narratives exist in a form that can be considered "authentic," many brief tales survive, which contain some common characteristics.

Many stories, for instance, involve a cosmic egg that breaks open and lets out a primeval serpent, typically a python. The world and every living thing in it are made from the body of the snake, so common on the African continent. In many other world myths, snakes have often played some fascinating—and contradictory—roles. Dangerous yet intriguing, they shed their skins, seemingly able to take on new life, an idea found in countless stories, including *Gilgamesh*. Their phallic connection only adds to this view of snakes as a mystical life-force. On the other hand, they are silently deadly and are often the supreme symbol of disorder and evil, whether in the Bible or ancient Egypt, where the serpent Apep tries to kill the sun each night. But in African myth, the concept of snake as a life-force predominates. According to the Creation story of the Fon of Dahomey, the serpent Aido-Hwedo serves the Creator goddess Mawu, daughter of the older, remote Creator Nana-Buluku. The rivers of the world wind around like the serpent's body, and the mountains are formed by great piles of the serpent's excrement. Mawu makes the serpent lie down in the waters surrounding the earth in a perfect circle with his tail in his mouth—a widely shared symbol of eternity. Sometimes he shifts, which explains earthquakes. And someday, when he swallows his tail, the world will come to an end.

Another primordial snake in southern and central African myths is Chinawezi. Called "the mother of all things," this female serpent shares the world with her husband, Nkuba, who sits up in the sky and waters the earth with the beneficial rain of his urine. Chinawezi rules the earth, and whenever the thunder rumbles in the sky, it is believed that she replies by making the rivers swell.

In many traditions, a supreme god begins the work of making the cosmos but then leaves others in charge, or leaves earth altogether in annoyance with human behavior. In still other Creation tales, the world already exists, and it is only the creation of humanity that is of interest. Some of these stories reflect the influence of Christian missionaries as the old stories were merged with the biblical tradition. The Efe of Zaire, for instance, have a story in which the female moon helps the supreme Creator make the first man, Baatsi, from clay covered with skin and filled with blood. When the Creator makes a female companion for Baatsi, the couple are instructed to make children but are warned not to eat from the "Tahu tree." They obey, and for many years everyone lives an idyllic existence until they get old and tired, and simply go straight to heaven without dying. But later on, when a pregnant woman has a craving for Tahu fruit and convinces her husband to pick some for her, the Creator decides that men and women must suffer the punishment of death.

Another Eve-like story, from the Dinka of southern Sudan, is about Abuk, the first woman. In the beginning, the High God allows the first man and woman to plant a grain of millet each day. When Abuk greedily decides to plant more, she accidentally whacks the High God on the toe with her tool, making him so angry he retreats to the sky and cuts the rope that links heaven and earth. Since then, humans have had to work hard to grow food, and suffer sickness and death.

Some of the other most prominent African Creation accounts, which also come down in fragments, are included in the first part of this "Who's Who" directory, which lists the god's principal tribal association and location. The second part of the list includes the other most significant group of African gods, the "tricksters," who are responsible for a mixture of good things and amusement, but more often bring mischief, chaos, and disaster.

WHO'S WHO OF AFRICAN DEITIES

Creator Gods

Amma (the Dogon of Mali) In Mali (western Africa), the Dogon revere a single god, Amma, the chief creator of all things. In one of their Creation myths, Amma exists at the beginning of time as a great egg that contains all the elements of creation—fire, earth, water, and air. In a series of great explosions, these all combine to make life.

In another version, Amma is a divine potter who casts the sun, moon, and stars out of clay that he flings into the sky. When the heavens are complete, he forms a woman—earth—and produces a jackal monster and two serpentlike twins with her. The twins invent speech and cover the bare earth with vegetation. Amma then couples with the earth, producing another set of twins, who become the ancestors of the Dogon.

Bumba (the Bushongo of the Congo) The chief Creator god of the Bushongo (central Africa), Bumba vomits the earth, sun, moon, and stars into existence. These are followed by the animals, from which all life descends. Described in some traditions as "white," Bumba gives fire to a man named Kerikeri, who charges a high price for embers to make fires for cooking. The king's daughter entices Kerikeri into marrying her, so she can learn the secret of fire from him. One night, she pretends to be cold and watches as Kerikeri builds the fire. After learning his secret, she deserts him. It is another story underscoring the widely shared African distaste for selfishness.

A similar Creation story of the Kuba of the rain forests of Zaire (central Africa) is about Mbombo, a spirit who, during the dark hours of the first day of Creation, has sharp pains in his stomach and vomits, producing the sun, moon, and stars. As the sun shines, the primordial water recedes, and the hills and plains of the earth are revealed. In a second convulsion, he sends forth a stream of vomit that produces the rest of Creation, including all of the animals and the first man and woman.

Cagn (San of the Kalahari) The chief creator of the San (once called Bushmen) of the Kalahari Desert region in southern Africa, Cagn is a wizard of great power, whose magical strength resides in one of his teeth. A shape-shifter, he can assume the forms of different animals, including a praying mantis and an antelope. He also has a pair of sandals that can turn into attack dogs. At various times, Cagn is eaten by ants and by an ogre, but he always comes back to life when his bones rejoin.

The San have a myth about another creator, **Dxui**, who takes the form of a different flower or plant every day and becomes a man at night, until he creates all the plants and flowers that exist. Afterwards, Dxui becomes a fly, water, and a bird, until he is finally transformed into a lizard, which, the San believe, is the oldest creature of all.

Chuku (Ibo of Eastern Nigeria) The Ibo (or Igbo) believe that the supreme creator, the benevolent Chuku ("great spirit"), creates an earthly paradise in which there is neither evil nor death. In a very typically "confused message" story, Chuku sends a dog to earth to tell people that those who die accidentally can be brought back to life if they are laid on the ground and sprinkled with ashes. But the dog carrying this message is too slow in delivering it, so Chuku next dispatches a sheep with the same message. But the sheep stops to eat along the way, and by the time it finally reaches mankind, the message is confused—the sheep tells the people to bury their dead. Because of this foolish sheep, death comes into the world.

Imana (the Banyarwanda people of Rwanda) In Rwanda (central Africa), the omniscient creator Imana has very long arms and is benevolent to mankind, but likes to keep his distance. One legend describes how Imana is hunting down Death in order to rid the world of it. But Death begs an old woman for protection, and she agrees to hide Death under her skirt. For this, Imana decides that Death should live with mankind, after all.

Kalumba (the Luba of Zaire) The creator of the Luba tribe (central Africa), Kalumba creates mankind and then wants to protect people from death and disease. He sends a goat and a dog to guard the road on which Life and Death are traveling. The animals have been instructed to only allow Life to pass, but they argue and split up. While the dog sleeps, Death is able to sneak past. Then, while the goat is on guard, Life gets turned away, so people cannot be saved from Death.

Leza (the Kaonde of southern Africa) Leza is a supreme god who rules the sky, sits on the backs of all people, and is supposedly growing old, so he cannot hear prayers as well as he once did. In one tale of the Kaonde, Leza gives a bird three calabash gourds to deliver to humans. Inside two of the gourds are the seeds to grow food, but the third is not to be opened. Like Pandora, the bird can't restrain its curiosity and looks inside the gourd, releasing all the evils of the world, which are contained inside. Leza and the bird are unable to recapture the evils once they are set loose.

Mawu and Liza (Fon of Benin) This pair of twin creator gods may have belonged to another tribe that was conquered by the Fon (western Africa), who possessed superior iron-working skills and probably had superior weapons. The defeated tribe's gods were then absorbed by the Fon. Mawu and Liza are born from an older creator god, Nana-Buluku, who is a sexless primeval creator, and Aido-Hwedo, the rainbow-colored snake who holds up the earth. The male god Liza is associated with the sun, power, the daytime, work, and strength; the goddess Mawu is associated with the moon, nighttime, fertility, motherhood, and joy. This divine pair shape the universe from preexisting material and then create all the other gods in the sky and on earth.

In one of the Fon Creation stories, these two come together during eclipses to create the other gods. They have a set of twins, Sagbata and Sogo, who, like most mythic twins, get into a dispute over which of them will rule. When the elder twin, Sagbata, is given precedence over the younger twin, Sogo, Sagbata angrily stops the rain, and soon all of creation is starving and thirsty.

On the second day of Creation, they send down their son, Gua (or Gu), the god of thunder, blacksmiths, and farmers, to help mankind. Gua does not anticipate that his tools will later be used for warlike purposes. In a separate version of this myth, Gua helps make the first people out of divine excrement.

'Ngai (Masai of southeastern Africa) A creator god of the cattle-herding Masai, 'Ngai gives every man a guardian spirit to ward off danger and carry him away at death. The good go to a rich pasture land, while the evil are carried off to a desert.

In the beginning of Creation, there is only one man on earth, Kintu. When 'Ngai's daughter Nambi sees Kintu, she falls in love with him, and they marry after he passes a series of challenges. Promising not to return to the sky, they go to earth with plants and animals in Nambi's dowry. But Nambi forgets to bring along grain to feed her chickens, and when she returns to the sky to get some, she meets Death, who follows Nambi home and then kills the couple's children. Death remains on earth after that. As in many African myths, the connection between heaven and earth is destroyed by human error or foolishness.

Nyame (Ashanti of Ghana) Supreme god of heaven and earth, as well as the sun and moon, Nyame is Creator of all the realms—heaven, earth, and the underworld. Nyame gives each soul its destiny at birth, and washes it in a golden bath. But Nyame is one of those gods for whom living with humans gets to be too much of a nuisance. After an old woman preparing yams keeps hitting Nyame with her pole, he goes away to seek a more peaceful home in the sky.

Unkulunkulu (Zulu, Xhosa) Known as "Old, Old One," Unkulunkulu is both the creator, a god of earth who has nothing to do with the heavens, and the first man. According to the Zulu (southern Africa) Creation myth, he evolves alone in the emptiness, and, once he comes into being, creates the first men out of grass. Unkulunkulu orders a chameleon to tell men that they will be immortal. But the creature lingers so long that the god angrily sends a lizard with the opposite message, and the lizard arrives with the news of death first.

To balance man's mortality, Old, Old One teaches humans about fertility rites, marriage, healing, and other basics of civilization. He also provides the dead with a dwelling place in the sky, and the stars are thought to be the eyes of the dead looking down upon the world.

Tricksters and Animal Gods

No matter what he is called, people everywhere love a trickster. To Shakespeare, he is the playful sprite Puck, who makes trouble in *A Midsummer Night's Dream*, or the whimsical spirit Ariel working his mischief in *The Tempest*. To generations of children, he is the magical boy Peter Pan who never grows up and knows how to fly, or the "wascally wabbit" Bugs Bunny, who constantly bedevils Elmer Fudd. In silent movies, he is Charlie Chaplin, sticking a wrench in the cogs of *Modern Times* to outwit the high and mighty. In *Star Wars*, he is Han Solo, the likable rogue out for himself. More recently, he is *Seinfeld*'s Kramer, who can create manic upheaval and disorder in less than thirty minutes.

Described as the "sacred clown," the trickster can be found in every mythology. Looking to put over a con, cause chaos, or get something for nothing, the trickster is a lovable loner who is almost always outside the ring of "civilized" behavior. As Jungian authority Dr. Joseph Henderson writes in *Man and His Symbols*, "Trickster corresponds to the earliest and least developed period of life. Trickster is a figure whose physical appetites dominate his behavior; he has the mentality of an infant. Lacking any purpose beyond the gratification of his primary needs, he is cruel, cynical, and unfeeling. . . . This figure, which at the outset assumes the form of an animal, passes from one mischievous exploit to another. But, as he does so, a change comes over him. At the end of his rogue's progress, he is beginning to take on the physical likeness of a grown man."

While the trickster's mischief can sometimes benefit humans—Prometheus in Greek mythology, for instance, tricks Zeus and brings fire to mankind—more typically, his amusing diversions bring discord and disorder to the world, making him an unwelcome member of the community. In African myths and legends—as well as in the mythology of the Native Americans—the trickster is an especially vivid character, most often appearing as an animal and always a male. Perhaps the trickster began simply as a way to explain the sudden, unexplained little mysteries of life—the food missing from the table, the muddied well-water, the vegetables filched from the garden—as well as the bigger anxieties, like the hint of a stranger in the marital bed, or the unexplained disappearance of a child.

Animals such as the chameleon, praying mantis, hare, tortoise, and spider take part in every area of African legend, from the Creation to the coming of death to humans. But probably the most common role of animals in African myth is that of the trickster. This list includes some of the most significant of them.

Anansi Perhaps the most famous character in African myth is "Mr. Spider," who is called Anansi in West Africa (and **Turé** in the Congo). Known for his cleverness, Anansi is a Creator god in some traditions, including the Ashanti, while in others, he is a man who gets kicked into a thousand pieces and becomes a spider because of his cunning tricks. A scoundrel and shape-shifter known for assuming disguises, Anansi is able to dupe other animals and even humans.

In one popular tale, Anansi makes a rather curious request—he says he wants to own all tribal myths that belong to the Ashanti's Creator sky god, Nyame.

The sky god tells Anansi that to get the stories, he must capture three things: hornets, the great python, and the leopard. Setting about his challenges, Anansi cuts a small hole in a gourd, throws some water on himself, sits inside the gourd, escapes, and tells the hornets to get in so they will not get wet. Once the hornets are inside the gourd, Anansi plugs up the hole with some grass, and takes the hornet-filled gourd to Nyame.

Next, Anansi cuts down a long bamboo pole and some strong vines. When he comes upon the python, he tells the snake that he has been arguing with his wife about what is longer—the python or the pole. The vain python allows himself to be measured by the clever spider, who then ties the serpent to the pole with the vines. Now caught, the python is delivered to Nyame.

Only the leopard is left. Digging a pit and covering it with brush, Anansi next traps the leopard, who is eventually strung up in the air by rope and killed. When Nyame sees the leopard's body, he is so impressed that he gives all his stories to Anansi, and they became known as the "spider's stories."

When anyone wants to tell one of the sky god's stories, he must pay homage to Anansi, who owns them all. Today, we would call this "copyright protection."

Eshu (Yoruba of Nigeria) Unlike many other African tricksters, Eshu is a god, not an animal. Capable of shape-shifting, and being both large and small at the same time, Eshu confuses men and drives them to madness, but also acts as a go-between for the mortals and the gods. The bringer of chaos and cause of all arguments, Eshu once persuaded the sun and moon to trade places, causing universal disorder. He also got the high god of humans to leave earth for heaven. For his tricks, Eshu is ordered to become the messenger who links heaven and earth and reports every day on what is happening on earth.

In a tale with several variations, Eshu walks between two neighbors, wearing a hat with different colors on each side. The neighbors eventually argue over what color the hat is, and come to blows. When their dispute lands in court, Eshu resolves the argument and teaches people that the way in which one sees the world can alter his perception of reality. In other versions of this tale, Eshu is far less benevolent, and the argument over the colored hat leads to complete annihilation of the tribe, which only amuses Eshu, who says, "Bringing strife is my greatest joy."

Among the Fon of Benin, which neighbors Nigeria, the trickster god Eshu appears as **Legba**, an attendant of the supreme god. Legba's job is to do all the harmful things to people that god wants done. When Legba tires of this role, he asks god why he must always do the dirty work and get all the blame. The high god tells him that the ruler of a kingdom ought to get credit for all the good things while his servants take the rap for all the evil. Talk about archetypal!

In one story, Legba steals the god's sandals, puts them on, and goes into the yam garden and steals all the yams. This time, the people are angry at god for stealing their yams. When the god realizes that Legba has tricked him, the deity decides to leave the world and instructs Legba to come to the sky every night and report on what happens on earth.

In another story, Legba asks an old woman to throw her dirty washing water into the sky. God is annoyed by the dirty water's constantly being thrown in his face, so he moves away, where he can't be bothered so easily. Again, he leaves Legba behind to report, which is why there is a shrine dedicated to Legba in many African houses and villages.

Legba's counterbalance is **Fa** (or **Ifa**), the god of fate and destiny, who teaches healing and prophecy. To the Fon, everything is fated to happen, nothing is left to chance, and Fa represents every person's fate. Divination or magic can help one discover their Fa. Whenever beginning work or starting a business, it is customary for the Fon to make an offering of food to Fa, but first give a taste to Eshu, to ensure that things go smoothly. The religions of Benin later influenced voodoo, one of the principal offshoots of the convergence of traditional African religions and Christianity in the Caribbean (see below).

Hare Alongside Spider, Africa's most popular animal trickster is Hare, about whom stories are told all over the continent. One typical story tells how Hare challenges an elephant and a hippopotamus to a tug-of-war. But instead of tugging, Hare ties the ends of the rope to each of the animals. As they pull against each other, Hare's land is plowed, which is exactly the job Hare was supposed to do for his wife.

Another story tells how Hare mixes up a message he has been given to deliver, and loses immortality for humans. When the moon sends Hare to tell people that they will die and then rise again, just like the moon, Hare confuses the message and tells people only that they will die. When the moon finds out what Hare has said, she beats him on the nose with a stick. Since that day Hare's nose has been split.

The Hare stories made the transatlantic crossing with the many Africans taken as slaves to the Americas. Mingling with many similar Native American tales of trickster rabbits, the Hare stories became best known in America as the Br'er (short for "Brother") Rabbit stories.

Having heard these stories told on a plantation, Joel Chandler Harris (1848–1908), a writer for the *Atlanta Constitution*, later collected them in a book called *Uncle Remus: His Songs and His Sayings* (1881). The character of Uncle Remus is a former slave who becomes a beloved servant of a Southern household and entertains the family's young son by telling him traditional animal fables, using a Southern African-American dialect. Besides Br'er Rabbit, the best-known characters of the stories are Br'er Fox, Br'er Bear, and Br'er Wolf. Most folklorists today agree that the Br'er Rabbit tales are thinly veiled racial allegories.* More than just a trickster, Br'er Rabbit represents the clever slave who could outwit his master.

Yurugu (Dogon of Mali) The child of the Dogon Creator, Amma, and the earth is Yurugu, a rebellious god and trickster. While in the

*The stories also formed the basis for the Disney film *Song of the South*, much criticized for its stereotyped portrayal of African-Americans. One of the few Disney features never released on video, the film did inspire the popular Disney theme-park ride "Splash Mountain."

cosmic egg, Yurugu steals the yolk, because he thinks his sister—or mother, in some versions—is inside, and he wants to mate with her. His incestuous behavior brings mischief and disorder into the world and makes some of the world's land arid. Amma turns Yurugu into a jackal, and he sires the many evil spirits of the bush.

How did a suicidal king become a god and end up in the Supreme Court?

Along with the trickster tales of Hare and Anansi the Spider, a great many other gods made the terrible transatlantic crossing known as the Middle Passage from Africa to the New World. Once settled in the fertile grounds of the Caribbean and the Americas, these African gods didn't just disappear. Myth—and belief—are hard things to break. Many of the gods and traditions of the Yoruba and Fon, in particular, crossed the Atlantic with the people taken from West Africa, and found a home in the New World. Among the many gods who were given new lives and new meaning was Shango, the powerful storm god of Yoruba.

Possibly based on an actual mortal king, Shango was famed for his abilities as both a warrior and a powerful magician. While dabbling in magic, Shango causes lightning to strike his palace, killing some of his many wives and children. Overcome by grief, he hangs himself. When his enemies scorn the dead king, they are destroyed by storms, and Shango is declared to be a god who controls the thunder and lightning. That's one version. In another telling of his legend, Shango is an oppressive ruler, and when his people rebel, he is exiled to the forest, where he hangs himself from a tree. Those who remain loyal to Shango refuse to believe that he committed suicide, and say he has gone to heaven to become god of rain and thunder, symbolized by a twin-headed ax. No matter which version of his life and death his believers accept, Shango is always revered as a great source of magic and a sexual dynamo.

When Shango's devoted followers were taken as slaves to the Americas, they continued to worship him, along with most of the other gods of the Yoruban and West African pantheon. Many of these gods emerged in new religions that fused African traditions with the Christianity that the slaves were forced to accept along with their chains. Few aspects of

African-American or Afro-Caribbean culture have been more mythologized or grossly stereotyped—especially by Hollywood—than the traditions that emerged as voodoo and Santeria.

- **"Voodoo"**

  Although commonly called voodoo, this New World religion traces its roots to the African traditional religion Vodun (also Vodoun, Voudou), a word meaning "spirit." Vodun was recognized as the official religion of Benin in 1996, is practiced by many in Haiti, and can be found in many large cities. One estimate is that 60 million people worldwide worship Vodun. Its followers, called Voduns, are concentrated in Benin, Ghana, Haiti, and in the United States, largely in the American South and wherever Haitian refugees have settled. Also practiced by the West African Yorubans, Vodun may have roots that stretch back thousands of years.

  During the colonial slave-trading era, slaves brought Vodun with them to Haiti and other islands in the West Indies, where, upon arriving, they were baptized as Catholics, and slave masters and priests tried to suppress the African Vodun belief. Its priests were killed or imprisoned, which forced the slaves to create underground societies to secretly worship their gods and venerate their ancestors. While attending Mass, as required by their masters, slaves simply continued to follow their original faith. An influential 1884 book called *Haiti, or the Black Republic*, by S. St. John, described Vodun as an "evil religion," falsely alleging that it included human sacrifice and cannibalism. Unfortunately, the image has stuck.

  Vodun has many traditions based on Yoruban religion, including the belief that Olorun, the chief god, is remote and unknowable. Olorun authorizes a lesser god, named Obtala, to create earth and its life forms. The pantheon of Vodun spirits called "Loa" includes Aida-Wedo (the rainbow spirit based on the creation serpent Aido-Hwedo), and Shango (also known as Sango).

- **Santeria**

  Also called Le Regia Lucumi or the Rule of Osha, Santeria originated in Cuba as a combination of West African Yoruba religion and Catholicism. There, as in Haiti and elsewhere, slaves were forced to follow Roman Catholic practices, which contradicted their native beliefs. But finding parallels between their own religion and Catholicism, and in order to please their slave masters while disguising the worship of their own gods, they, too, created a secret religion. Santeria uses Catholic saints and personalities as "fronts" for the traditional African god and his spiritual emissaries, the orishas of the original Yoruba religion. Santeria spread quickly among these West African slaves, and even when the slave trade was abolished, Santeria flourished, its African-based religious traditions continuing to evolve and fuse with Christian ideas, native Cuban traditions, and, later, Enlightenment ideas brought from France.

Santeria has no sacred texts, and has been passed on orally to initiates for hundreds of years. Today it continues in small numbers in many countries, including the United States, where it is still practiced—in New York and Florida, in particular.

Like voodoo, much of Santeria corresponds to Yoruban religious traditions and mythical stories. The supreme god is Olodumare (or Olorun), who is the source of all energy in the universe, and is equated with Christianity's Jesus Christ. Olorun's emissaries, the orishas, are equated with specific Roman Catholic saints. Just as Legba is the messenger god in Africa, Legba (or Elegba) of Santeria acts as the intermediary between humans and the orishas. He is equated with the Catholic St. Anthony; nothing can be done without his intercession. Shango, who rules thunder and lightning, is called Chango in Santeria and is linked with Catholicism's St. Barbara. Demoted from the litany of saints in a 1969 reform of Roman Catholic liturgy that downgraded her along with many other notable martyrs, Barbara was said to have been beheaded by her own father for her Christian faith. When her father was killed by lightning, Barbara became associated with the force of lightning and with death that falls from the sky. That provides the connection to Chango, who is also a god of thunder, lightning, and martial power.

For five hundred years, the traditions of Santeria—including a set of Eleven Commandments, roughly equivalent to the biblical Ten Commandments, with the addition of a prohibition against cannibalism—have been maintained by its followers. These traditions include a belief in magical spells and trance possessions, which both play an integral part in Santeria. The trance possession occurs at shamanistic drumming parties, during which dancers try to attain a sacred state of consciousness and ecstasy.

While Santeria and voodoo share West African mythical and religious roots, there are differences between them as they now exist. Principally, the Santerians believe that Catholic saints and orishas are the same spirits, while voodoo believes that the two groups are distinct, and reveres both.

Finally, both religions have attracted attention because of animal sacrifice, which is probably the most controversial and publicized aspect of Santeria. The sacrificial animal, according to Santeria tradition, must be killed quickly and painlessly, and the meat eaten by participants in the service. In the early 1990s, the city of Hialeah, Florida, attempted to halt such practices, but a local branch of the Santeria church sued and—with the support of mainstream churches and Jewish organizations—won their case in the U.S. Supreme Court. In deciding the case of *Church of the Lukumi Babalu Aye v. City of Hialeah* in June 1993, Justice Anthony Kennedy wrote, "Though the practice of animal sacrifice may seem abhorrent to some, religious belief need not be acceptable, logical, consistent, or comprehensible to others in order to merit First Amendment protection."

CHAPTER NINE

SACRED HOOPS

The Myths of the Americas and Pacific Islands

For the forming of the earth they said "Earth."
It arose suddenly, just like a cloud, like a mist,
now forming, unfolding . . .

—Popol Vuh

Screaming the night away
With his great wing feathers
Swooping the darkness up;
I hear the Eagle bird
Pulling the blanket back
Off from the eastern sky.

—invitation song of
the Hodenosaunees

From the beginning of creation,
We were placed here . . . we are
This holy land . . .

—from a Navajo prayer

While I stood there I saw more than I can tell and I understood more than I saw; for I was seeing in a sacred manner the shapes of all things in the spirit, and the shape of all shapes as they must live together like one being. And I saw that the sacred hoop of my people was one of many hoops that made one circle, wide as daylight and as starlight, and in the center grew one mighty flowering tree to shelter all the children of one mother and one father. And I saw that it was holy.

—*Black Elk Speaks* (1932),
as told through John G. Neihardt

In the 1990s, Indian religions are a hot item. It is the outward symbolic form that is most popular. Many people, Indian and non-Indian, have taken a few principles to heart, mostly those beliefs that require little in the way of changing one's lifestyle. Tribal religions have been trivialized beyond redemption by people sincerely wishing to learn about them. In isolated places on the reservations, however, a gathering of people is taking place and much of the substance of the old way of life is starting to emerge.

—Vine Deloria Jr., *God Is Red* (2003)

How did Native American myth go up in smoke?

Is there an "American" mythology?

What is the Popol Vuh?

Who were the Mayans who produced the Popol Vuh?

Which gods like a good ball game?

Who's Who of Mayan Gods

What sets Mesoamerican myth apart?

Did the Aztecs really think the Spanish were gods?

What is the "Day of the Dead"?

Who's Who of Aztec Gods

Was the "lost city" of Machu Picchu really a "sacred place"?

Did the Incas have a foundation myth?

Who's Who of Incan Gods

Is there a "North American" mythology?

Who's Who of North American Native Gods

Which goddess gets her own "planet"?

What famous poem contributed to the "myth" of the Native Americans?

Do Native American myths still matter?

Which mythic character created the Pacific Islands?

What is Dreamtime?

MYTHICAL MILESTONES

The Americas

c. 12,500 years ago Monte Verde sites in Chile include dwellings and stone tools; earliest evidence so far for people in the New World.

c. 11,500 years ago "Clovis culture": the earliest evidence of human habitation in North America, based on spear points found in Clovis, New Mexico, first discovered in 1932. Earlier dates have been suggested for the Meadowcroft Rock Shelter near Pittsburgh, Pennsylvania, and other sites in Virginia and the Carolinas, but they remain controversial.

C. 9,200 years ago "Kennewick man": the evidence of oldest known skeletal remains in North America.

Before the Common Era (BCE)

c. 5000 Corn cultivation begins in Central America.

c. 4750 First evidence of animal domestication in Central America.

c. 4500 Corn cultivated in eastern North America.

c. 3500 Cotton cultivation in Central America; used to make fishing nets and textiles.

c. 2600 Large temple complexes built along the Andean coast of South America.

c. 2500 Large permanent villages appear in South America.

c. 2200 Earliest known pottery in South America.

c. 1750 Large ceremonial centers built in Peru.

c. 1500 Earliest evidence of metalworking in Peru.

c. 1200 Olmec, the first major pre-Columbian civilization, emerges in Yucatan lowlands. Olmec civilization is destroyed around 400 BCE.

c. 1000 Adena culture develops in middle Ohio River valley. The people in this village culture are famous for their large burial mounds, which begin to appear around 700 BCE.

c. 850 Chavin culture, based in Peru, with worship of part-human, part-animal beings, reaches its height. Grave goods include copper jewelry, decorated human skulls, and pipes for early tobacco use.

c. 800 Mayans begin to move from Central America into southern Mexico.

c. 400 Beginnings of Moche civilization in northern Peru.

c. 200 Nazca culture emerges in Peru; famed for "Nazca lines"—geometric and figurative designs etched into the surface of the Peruvian desert and attributed to aliens in pseudoscientific circles. Most likely, the lines were offerings made to the gods of sky and mountains.

c. 150 Great Serpent Mound in Ohio: 1,312 foot (405 meters) snake-like earthen effigy.

c. 150 Mayan "Golden Age" begins in Mesoamerica.

50 Teotihuacán in the Valley of Mexico is largest city in America.

Common Era

100 Emergence of Anasazi in southwestern North America.

Pyramids of the Sun and Moon constructed in Teotihuacán, Mexico, by unnamed civilization.

c. 420 Moche culture (Peru). Temple of the Sun constructed with 50 million bricks.

600 City of Palenque (Chiapas, Mexico) is constructed.

c. 700 City of Teotihuacán burns and is abandoned.

790 Decline of Maya civilization begins as many sites are abandoned.

c. 800 First use of bows and arrows in the Mississippi Valley.

900 Rise of Toltecs, warrior people from central Mexico, as Mayan empire collapses; they dominate central Mexico for the next 300 years.

987 Toltec priests are expelled from city of Tula (modern Hidalgo, Mexico) by a rival cult that favors human sacrifice.

990 Exiled Toltecs take over Maya city of Chichén Itzá on Yucatán peninsula.

c. 1000 Viking voyages to Newfoundland in North America; despite brief settlement, they leave no lasting impact on Native American culture or history.

Incas found Cuzco (Peru).

c. 1100 Fortified cliff dwellings of the Anasazi people are first built in southwestern North America.

c. 1175 Toltec empire destroyed by famine, fire, and invasion.

1200 Entry of people called Mexica (generally known as Aztecs) into Valley of Mexico. Originally a farming people from western Mexico who became mercenary warriors, they migrate to Valley of Mexico and settle on two marshy islands in Lake Texcoco. Begin to construct city of Tenochtitlán (site of present-day Mexico City) on one of the islands.

Toltec-Mayan city of Chichén Itzá is abandoned.

c. 1300 Anasazi pueblo villages in American Southwest are deserted, possibly due to climate changes.

1410 Inca empire of Peru expands.

1428 Aztec Empire expands.

1440 Moctezuma I is ruler of Aztecs. (Moctezuma's name is also spelled Montezuma or Motecuhzoma)

1487 Inauguration of great pyramid temple at Tenochtitlán; according to traditional acccounts, 20,000 people are ritually sacrificed there to "celebrate" the temple's completion.

1492 Christopher Columbus, on his first voyage in search of westward route to Asia, lands in the Bahamas, Cuba, and Hispaniola (Santo Domingo).

1496 Columbus establishes first Spanish settlement in Western Hemisphere.

1500 Portuguese reach Brazil and claim it for Portugal.

1502 Beginning of reign of Moctezuma II.

African slaves introduced to Caribbean.

1507 Waldseemüller's world map names newfound lands in honor of Amerigo Vespucci.

1508 Spanish settlers on Hispaniola enslave natives.

1509 Spanish settlement of Central America begins.

San Juan (Puerto Rico) is founded.

1513 Ponce de Leon claims Florida for Spain.

1514 Spanish force the conversion of natives to Christianity under the threat of death.

Spanish priest Bartolomé de las Casas begins to record the depraved behavior of Spanish colonists toward the natives.

1519 Hernando Cortés lands at Veracruz with 500 men; marches to the Aztec capital. Moctezuma II surrenders without a fight, is held captive and dies, probably executed by the Spanish, in 1520. The Spanish are later driven out by Aztec leader Cuauhtémoc. In 1521, Cortés returns with Indian allies and retakes Tenochtitlán following a smallpox epidemic that devastates the Aztecs. The Spanish level the city and begin to build Mexico City on the ruins. In 1522, Cortés becomes governor of New Spain. Cuauhtémoc, the last Aztec king, is hanged in 1524 on a charge of treason.

1526 Dominican monks arrive in Mexico.

1530 Portuguese begin to colonize Brazil.

1533 Francisco Pizarro captures Inca chief Atahualpa; orders his execution.

1541 Jacques Cartier founds a French colony at Quebec in Canada.

Pizarro is assassinated by rival Spaniards.

1545 Discovery of vast silver mine in Potosí (Peru); by the 1590s, Spain is exporting 10 million ounces of silver per year from the New World.

1550 First Jesuits reach Brazil.

1552 Bartolomé de Las Casas's scathing account of treatment of natives, *History of the Indies*, is published.

1570 Iroquois in northern North America form a league of tribes known as the Iroquois Confederacy.

1607 Foundation of first permanent English colony at Jamestown, Virginia. First African slaves arrive in 1619.

1620 Pilgrims arrive at Plymouth, Massachusetts.

BREAKING NEWS FROM THE NEW WORLD . . .

- In the Central American rain forests of Guatemala, in 2004, archaeologists uncover a royal palace beneath the thick tropical canopy. Inside a tomb within the palace ruins, resting on a stone platform, they find the body of a Mayan queen who reigned more than twelve hundred years ago. Her remains are surrounded by pearls and crown jewels, along with masterpieces of carved jade and artifacts that throw new light on an ancient people about whom there are still mysteries. The researchers who make this remarkable find say it may unlock many secrets of a magnificent civilization. (*New York Times*, May 11, 2004.)

- A tomb near Mexico's 2,000-year-old Pyramid of the Moon in Teotihuacán is uncovered in 2004, yielding the remains of ten headless human bodies, most likely sacrificial victims. This extraordinary, if grisly, discovery comes after some 200 years of excavations at the site of the first major city in the Americas. Located about 35 miles northeast of Mexico City, and home to an estimated 200,000 people in 500 CE, Teotihuacán mysteriously collapsed about 200 years later. With its massive Pyramids of the Moon and the Sun, it was called "the place where men became gods" by the Aztecs when they rose to power in Mexico in about 1400 CE. (Reuters, December 2, 2004.)

- In a case pitting scientists against Native Americans, a federal court rules in 2004 that the skeletal remains of "Kennewick man," discovered in Washington state in 1996 and the oldest human remains yet found in North America, can be studied for scientific purposes. Tribes from three states in the American Northwest had sued to prevent any further investigation of the 9,200-year-old skeleton under a law that requires the reburial of any Native American ancestral remains. The researchers hope that more extensive testing and study of "Kennewick man"—who was shown to be unrelated by DNA to any of the tribes—will shed more light on long-standing mysteries over who came to the Americas and when they arrived (*New York Times*, July 20, 2004).

These headline-making stories all point to the long, rich, yet still grossly misunderstood history and mythic traditions of the ancient Americas. Each new scientific advance and archaeological find makes it increasingly clear that our image of Native America has been tainted by the antiquated, "cowboys and Indians," Hollywood version. Or that we romanticize Native Americans as "noble savages" living in Eden-like spiritual and ecological harmony. Or, more recently, they have been depicted mostly as wealthy casino operators running gambling meccas on tribal lands. Needless to say, none of these views is accurate or complete, in part because there were—and are—so many different Native Americans.

Just as the people of Africa were a diverse group, the Native Americans were also a multicolored, many-voiced lot. Set apart by the Atlantic and Pacific Oceans and living on two large continents that span the Western Hemisphere, the native peoples of the Americas ranged from the Aztecs of Mexico, the Mayas of Central America, and the Incas of South America, to the vastly different tribes of North America. These tribes included the Sioux and other people of the Great Plains; the Navajos and Hopis of the Southwest; the "Five Civilized Tribes" of the Southwest; the "Iroquois Confederacy" of the Northeast; and the Aleuts and Inuits of the Arctic regions—to name just a few of the hundreds of Native American groups.*

The mythologies of the Native Americans were as rich and diverse as the people themselves. Full of nature deities, mischievous animal tricksters, heroic braves, and dueling twins, they all included a great, world-encompassing Creation story. This thinking led to a deep reverence for

*What to call these people who greeted Columbus? Of course, he called them *los indios*, thinking he arrived in India or the East Indies, and the name "Indian" has stuck, although not always happily for many of those so identified. The term still provokes disagreement in scholarly, legal, and other circles. For the purposes of this book, the term "Native American," while admittedly imprecise, is used. It is preferred by many living Native Americans, and is also used to avoid confusion with the "other" Indians of chapter 6. A "correspondent" on the very amusing *Daily Show* once clarified this confusion by asking a self-identified "Indian" if they meant "Gandhi Indian or Sitting Bull Indian." Using "Native American" seems a graceful and polite way of avoiding that rather dubious distinction.

nature and the concept of a benign Earth Mother. As Native American historian Vine Deloria puts it in his provocative book *God Is Red*, "For many Indian tribal religions the whole of creation was good, and because the creation event did not include a 'fall,' the meaning of the creation was that all parts of it functioned together to sustain it."

Presiding over most of these traditions—which were all preliterate except for the Mayas and Aztecs—was the shaman. This powerful figure supervised group chanting, healing, spirit communication, sweat lodges, and the pipe rituals aimed at connecting with the "Great Mystery." In the Mesoamerican civilizations, shamans formed a priestly class—like the Celtic Druids—that presided over blood rituals and human sacrifices. Not intended for squeamish audiences, these sacrifices included tearing or cutting out the still-beating heart from a victim's chest to appease the gods. This grotesque cruelty was sometimes performed on infants while their mothers looked on.

One of the great scientific and academic debates simmering today has to do with who the Native Americans were, how they got to the Americas in the first place, and how long they have been here. Usually, when you hear the expression "Early American," it refers to colonial-era antiques and life in the seventeenth and eighteenth centuries. But those "early Americans" were Johnny-come-latelies compared with the true "early Americans," who lived in the Americas thousands of years earlier. What you probably learned about their history in school goes something like this:

Near the end of the last great Ice Age, some 12,000 years ago, groups of nomadic hunter-gatherers crossed the 1,000-mile-wide land bridge connecting Siberia to what is now Alaska and Canada. This crossing took place before the great North American ice sheets melted and raised sea levels by some 300 feet, inundating the grassy steppe that allowed people and animals to move between Asia and North America. Probably pursuing large game such as mastodons, these really-early Americans spread out over the two continents and gradually diversified into the tens of millions of people who were present when the Spanish arrived in 1492.

But an array of new research from the worlds of archaeology, biology, and linguistics has shaken that notion to its permafrost foundation. It is now more widely accepted that the real "Pilgrims" may have come to the Americas 20,000 to 30,000 years ago, probably in successive waves of migrations carried out over a long time span, from Siberia, Mongolia, and other parts of Central Asia. Kennewick man, for instance, has been related—based on DNA evidence—to the Ainu, the prehistoric people who first inhabited Japan. Instead of strolling across the land bridge now covered by the Bering Straits, in one great prehistoric walkathon, perhaps Kennewick man and some of the other first Americans came in small, skin-covered boats, hugging the Pacific coastline down to the southernmost parts of South America. Gradually, these ancient people spread out across both continents, a process that continued for a long time.

The likelihood of many waves of immigrants would help explain the tremendous diversity of tribes, languages, myths, and civilizations the Europeans encountered when they arrived in the 1500s. Well before the beginning of the Common Era, tribes had spread out across North America, and impressive civilizations had begun to emerge in Central America and Mexico. These civilizations included cities that were larger, cleaner, and more organized than most European cities of the same period. Their inhabitants wrote hieroglyphic books and erected temples and pyramids as sophisticated staging areas for their religious rites. When the Spanish conquistadors arrived, these extraordinary accomplishments would be relentlessly and mercilessly plundered, and the Mesoamerican city-dwellers would experience death and destruction on an unimaginable scale.

To some degree, myth may have helped make that destruction possible. While some scholars now question the notion, many historians have long held that one reason a fairly small group of Spaniards was able to subjugate populations numbering in the millions lay in the story of an Aztec god named Quetzalcoatl. According to Aztec myth, Quetzalcoatl had departed from the Aztec people with the messianic promise of returning to usher in a new Golden Age. Supposedly, the notion that these white Spanish explorers might be the returning Aztec messiah helped win the Aztecs' initial welcome, if not their hearts and minds. The Spanish then used treachery, technology, and brutality as the real means to conquest. European diseases, against which these Native Americans had no natural immunities, did the rest of the dirty work.

Far more significant, myth played a role in crushing religious traditions, which contained striking similarities to Catholicism. Powerful images of death, penitence, self-mortification, blood sacrifice, and a dying-and-reborn god—many of them central to Catholicism—pervaded the Aztec, Mayan, and Incan traditions. These parallels were not lost on the Spanish soldiers and the priests who followed them. Professing concern for the Mesoamerican soul, the Spanish co-opted native beliefs and set about their real goal—acquiring massive tracts of land and seizing as much gold and silver as they could for Spain's royal coffers.

The path to exploitation in North America was a variation on a theme. Unlike the fairly swift, relentless Spanish conquest of the natives of Mexico and South America, the Europeans moved more slowly in North America. They had to. First, there were hundreds of tribes and groups to conquer, across a huge and largely uncharted space. While some of these people lived in sophisticated, organized settlements, others occupied remote places or were on the move, following the buffalo and better weather. The Europeans had to work longer at dropping a moving target that quickly learned the value of fighting back and was often quite good at it.

But in the end, the song remained the same. What took place all over North America was a grotesque *Groundhog Day*, in which identical awful things happened over and over, minus a happy ending. The pattern was this simple: The Europeans arrived and were welcomed and often aided by the natives. Once the conquerors had promised to keep the peace, they aggressively expanded, broke treaties, declared war, pitted one tribe against another, and unwittingly (for the most part) introduced diseases that nearly exterminated the native people, along with their language, mythic traditions, and sacred beliefs. In the four hundred years between Columbus's arrival and the massacre at Wounded Knee in 1890, upward of 90 percent of an estimated 40 to 100 million people were wiped out in one of the largest "ethnic cleansings" the world has ever witnessed—all in the name of progress, "civilization," and the God of Christianity.

The story of what was lost is still being written as new discoveries are made each day. A palace with a Mayan princess is found. Sacrificial victims in an ancient tomb are unearthed. The skeletal remains of an early man are opened up for study. And a new generation of scholars, eager to retell this tale from the "loser's" point of view, continues to stir up the long-accepted histories and theories of what happened to the "primitive" people of the New World. As a result, each day new light is shed on the vibrant mythic traditions of the Americas.

Mythic Voices

They must be good servants and very intelligent, because I see that they repeat very quickly what I told them, and it is my conviction that they would easily become Christians, for they seem not to have any sect. If it please our Lord, I will take six of them that they may learn to speak. The people are totally unacquainted with arms, as your Highnesses will see by observing the seven which I have caused to be taken in. With fifty men all can be kept in subjection, and made to do whatever you desire.

—Christopher Columbus,
*from his diary, October 12, 1492**

The tribe has no belief in God that amounts to anything; for they believe in a god they call Cudouagny, and maintain that he often holds intercourse with them and tells what the weather will be like. They also say that when he gets angry with them, he throws dust in their eyes. They believe furthermore that when they die they go to the stars and descend on the horizon like the stars. . . . After they had explained these things to us, we showed them their error and informed them that Cudouagny was a wicked spirit who deceived them, and that there is but one God, Who is in Heaven, Who gives everything we need and is the Creator of all things and that in Him alone we should

*The original native cultures of the Caribbean Islands were so devastated that there is very little information with which to assess them and their myths. Since the slave era, with its huge influx of Africans, these areas were dominated by the fusion religions discussed in chapter 8.

believe. Also that one must receive baptism or perish in hell. . . .

—*French explorer* JACQUES CARTIER *(1491–1557), describing the Hurons of eastern Canada*

How did Native American myth go up in smoke?

What is "civilized"? What is "savage"? To the "civilized" Europeans who came to the Americas in the 1500s, the answer was simple. "Civilized" meant clothes-wearing, literate, European Christians. The word "savage" meant "Indian." Led by medicine men smoking pipes, having visions, curing with herbs, and rejecting the white man's "salvation," the native tribes were, in the European view, doomed souls. Unfortunately, that view dominated from the sixteenth century on.

Maybe that is why the public today is still largely in the dark about Native American mythology and beliefs. The Europeans—and later, Americans—didn't just crush the Native American "savages." They composed the diaries and letters, painted the artwork, took the photographs, and wrote the first histories that either ignored or demeaned a conquered people. They suppressed native mythologies and languages to near-extinction, allowing them to go up in the smoke of burning villages. Church schools, missionaries, and government agencies like the notorious Indian Affairs Bureau added to the catastrophe, forcing native children to accept "Anglo" names and denying them the right to speak their mother tongues or learn their ancestral sacred stories. Sacred native places were built over and renamed—a process still going on as a Wal-Mart outlet goes up near Teotihuacán in Mexico, or an astronomical observatory is placed on a mountain sacred to the San Carlos Apaches in the state of Arizona.* People who shrug this off might see the issue dif-

*That one of the participants in the construction of this telescope was the Vatican Observatory is, of course, a delicious irony. Four hundred years earlier, the same Vatican was putting Galileo on trial for looking through a telescope. Pope John Paul II had also spoken in Arizona, in 1989, of the importance of native people's maintaining their customs. The story of the observatory is told in *Sacred Lands of Indian America*.

ferently if Native Americans secured rights to build a gambling casino atop Arlington National Cemetery.

But there is more to the story. As the authors of *The Encyclopedia of Native American Religions* write, "Indian country in North America is still home to hundreds of religious traditions that have endured, despite a long history of persecution and suppression by government and missionaries. . . . Native American sacred beliefs are as dignified, profound, viable, and richly faceted as other religions practiced throughout the world. Native sacred knowledge has not been destroyed or lost but in fact lives on as the heart of Native American cultural existence today."

What is known of these sacred traditions today comes largely through efforts during the past century to interview native survivors who preserved an oral tradition. These survivors included Black Elk, whose 1932 memoir, *Black Elk Speaks*, records the visionary recollections of an Oglala Sioux holy man who witnessed the nineteenth-century spiritual revival called the "Ghost Dance" and the massacre at Wounded Knee. Memories such as his have been added to the handful of written sources that do exist, such as the Popol Vuh, a re-creation of sacred Mayan writings discovered 300 years ago in a Guatemalan church, and a few hieroglyphic books from the Aztecs and Mayas. From the Spanish colonial era, there is also a large library of works about native beliefs, but these are somewhat suspect, given their source—often priests, or natives who may have wanted to curry favor with their Spanish masters, or deceive them.

More recently, the effort to preserve Native American traditions has been invigorated by a generation of scholars far more sensitive to their subject. In September 2004, more than 20,000 members of some 500 Native American tribes gathered in Washington, D.C., to celebrate the opening of the National Museum of the American Indian. Built at a cost of more than $219 million, the museum housed the Smithsonian Institute's hundreds of thousands of Native American objects.* There is also a revival of interest in tradition among young Native Americans, who hope to save something of their past as a complement to their height-

*Expected to attract 600,000 visitors each year, the museum is not a cause for celebration among some Native Americans, including those who wanted to call it the "Museum of the American Indian Holocaust."

ened political activism in both the United States and Latin America. Award-winning poets, short-story writers, and novelists such as Leslie Marmon Silko, Louise Erdrich, and Sherman Alexie have joined in the rescue effort with creative works such as *Ceremony*, *Love Medicine*, and *Tonto and the Lone Ranger Fistfight in Heaven*, which explore tribal mythic traditions and their impact on contemporary Native Americans. Finally, archaeology and other scientific research have also added immensely to a picture that has been rescued from the ashes of the Native American holocaust.

MYTHIC VOICES

You said that we know not the Lord of the Close Vicinity, or to Whom the Heavens and earth belong. You said that our gods are not true gods. New words are these that you speak; because of them we are disturbed, because of them we are troubled. For our ancestors before us, who lived upon the earth, were unaccustomed to speak thus. From them we have inherited our pattern of life, which in truth they did hold; in reverence they held, they honored our gods.

—*Aztec scholars to the first Franciscans in Mexico City, 1524*

Is there an "American" mythology?

Vast differences distinguish the many cultures and people once lumped together as "Indians." The Cherokee farmers of the Southeast were very different from the Great Plains Sioux who followed the buffalo herds that sustained them. The democratic Hodenosaunee (or Iroquois) of the long houses in the Northeast had little in common with the Pueblos in their adobe "apartment buildings" in the Southwest. And none of these native people could be confused with the city dwellers of Mexico and Central America, or the resilient oceangoing fishermen of the Northwest and Arctic. But there are common patterns among the beliefs of many

Native Americans, which, some scholars think, may stretch all the way back to shared prehistoric origins in Asia. These characteristics include:

- A "Great Spirit." A supreme god with ultimate power who both has created and oversees the universe is a common feature in Native American mythology. Often male and usually related to the sun, the great spirit goes by many names, as David Leeming and Jake Page point out in *The Mythology of Native North America*. "Usually this supreme god—the Great Spirit, the Great Mystery, Father Sky, Old Man, Earthmaker, or one of several other names—is the prime creator."

  To the Huron of the Northeast woodlands, the Creator god is Airsekui, to whom the tribe offers the first of its fruits and meats each harvest and slaughter time. To the Incas of South America, he was Inti, the godhead who also founded the Inca dynasty. To the Sioux, Osage, and other tribes of the midwestern plains, the Creator god is more of a force than a personalized deity, and is called Wakonda or Wakan Tanka. Wakonda is the force behind all life and creation, wisdom, knowledge, and power. Sometimes envisioned as a large bird, Wakonda sustains the world and gives authority to the medicine men. For Algonquian tribes (who range across North America from the East Coast through the Great Lakes to the Rockies), the high god is Kitchi Manitou or "great mystery," a divine energy that created the world by thinking of it, exists in all things, and can be sought—in almost Eastern mystical terms—to achieve selfhood.

- An Earth Mother. Source of all fertility, the Earth Mother is a popular deity in the Americas, who is nearly always a nurturing force. In the myth of the Cherokee—originally from the southeastern United States and forced west to Oklahoma in the notorious "removals" of the 1830s—their Earth Mother is the goddess known as Grandmother Sun. To the Hopi of the Southwest, she is Spider Woman or Kokyanwuuti, the goddess of Creation who teaches the people how to weave and make pottery.

  Often these Earth Mothers or great goddesses have twin children or grandchildren—another common Native American theme—who are frequently tricksters. One Earth Mother of twins is the main goddess of the Navajos of the Southwest, who call themselves Diné ("the people"). Born from a piece of turquoise and made pregnant by the sun god, their Earth Mother is known as Changing Woman, or Estsanatlehi. She is a miraculous birth-giver, whose twin children—Monster Slayer and Born for Water—make the world safe for the Navajos. Changing Woman creates people from a mixture of corn dust and skin from her breasts. Growing old and young in a never-ending cycle, Changing Woman lives on an island in the west, from which she sends life-giving rain and fresh winds to keep the people alive. One of the most important rites among the Navajo is the female puberty ritual, a four-day ceremony in which a girl becomes a woman and gains the healing power granted by Changing Woman.

- "Earth Diver" Creation Stories. Probably the most prevalent and archetypal Native American Creation story, especially in North America, features an animal—typically a beaver, beetle, duck, or turtle—who plunges into the waters covering the earth and returns to the surface with bits of mud or soil, from which the Creator then makes the earth.

  To the Yuchi and Creek of Alabama and Georgia, the earth diver is Crawfish, who goes to the bottom of the water where the Mud People live. Angry that Crawfish comes and goes, stealing their mud and constantly stirring up the water, the Mud People try to stop him, but he moves too fast. Buzzard soars over the mud and dries it out with his wings, making the mountains and valleys. Finally, great mother (the Sun) gives light to the world and drips her menstrual blood on earth, giving birth to the first people.

  For the Seneca of the Northeast, the earth divers Toad and Turtle work together to create land. They do this after Star Woman, the daughter of the Sky Chief, falls through a hole in the sky. Caught by birds, she rests on Turtle's back until Toad brings up enough soil from beneath the water to create the earth for her to live on.

- Tricksters. Like Africa's mythology, Native American traditions show a special fondness for the malevolent and often aggressively oversexed trickster—animal gods such as Coyote and Hare, or a man-animal like Iktomi, the Spiderman of the Lakota Sioux. The Aztecs of Mexico have a lusty collection of tricksters called Centzon Totochtin, or "Four Hundred Rabbits." And classic Maya pottery also depicts a rabbit stealing—in true trickster fashion—an unidentified old god's hat and clothes. Some of these tricksters were the inspiration for two modern American cartoon icons—Wile E. Coyote of *Road Runner* fame and a "wabbit" named Bugs, whose animated antics are far less malicious and X-rated than those of their ancient ancestors.

Sometimes the tricksters are cunning "culture heroes," like the Mayan twins of the Popol Vuh who confound death with their amazing abilities. Many scholars believe that the now-ubiquitous Kokopelli—the hunchbacked flute player depicted on prehistoric Anasazi rock art in the Southwest—was a combination trickster-fertility god, similar in many respects to the Greek Pan. Kokopelli may have even been based on legends of an actual trader from Central America who made his way to the South and left a lasting impression. Other Native American tricksters are smart, brave, and resourceful—but they can also be vindictive, spiteful, and selfish. In some accounts, tricksters create man, steal fire from heaven, survive floods, and defeat monsters.

Summarizing tricksters, Native American myth expert Richard Erdoes writes, "Always hungry for another meal swiped from someone else's kitchen, always ready to lure someone else's wife into bed, always trying to get something for nothing, shifting shapes (and even sex), getting caught in the act, ever scheming, never remorseful." Tricksters are, he adds, "clever and foolish at the same time, smart-asses who outsmart themselves."

- The Shaman. A figure revered in many worldwide traditions, the shaman, or "medicine man,"* plays a central role in many tribes throughout the Americas. Widely thought to be a carryover from the ancient Siberian beginnings of Native America, and sometimes related to the trickster, the shaman in Native American tradition is the person considered to have magical powers that come from a direct contact with the supernatural, usually through ecstatic trances

*The word "medicine man" was first applied to many Native American religious leaders by the English in the seventeenth century, according to *The Encyclopedia of Native American Religions*.

or dream visions. Shamans were often healers who used a combination of herbal remedies and "spiritual" healing—traditions that continue today across the Americas.

American tribal names for the shaman vary, and it is not a term used by American tribal people. For the Arctic Inuit, the shaman or medicine man is Angakoq, who is the repository of lore and magic and the actual connection to the spirit world. To the Oglala Sioux, the shaman is a *wichasha wakon*, a holy man like Black Elk, who, at the age of nine, had a powerful spiritual vision. Serving as tribal priests, diviners, and healers, the shamans underwent training that usually included a "vision quest," in which the initiate sought to communicate with the spirit world. The shaman's apprenticeship might last anywhere from a few days to many years, as novices had to experience extreme hardships to learn how to control "spirit helpers." In Peru, centuries after Catholicism was established, the Church continued to "investigate" what it called *idolatrias* ("idolatries"), which involved curers and diviners who persisted in the traditional worship of sacred Incan places in the mountains. Like many Native American tribes and cultures, the Incas regarded many places and things as *huaca* ("sacred"). These included springs, stones, caves, and mountain peaks, each of which had its own spirits.

- The Totem. Somewhat unique to Native Americans, but similar in some respects to the African fetish, the totem is a symbol of a tribe, clan, or family. But it is also an object imbued with spirit power. As writer Jonathan Forty describes it in *Mythology: A Visual Encyclopedia*, "The totem was a coat of arms, an altar, shrine, flag, and a family tree all rolled into one."

  Although "totem" is most often associated with the great carved poles that lined the village streets of the Northwest and Alaskan tribes, the word comes from the Chippewa (or Ojibwa) of the Great Lakes area. In the broader sense, "totem" is a powerful symbol that united people who sometimes occupied vast territories. In discussing the importance of the totem to people in what he calls the "primal world," religion authority Huston Smith writes in *Illustrated World's Religions*, "To be separated from the tribe threatens them with death, not only physically but psychologically as well. The tribe, in turn, is embedded in nature so solidly that the line between the two is not easy to establish. In the case of totemism, it cannot really be said to exist. Totemism binds a human tribe to an animal species in a common life. The totem animal guards the tribe, which, in return, respects it and refuses to injure it, for they are 'of one flesh.'" That idea is so powerful that a clan might have rules against killing or eating the species to which their totem—a bird, fish, animal, plant, or other natural object—belongs. (Maybe you have a "Jesus fish" on your car? Or wear your favorite team's tiger, wildcat, or cardinal on your cap? Or maybe you pledge allegiance to a flag with an eagle on top? Totem, totem, totem.)

In the Pacific Northwest, highly skilled artisans carved the family and clan emblems on the elaborate cedar totem poles that eventually came to be viewed as "status symbols." Captain Cook, the English explorer, saw these totems during his travels in the Pacific Northwest and noted them in his journals in the 1700s. The famed photographer of the Native Americans Edward S. Curtis first took pictures of them in the late nineteenth century as part of an expedition to Alaska led by railroad magnate E. H. Harriman, who stripped entire villages of their totems and other sacred objects. But the early history of totem poles is otherwise obscure, except for legends hinting that they go very far back in time.

In a flagrant example of attempted "culturecide," in 1884, the Canadian government outlawed the large ceremonial gatherings called the "potlatch," at which totem poles were raised. Many native children were then sent to government schools, and totem pole carving nearly died as an art form. There has been a revival in recent decades among a younger generation of artists who want to preserve the old ways.

By the way, the "low man on the totem pole" usually wasn't. In fact, the bottom figure was often created by the best carver, who wanted his work to be most visible.

THE MYTHS OF THE MAYAS

The Mayas once occupied an area that today consists of the Mexican states of Campeche, Yucatán, and Quintana Roo; part of the states of

Tabasco and Chiapas; and most of Guatemala, Belize, and parts of El Salvador and Honduras. Recent discoveries show that the Mayan civilization began to reach its peak as early as 150 BCE and grew vibrantly until 900 CE. By then, most of the Mayas had moved to areas to the north and south, including Yucatán in Mexico and the highlands of southern Guatemala, where they continued to prosper until Spain conquered most of their territory in the mid-1500s. Descendants of the Maya still live in Mexico and Guatemala—where they are among the world's poorest people. They speak Mayan languages and retain many of the religious customs of their ancestors.

MYTHIC VOICES

> **There is not yet one person, one animal, bird, fish, crab, tree, rock, hollow, canyon, meadow, forest. Only the sky alone is there; the face of the earth is not clear. Only the sea alone is pooled under all the sky; there is nothing whatever gathered together. It is at rest; not a single thing stirs. It is held back, kept at rest under the sky.**
>
> **Whatever there is that might be is simply not there: only the pooled water, only the calm sea, only it alone is pooled, Whatever there might be is simply not there: only murmurs, ripples, in the dark, in the night. Only the Maker, Modeler alone, Sovereign Plumed Serpent, the Bearers, Begetters are in the water, a glittering light. They are there, they are enclosed in quetzal feather, in blue-green.**
>
> —*from Popol Vuh, translated by Dennis Tedlock*

What is the Popol Vuh?

The Holy Bible of Judaism and Christianity is a book of Creation, a list of divinely ordained rules and rituals and a foundation history of the Hebrew people, which includes a list of ancient Israel's many legendary and real kings. Perhaps most important, the Bible is believed to be the word of God.

The Mayan Popol Vuh is a book of Creation, a list of divinely ordered rules and rituals and a foundation history of the Mayan people, which gives the Mayan kings a heavenly mandate and links them to a list of legendary rulers. Perhaps most important, the Popol Vuh was believed to be the word of the gods.

So, the Bible and the Popol Vuh have some things in common. Both books were the sacred texts at the core of their culture's religious traditions. Both were written down by scribes only after centuries of oral transmission. Both contain poetic accounts of Creation and grim stories of death and destruction. Yet, most people have never heard of the Popol Vuh. There aren't Popol Vuh study courses offered in most local colleges. Or pamphlets with scriptural excerpts from the Popol Vuh. Or concordances published for easy referencing of the Popol Vuh. To a significant degree, the book is a well-kept secret.

Once again, we have the Spanish conquerors to thank. After their arrival in the Americas in the 1500s, the Spanish—as many conquerors do—prohibited the use of the Mayan and other native languages and began to enforce the use of Spanish and Latin as the common vernacular. Of course, Catholicism became the official—and, presumably, only—religion wherever the Spanish went. And wholesale "culturecide" began to take place. In an introduction to his English translation of the Popol Vuh, scholar Dennis Tedlock describes the Spanish approach to destroying a culture: "Backed by means of persuasion that included gunpowder, instruments of torture, and the threat of eternal damnation, the invaders established a monopoly on virtually all major forms of visible public expression, whether in drama, architecture, sculpture, painting, or writing. In the highlands, when they realized that textile designs carried complex messages, they even attempted to ban the wearing of Mayan style clothing." (Oh, those terrible Spanish. But just remember, in America you can be thrown out of a shopping mall for wearing a T-shirt with a message the authorities don't like. And the French banned the head scarves worn by Muslim girls in public schools. Clothing is, and always has been, a form of spiritual, cultural, and political expression.)

During the mid-sixteenth century, working secretly and anonymously, Mayan priests and clerks who had been taught Latin translated copies of the old hieroglyphic Mayan books into Latin. They also began to blend Catholicism in with their own religious beliefs, merging the two much the way the African practitioners of Santeria and voodoo did in the Caribbean. Around 1700, a Latinized version of ancient Mayan texts was discovered by a Franciscan priest in a Guatemalan town. Instead of destroying the book—which is what happened to most ancient Mayan and Aztec hieroglyphic writings—the priest translated it into Spanish and added the names of some of the Spanish governors of Guatemala to the lists of Mayan kings. Maybe the priest thought this addendum would keep him out of ecclesiastical hot waters if his heresy was ever discovered.*

This, then, is how the Popol Vuh survived. Coming down from Mayan scribes who valued the "ancient word" over the "preaching of God" forced on them by the Spanish, the Popol Vuh is a rare and important source, although one that is clearly filtered through the Spanish colonial era.

Divided into five parts and a little over one hundred pages long in English, the Popol Vuh begins with a Creation account in a world that has only an empty sky above and a sea below. Central to this Creation narrative are two groups of gods, one from the sea and one from the sky, who decide to create the earth, plants, and people. The role of the people, interestingly, is to praise the gods and provide them with offerings. The first people the gods make have no arms and can only chatter and howl—so they become the first animals. A second try produces a being made of mud, which cannot walk or reproduce and which dissolves into nothing. After consulting a wise old divine couple, the gods make a third attempt and create people out of wood. But the results are only slightly improved. The wooden people can speak and reproduce, but they prove to be very poor at praying and providing the requisite offerings.† The god Huracan—a name appropriated by the Spanish and transformed into the word "hurricane"—decides to do away with the wooden people with a flood, and he sends a gigantic rainstorm along with terrible monsters to

*That copy of the Popol Vuh is now the property of Chicago's Newberry Library. Four other Mayan books written in hieroglyphics survive. Each is known as a "Codex," and they are in various libraries around the world.

†This creation account parallels both Greek and Chinese Creation stories, which also feature several generations of flawed attempts at making humans.

attack them. The people are destroyed, but some manage to survive in the jungles and become the ancestors of the monkeys.

After this round of botched human creation, the Popol Vuh shifts to a long, complex, and admittedly bizarre narrative account of the Maya's two sets of semidivine national heroes, the twins One Hunahpu and Seven Hunahpu, and another set of twins, named Hunahpu and Ixbalanque—the slayers of demons who defeat the gods of the underworld. The extraordinary adventures of these two sets of twins have all the makings of a modern hit video game. A heroic quest into a multilevel underworld filled with demons with names like Scab Stripper, blowguns, monsters, deadly bats, and frequent decapitations all occupy the center of the elaborate Mayan tales described in the Popol Vuh. (See page 459, *Which gods like a good ball game?*)

Who were the Mayas who produced the Popol Vuh?

Peeking inside the Popol Vuh raises the question—how did the civilization that produced this extraordinary book evolve?

The simple answer is "farming." Civilization in Mesoamerica—the area now known as southern Mexico and much of Central America—began when people shifted from hunting-gathering to farming. Creating small villages in clearings in the rain forest, the first farmers of Mesoamerica raised tomatoes, peanuts, avocados, tobacco, beans, squashes—many plants and foods unknown in Europe until Columbus carried them back. But these farmers' most important crop was maize, which evolved from a tamed wild grass. Commonly known as corn, maize would become the staple of the Mesoamerican diet, feed its herds and support large populations.* By about 1700 BCE, improved farming techniques were producing maize in surplus quantities—an economic necessity for developing a more advanced civilization.

The first major Mesoamerican civilization, the Olmecs, had settled along the Gulf Coast of Mexico, in what is now Veracruz and Tabasco,

*Europeans knew nothing of corn until Columbus brought some corn seeds from Cuba back to Spain. By the late 1500s, corn had become a well-established crop in Africa, Asia, southern Europe, and the Middle East.

around 1500 BCE. Clearly poised to achieve great things, the Olmecs created, within a few hundred years, a fairly sophisticated society with temples, pyramids, a single ruler, and a powerful religion-based culture that spread throughout Mesoamerica. One of their distinguishing accomplishments were the great carved heads of supernatural beings and animal deities produced out of large basalt stones that weighed up to 36,000 pounds (16,300 kilograms). These massive stone blocks had to be transported more than 50 miles (80 kilometers) through difficult terrain, without benefit of the wheel or large draft animals, and were most likely rafted on rivers. Discovered at Olmec sites at La Venta in Tabasco and San Juan Lorenzo in Veracruz, five of these colossal heads can now be seen at an outdoor park in La Venta, and others are displayed in museums, including the National Museum of Anthropology and History in Mexico City and the Veracruz Museum of Anthropology in Xalapa.

The Olmec religion, sculpture, and other arts would significantly influence all the later Mesoamerican groups, including the next great Mesoamerican civilization, the Mayas. Eventually producing cities with towering pyramids and broad plazas centered mostly in Guatemala and southern Mexico, Mayan civilization, new research shows, began to reach a level of complexity as early as 150 BCE—much earlier than previously thought—and reached the peak of its development about 200 CE, then continued to flourish in a "classic period" for hundreds of years, declining around 900 CE.

Although not an empire in the usual sense, the Mayas formed a loose collection of about ninety city-states, with several different languages. But theirs was one of the first cultures in the Americas to develop an advanced form of writing, a hieroglyphic picture-language that was used to record the sacred texts. The Mayas also made great strides in astronomy and mathematics, developed an accurate yearly calendar, and produced remarkable architecture, painting, pottery, and sculpture. By about 900 CE, for reasons unclear, most of the Mayas abandoned the cities of the Guatemalan lowlands. Invasion and changing climates may have been the cause, but some of the Mayas moved south and others north, to the Yucatán Peninsula. There, between 900 and 1200 CE, the city of Chichén Itzá grew into the largest and most powerful Mayan city.

Like all Mayan cities, Chichén Itzá was a religious center with a temple where a priestly class resided and served the needs of surrounding rural populations. Each day, priests performed the daily sacrifices and rituals. Farmers from nearby villages would come to the city to attend regular festivals that included dancing, competitions like the ball game (see below), dramas, prayers, and sacrifice. Sometimes these sacrifices involved ordinary foods. But human sacrifices were also demanded to appease the gods.

Governed by a council of nobles, Chichén Itzá dominated Yucatán with a combination of military strength and control over important trade routes until it declined around 1200. For the next two hundred years, the Yucatán was divided by civil wars, and the Mayas later merged with the Toltecs, a warrior group that moved in from northern Mexico. In the early 1500s, the Spanish invaded the Mayan territories, and by 1550 had overcome almost all the Mayas, enslaving those who survived on the large-scale plantations they began to build.

Which gods like a good ball game?

Finally, a myth that every red-blooded sports fan can love! It involves a ball game. Today, people often speak of the mythic accomplishments of certain athletes—like Babe Ruth's prodigious home runs promised to a sick boy. But great athletes were actually a part of the Mayan mythology. Along with other Mesoamericans, the Mayas passionately played a sport that takes center court—literally—in the Popol Vuh.

The Mayan sport, simply called "the ball game," was more than just a game. Combining ritual elements with Super Bowl–level excitement, the sport was played on a ball court with two walls. The largest such court found in the ancient Mayan world is at Chichén Itzá and measures 140 by 35 meters (approximately 153 by 38 yards, or longer and narrower than a typical international soccer field or American football field). The courts featured two steeply sloping parallel stone walls inset with round disks or rings set high on the walls at right angles. Two teams competed in a contest to pass a rubber ball through such a ring. Other versions of the game included markers that could also be hit to score points. The game was probably invented by the Olmecs, who were the first to cultivate the rubber tree and whose name came from Aztec and meant "the people who use rubber."

Sounds easy—like basketball. But the tricky part was that the ball couldn't touch the ground, and had to be hit off the walls, using only the elbows, knees, or hips. A single score—or the ball touching the ground—usually ended the match. So, winning must have been difficult—but losing was even harder. "Sudden death" in this ball game could be literal, since the leader of the losing team sometimes became a sacrificial victim. It didn't happen at every ball game but often enough, as human sacrifice was essential in the Mesoamerican religions. And modern professional coaches think they are under a lot of pressure!

The central importance of the ball game is underscored in a story in part 3 of the Popol Vuh, in which two different sets of heroes are very good "ball players," who play an "under-World Series" with the gods of death.

The first set of twins, One Hunahpu and Seven Hunahpu (their names are actually dates from the elaborate Mayan calendar), are playing ball one day, but the noise of their game annoys the lords of Xibalba (hell), One Death and Seven Death, who invite the brothers to the underworld for a game. Among the other lords of the Mayan underworld are the charmingly named Scab Stripper, Blood Gatherer, Demon of Pus, and Demon of Jaundice. When the twins arrive and play the game, the lords of hell flat out cheat, kill the brothers, and decapitate them—losing heads is a recurring theme in the Popol Vuh.

One twin's head is placed in a tree as a warning not to mess around with the lords of hell. But Blood Moon, the daughter of Blood Gatherer, is fascinated by the head and is even more surprised when it speaks to her. The head tells the girl to put out her hand, and then spits on it. This makes Blood Moon pregnant. Her father is so angry that he calls for her sacrifice. But she conspires with the messengers sent to sacrifice her to use a false heart, just as in the story of Sleeping Beauty, when the hunter who is supposed to kill her offers an animal heart as proof instead. Blood Moon seeks out her mother-in-law on earth, where she gives birth to the hero twins, Hunahpu and Ixbalanque.

It is clear from the start that these twins are magical, because they grow up fast, are unusually expert hunters, and can perform all sorts of miracles, like killing monsters. One day they discover their father's ball-playing equipment and decide to play. Like their father and uncle before them, they disturb the gods of Xibalba and are summoned to the underworld and given a series of challenges by the lords of hell. Unlike their ancestors, however, these twins are able to outwit the lords of hell. Supreme tricksters, they meet every challenge put to them. At last, they are placed in a Bat House. A bat swoops down and beheads Hunahpu, and his head is used as the ball in the next ball game. But Ixbalanque switches his brother's head with a squash, and Hunahpu is restored to life. When the lords of death realize they have been tricked yet again, they decide to burn the twins.

With the help of a magician, the twins willingly jump into a pit of fire and are resurrected five days later. They then begin to travel the land as magicians. Hearing of their wonderful magical skills, the lords of hell order a command performance, and the twins amaze them with a series of dismemberments and decapitations of animals and themselves—from which they are able to recover! Seeing this miracle, the lords of hell want to become part of the act and ask the twins to kill and then restore them. The heroes readily agree, but then don't bring the lords of hell back to life. In this way, death is defeated, giving hope to mankind. For this great service, the twins are rewarded by being made the sun and the moon.

As the Popol Vuh, with its streams of blood and frequent decapitations, proves, Mayan myth wasn't all games. Obsessed with images of death, Mayan religion included sacrifice. In their cities, built essentially for ceremonial purposes, the Mayas constructed limestone pyramids topped with small temples where priests performed the gory rituals. The gods, whose help was required to continue the cycles of nature and to ensure fertility, demanded nourishment. To obtain the help of the gods, the Mayas fasted, prayed, and offered sacrifical deer, dogs, and turkeys.

The Mayas frequently offered their own blood as well, and sometimes their blood sacrifice involved a priest or noble person piercing the tongue, penis, ears, lips, or other body parts, which they spattered on pieces of bark paper or collected in bowls. Some occasions called for the living heart of a victim to be cut out in sacrifices performed at the top of the pyramid. In a culture that believed the world had been created five times and destroyed four, and would be destroyed again, this was part of the balance. For most people, death meant Xibalba, or hell. Heaven was reserved for those who had died in childbirth or battle, or those who were hanged or were offered as sacrifice. The ideas of penitence, fasting, abstinence, a world-ending flood, and a tortured, dying god were all part of the Mayan traditions—which made them fertile ground for the Catholic religion.

WHO'S WHO OF MAYAN GODS

The Mayan pantheon was quite vast. These are among its most significant gods.

Ah Puch Depicted with a skeleton's exposed ribs and the face of death—or as a bloated corpse— Ah Puch is unmistakable. He is the "lord of death," who visits the homes of the sick and dying to snatch them away to the kingdom of the dead. A later name for him is Cizin, "the flatulent one." Ah Puch is apparently still feared today by modern descendants of the Mayas in Guatemala, who call him Yum Cimil.

Chac Portrayed as a warrior whose tears brought rainwater to earth, Chac is a rain god, an agricultural and fertility deity, and one of the longest continuously worshipped gods of Mesoamerica. Responsible for bringing maize (corn) to the people by opening a stone in which the first maize plant was hidden, Chac is often worshipped as four separate but beneficial gods, one for each point of the compass. In the ritual that required the sacrifice of a human victim and the removal of his still-beating heart, the four men who assisted the priest were called *chacs*.

Hunab The remote Creator deity, Hunab renews the earth three times after flooding it. Once he repopulates earth with dwarfs, the second time with an obscure race, and finally with the Mayas, who are destined to be overcome by a fourth flood. Hunab may also have been the father of the chief god Itzamna.

Itzamna The greatest deity of the ancient Maya, Itzamna is lord of the heavens—the god of day, night, and moon, who brings writing, religious rituals, and civilization to the Mayas. Far from an awesome Zeus-like figure of power and glory, Itzamna is portrayed as a wizened, toothless old man. But don't be misled. Itzamna is also lord of medicine, with healing powers that allow him to banish fatal illness and raise the dead.

According to some scholars, Itzamna is never responsible for anything bad—unlike his wife, **Ixchel** or Lady Rainbow, who is loathsome and frightening. Depicted as an angry old woman with great power, Ixchel is the goddess of pregnancy, midwifery, and childbirth, and can tell the future. But she is also the storm goddess, who creates disastrous rains and floods—presumably the connection to her Rainbow epithet. She is often depicted wearing a skirt decorated with crossed bones, and a snake on her head. This snake is the Sky Serpent, which contains all the waters of heaven in its belly. In most artistic renderings, Ixchel holds a water jug, the vessel of doom, from which she can pour a destructive torrent at any time.

Ixtab An unusual goddess, Ixtab is often depicted hanging from a tree, partially decomposed, and is said to be the goddess of suicide, who takes the souls of those that die by hanging to eternal rest. The Mayas were preoccupied with death, especially violent death, and may have believed that suicide was an honorable way to enter the afterworld. Ixtab takes the souls of suicides, fallen warriors, sacrificial victims, and women who die in childbirth, to eternal rest.

Kinich-Ahau The ancient Mayan sun god Kinich-Ahau takes different forms in much the way Egypt's Re does. As Kinich-Ahau travels across the sky during the daytime, he appears old and young. During the nighttime, he is transformed into the jaguar god. The largest and most powerful Central American cat, the jaguar was feared and admired by the earliest people of Mexico and is one of the region's oldest gods. Jaguar also rules the underworld and is a symbol of power, fertility, and kingship. In order to show that they possessed these qualities, Mayan priests typically wore jaguar skins.

Pauahtun A god with four incarnations, Pauahtun stands at the four corners of the world holding up the sky. In spite of this very important job, he is thought of as a drunkard and the unpredictable god of thunder and wind.

THE MYTHS OF THE AZTECS

The name "Aztec" is widely but somewhat inaccurately applied to the people who settled in the Valley of Mexico sometime in the 1200s and founded the city of Tenochtitlán, on the site of present-day Mexico City, in 1325, according to Aztec traditions. A huge, oval basin about 7,500 feet (2,300 meters) above sea level, the valley is in the tropics but has a mild climate because of its altitude. Technically, all of the people speaking a language called "Nahuatl" in the Valley of Mexico are "Aztec," while the tribe that came to dominate the area was a group called the Tenochca, a division of the larger group called Mexica, a word the Spanish transformed into "Mexico." According to Aztec legend, the ancestors of the people who founded Tenochtitlán came to the Valley of Mexico from a place in the north called Aztlan, from which the name "Aztec" derives. By the early 1400s, they had come to dominate the region.

MYTHIC VOICES

> **They have a most horrid and abominable custom which truly ought to be punished and which until now we have seen in no other part, and this is that, whenever they wish to ask something of the idols, in order that their plea may find more acceptance, they take many girls and boys and even adults, and in the presence of these idols they open their chests while they are still alive and take out their hearts and entrails and burn them before the idols. . . . Certainly Our Lord God would be well pleased if by the hand of Your Royal Highnesses these people were initiated and instructed in our Holy Catholic Faith, and the devotion, trust, and hope which they have in these idols were transferred to the divine power of God.**
>
> —HERNANDO CORTÉS* *(1521)*

*The name of this Spanish explorer and conquistador is spelled in a variety of ways. This is the version in the *American Heritage Dictionary*.

The Spaniards made bets as to who would slit a man in two, or cut off his head at one blow; or they opened up his bowels. They tore the babies from their mother's breasts by their feet, and dashed their heads against the rocks. . . . They spitted the bodies of other babies, together with their mothers and all who were before them on their swords. . . . [They hanged Indians] by thirteens, in honor and reverence for our Redeemer and the twelve Apostles, they put wood underneath and, with fire, they burned the Indians alive. . . . I saw all the above things. . . . All these did my own eyes witness.

—FRAY BARTOLOMÉ DE LAS CASAS,
History of the Indies, *1552*

What sets Mesoamerican myth apart?

There is little question that what sets Mesoamerican myth apart from many others is its preoccupation with human sacrifice. Other civilizations throughout history clearly used human sacrifice, but nowhere else does it seem to occur quite on the scale it did in Mesoamerica. And nowhere in Mesoamerica was it more pronounced than among the Aztecs, a group originally known as the Tenochca.

In their foundation myth, the Tenochca were commanded by their god Huitzilopochtli to journey from their home base in the north to the Valley of Mexico. At first, they lived in the town of Culhuacan. But after they sacrificed a daughter of Culhuacan's king, the Tenochca were forced to move and start their own city, Tenochtitlán, on an island in the middle of Lake Texcoco. Becoming more powerful and skilled as warriors, they often served as mercenaries in the ongoing conflicts among other people in the area. By the mid-1400s, the Tenochca built a causeway that linked their island city to the mainland, and began to conquer the Valley of Mexico, emerging as a powerful city-state that controlled the region. Under Moctezuma I (also known as Montezuma), who ruled from 1440 to 1469, the Tenochcas conquered large areas to the east and south, and the name Aztec now commonly refers to this larger group who made up this empire. Moctezuma's successors expanded the empire until it reached what is now Guatemala, to the south, and the state of San Luis Potosí, about 225 miles north of Mexico City.

As they did, the Aztecs assimilated many of the gods, beliefs, and practices of the surrounding area into their own religion and myths, including the ancient gods of the mysterious ancient city of Teotihuacán, which they named "the place where men became gods," and the remnants of Toltecs, another warrior tribe that had conquered much of the Mayan territory in Yucatán. When Moctezuma II became emperor in 1502, the Aztec empire was at the height of its power, and hundreds of nearby conquered towns paid heavy taxes to the empire. By the time the Spanish arrived in the 1500s, Tenochtitlán may have had a population of 200,000 to 300,000, larger than any Spanish city of that time. But the Aztec rulers also had made plenty of enemies among the people they taxed and fought with so relentlessly. The Spanish were able to use that local antagonism to make allies among some of the natives eager to see Moctezuma brought down.

A highly militarized society with the kind of sharp class distinctions found in European feudalism, the Aztecs were divided into nobles, commoners, serfs, and slaves. While an emperor presided over all, a council of nobles commanded military units stationed in key locations throughout the empire. The military class included a hierarchy of knights and other ranks, whose main objective was to fight in what were called the "flowery wars" (*la guerra florida*). Don't let the name deceive you. These wars had nothing to do with gardening. The "flowery wars" resulted from an agreement between the Aztecs and other tribes to essentially hold mock battles in order to secure live prisoners for the sacrifices. On these set dates, the young members of the warrior class fought in order to prove themselves, and prisoners for sacrifice could be taken.

To the Aztecs, warfare was a religious duty aimed at taking prisoners to offer to the gods, and providing blood for the gods was a sacred duty. As a result, Aztec methods of combat were designed to capture prisoners rather than kill them. The chief Aztec weapon, a wooden club edged with sharp pieces of obsidian, was effective for disabling an opponent without finishing him off. For protection, warriors carried wooden shields and wore padded cotton armor. Clearly, these weapons and armor did not serve them very well against the steel swords, metal armor, firearms, and cannons of the Spanish. The Aztec and other native warriors, accustomed to taking prisoners in battle, were also unprepared at first to fight battles in which killing was the point.

The fruit of the flowery wars—the blood flowing from a wound was described as the "flower of war"—was offered up at great ceremonies, during which human hearts were proffered to Huitzilopochtli and the other major divinities. Believing that the world had already been destroyed four times, the Aztecs thought that this feeding of the gods would forestall the end of the universe. The grim "open heart surgery" was performed by priests who slashed open the chest of a living victim and tore out the heart. Like the Mayas, the Aztecs believed that the gods needed human hearts and blood to remain strong. Before their deaths, sacrificial victims, who symbolically represented the gods, were dressed in rich clothing, given servants, and treated with honor. Once dead, their souls flew immediately to Tonatiuhichan—the House of the Sun. This was the highest paradise, where dead warriors spent their eternal lives—and lived forever in happiness. By some accounts—not universally accepted—priests or worshippers sometimes ate portions of a victim's body, believing that the dead person's strength and bravery passed to anyone who ate the flesh. While most victims were prisoners of war, the Aztecs also sacrificed children to the god Tlaloc.

Some modern critics and historians dismiss the accounts of Aztec sacrifice as propaganda written by the Spanish invaders to justify their own brutality. But the vast majority of scholarly research and recent archaeology supports the view that the Aztecs had elevated human sacrifice to a ghastly cultural rite.

Did the Aztecs really think the Spanish were gods?

Back in grade school, if they were still teaching anything about the arrival of the Spanish in what would become Mexico, you may have heard this version of events. When the Spaniards arrived, riding horses then unknown in the Americas and wearing metal armor that made great noise, the "primitive" Aztecs unwittingly welcomed them, believing that Cortés was the returning god Quetzalcoatl. With a relatively small band of men, Cortés entered Tenochtitlán, took Moctezuma captive, and, in short order, captured the city and the Aztec empire, eventually destroying it.

The real history, as usual, is a little more complex. First, we should begin with the source of this story—or legend, as it might be called. Most accounts of the arrival of Cortés and the Spaniards, and specifically his encounter with Moctezuma, come from Cortés and other conquistadors, and, later, priests. That's like reading Captain John Smith's history of colonial-era Virginia, or accepting Hitler's view of the invasion of Czechoslovakia. It is hardly unbiased history.

In fact, Cortés landed on the east coast of Mexico in 1519, and marched inland to the Aztec capital. He and his men, who were not all trained soldiers, were joined by thousands of natives who had been conquered by the Aztecs and resented Aztec rule. Moctezuma II did not oppose the advancing Spaniards and did invite them in, but then they took him hostage. In 1520, the Aztecs rebelled and drove the Spaniards from the city. Moctezuma died that year, either executed or from wounds received early in the rebellion. Cortés reorganized his army and began a bloody attack on Tenochtitlán in May 1521. Moctezuma's successor, Cuauhtémoc, surrendered in August the same year. He was later hanged by the Spanish for "treason."

So, the question remains: did myth play any role in this fatal encounter? Most histories of the conquest state that Moctezuma believed Cortés represented the returning god Quetzalcoatl. But anthropologist and Latin studies expert Matthew Restall counts this as one of the *Seven Myths of the Spanish Conquest*, the title of his 2003 history. He is joined by other recent scholars who question this very old assumption. John H. Elliott, a British historian, suggests that this story of the role of Aztec myth in the conquest is layered with its own set of legends. First of all, Cortés himself never mentioned the Quetzalcoatl story in his own writings. Restall and Elliott believe that the stories of a returning god from the East only sprang up later—perhaps twenty years after the Spanish arrived. Elliott also dismisses Cortés's accounts of two speeches made by Moctezuma as the elaborate creation of the Spaniard, written for the consumption of the Spanish royal court. There are no contemporary records—in the writings of Cortés or later Aztec accounts—to confirm the idea that the Aztecs thought the Europeans were gods.

Whether or not the Aztecs actually believed that Cortés was the returning Quetzalcoatl remains an intriguing historical mystery. But it is certainly not what brought about the ultimate downfall of a mighty empire. In his landmark book about the role of disease in history, *Plagues and Peoples*, William H. McNeill writes: "Four months after the Aztecs had driven Cortés and his men from their city, an epidemic of small pox broke out among them, and the man who had organized the attack on Cortés was among those who died. . . . Such partiality could only be explained supernaturally, and there could be no doubt about which side of the struggle enjoyed divine favor. The religions, priesthoods, and way of life built around the old Indian gods could not survive such a demonstration of the superior power of the God the Spaniards worshipped."

What is the "Day of the Dead"?

When the Spanish arrived in Mexico and the rest of Mesoamerica around 1500, one of the native traditions they encountered was a monthlong ritual that seemed to mock death. Though the tradition had roots stretching back thousands of years, the Catholic priests saw it as "pagan" and did their best to eradicate it.

During this celebration, the Aztecs and other Mesoamericans displayed skulls, which symbolized the twin ideas of death and rebirth. The skulls were used to honor the dead, who were thought to come back and visit during the monthlong celebration, which was presided over by Mictecacihuatli, the goddess of the underworld known as "lady of the dead."

The Spanish considered the ritual barbarous and sacrilegious, an extension of the human sacrifices that they had eliminated. "Good Christians" simply didn't go around worshipping skulls or other body parts (unless, of course, they should happen to be the remains or "relics" of a dead saint!). In their attempts to convert the natives to Catholicism, the Spanish tried to stamp out this celebration, which fell in the Aztec solar calendar's ninth month—around August. When they could not eliminate it, the priests simply moved the celebration to coincide with the Catholic feast days, All Saints' Day and All Souls' Day (November 1 and 2).

It is another classic example of how myths are transformed from one culture to another. Just as the Church had succeeded in converting the Celtic Samhain—also a time when the dead walked the earth—into All Souls' Day (see chapter 5), the Aztec celebration of the dead was merged with Catholic tradition. But the ancient native roots of what is now known as Día de los Muertos ("day of the dead") didn't go away. Although increasingly commercialized into a "Hispanic Halloween" festival that extends well beyond October 31, the ancient traditions of this celebration of the dead are very much alive. Today, people in Mexico and Latin America, as well as many Hispanic Americans, don wooden skull masks and dance in honor of their deceased relatives. Wooden skulls are also placed on altars dedicated to the dead. Candy "skulls" of sugar and "Day of the Dead" cookies are widely sold. In many places, it is customary to make a trip to the cemetery for a graveside picnic comprising the deceased person's favorite foods. Gifts for the dead are also placed on graves.

WHO'S WHO OF AZTEC GODS

The Aztecs worshipped hundreds of divinities who were believed to rule all human activities and aspects of nature. This list includes some of the Aztecs' central deities.

Centeotl (**Cinteotl**) God of the all-important maize, Centeotl is a key fertility figure. Every April, people offer him their blood, which is dropped on reeds and displayed on front doors. Centeotl also performs penitence that ensures abundant crops for mankind. All the attributes connected to Centeotl—blood sacrifice, penitence, and an April festival—were connected by Catholic priests to Jesus and his springtime crucifixion and resurrection.

Coatlicue Known as the Lady of the Serpent Skirt, Coatlicue is the mother of the central god Huitzilopochtli as well as an earth serpent goddess. She wears a skirt of writhing snakes and a necklace of human hearts, and carries a skull pendant. Endowed with flabby breasts and clawed hands and feet, Coatlicue feeds on human corpses. But she is not totally without redeeming qualities. Since she is goddess of the fertility of the earth, she freely gives life-sustaining crops to humanity.

In a key Aztec myth, Coatlicue is magically impregnated in an "immaculate conception" when a ball or clump of feathers falls from the sky and lands on her breast. Thinking that their mother has disgraced herself by becoming pregnant, Coatlicue's 400 children plan to kill her to uphold the family honor. In some accounts, Coatlicue is killed; in others, she lives. Either way, she gives birth to Huitzilopochtli, who springs from her body fully formed and kills many of his half-siblings. The idea of the virgin birth of Jesus would have been completely acceptable to the people who embraced this myth of a god born from a pregnancy that came from a clump of heavenly feathers.

Huitzilopochtli While many Aztec deities were borrowed or transformed from other myths of Mesoamerica, Huitzilopochtli is "All-Aztec." Chief deity of the Aztecs, the god of war and the sun, Huitzilopochtli commands the Aztec warriors to create an empire, fight without mercy, and gather the captives necessary for sacrifice to the gods. Each night he undergoes a transformation, much like the Egyptian Re, becoming bones and returning to the world the next morning. His name means "blue hummingbird of the left," because dead warriors become hummingbirds and fly to the underworld. Appropriately, Huitzilopochtli is depicted as a blue man fully armed and decorated with hummingbird feathers.

Huitzilopochtli's birth is exceptional because he springs fully formed from his mother Coatlicue's body just as she is about to be killed by her 400 children. Huitzilopochtli kills his half-sister **Coyolxauhqui**, or Golden Bells, and tosses her head into the heavens, where it becomes the moon. With his mother the earth, his sister the moon, and his 400 brothers who comprise the stars of the Milky Way, Huitzilopochtli and his family make up the entire cosmos.

Mictlantecuhtli God of death, Mictlantecuhtli rules the silent kingdom of the dead known as Mictlan. Depicted as a skeleton wearing a pleated conical cap, Mictlantecuhtli figures in the Aztec story of the origin of people. Once the gods decide to repopulate the earth—after a flood!—they send the god Quetzalcoatl to the underworld to gather the bones of the dead who will be brought back to life. While Quetzalcoatl is carrying these bones, Mictlantecuhtli tries to trick him. Quetzalcoatl drops some of the bones and breaks them. When he gathers them up and returns to earth, the bones are sprinkled with the blood of the gods and are changed into men. Because some of the bones are broken, men come in different sizes.

Ometecuhtli (**Ometeotl**) The supreme creator god of the Aztecs, Ometecuhtli lives in the highest part of heaven and is known as the "dual lord" or "two-god." His name is fitting, since the "dual lord" takes a variety of forms, including a dual incarnation as a divine couple who are the parents of the four great Aztec gods: Huitzilopochtli, Xipe Totec ("the flayed lord"), Tezcatlipoca, and Quetzalcoatl.

Quetzalcoatl The dying and rising god, Quetzalcoatl is the great king and bringer of civilization. Known as the "plumed serpent," Quetzalcoatl is depicted as a combination of a snake with the feathers of the quetzal, a brilliantly colored bird whose feathers signal authority among the Maya (it is still the national bird of Guatemala). A semilegendary ruler with roots in the older Toltec and Mayan myths, he may have been based on a Toltec priest-king, although one of the Mayan Creation gods, **Gucumatz** (or Kukulkán), is also called "plumed serpent" in the Popol Vuh.

His wife or sister, **Chalchiuhtlicue**, is goddess of running water. She protects newborn children, marriage, and innocent love.

In the complex Aztec-calendar religion, which includes four eras of varying length from hundreds to thousands of years called "suns," Quetzalcoatl rules the second sun, which ends with hurricanes, and men being transformed into monkeys—a vestige of Mayan myth. The first sun is ruled by Tezcatlipoca, Quetzalcoatl's brother, and comes to an end when beasts consume the world. The third sun is ruled by Tlaloc, god of rain and fertility, and ends in fire. The fourth sun is ruled by Tlaloc's wife, Chalchiuhtlicue, and ends in the flood in which men are changed to fish. After the fourth sun, Quetzalcoatl makes his trip to the underworld to repopulate the earth. Humanity at present is in the fifth sun, ruled by the fire god **Xiuhtecuhtli**, which will end with earthquakes. This highly apocalyptic view of the world squared neatly with Catholic teachings.

Quetzalcoatl figures in an important myth, in which he argues with his brother Tezcatlipoca. There are two accounts of what happens next. Quetzalcoatl either sails away in a raft or immolates himself, in either case promising to return someday. This is the myth Cortés supposedly exploited in his conquest of the Aztecs, although the jury is now out on that one.

Tezcatlipoca The brother and sometimes adversary of Quetzalcoatl, Tezcatlipoca ("lord of the smoking mirror") is god of the summer sun and the harvest as well as drought, darkness, war, and death. His name derives from the mirrors made from obsidian, which sorcerers used to predict the future. A black stone, obsidian was also employed to make spear points, war axes, and, most important, sacrificial knives.

Tezcatlipoca is a fickle deity with a split personality, who can be cruel or kind. Taking pleasure in battle, he is thought to die each night and return to the world in the morning.

In Tenochtitlán, custom held that handsome young men were sometimes selected to impersonate Tezcatlipoca for a year, after which they were killed with an obsidian knife, their hearts removed and offered as sacrifices.

Tlaloc An ancient rain-and-fertility god adapted from the earlier Toltec people, Tlaloc is portrayed as a black man with tusklike jaguar's teeth, rings around his eyes, and a scroll emerging from his mouth. He controls rain, lightning, and wind, as well as afflictions such as leprosy. According to the grim accounts of Tlaloc, his ritual sacrifices in Tenochtitlán required infant subjects. If the mothers of the sacrificial infants wept, the worshippers believed rain for the crops was assured. The flesh of these sacrificial victims was then eaten by the priests and nobility. (Tlaloc corresponds to the Mayan **Chac**, who also demanded sacrifice.)

Tlaloc's consort is She of the Jade Skirt (Chalchiuhtlicue), the goddess of rivers and standing waters. She of the Jade Skirt also protects children. Perhaps she is associated with them because of the water that breaks before a woman gives birth.

Tlazolteotl Certainly one of the least appealing deities in any pantheon, Tlazolteotl is called "eater of the excrement" and is aptly know as the filth goddess. Associated with the consequences of lust and licentiousness, she is depicted squatting in a traditional birthing position. She is also linked with confession, purification, and penitence.

Xipe Totec The god of agriculture and penitential torture, Xipe Totec is "the flayed lord." According to myth, the flayed lord undergoes self-torture, which the Aztecs imitate by lacerating their bodies with cactus thorns and sharp-edged reeds. There may have been a connection between this ritual and corn, which loses its skin when the shoots begin to burst through. Or the link between new skin and spring growth. Xipe Totec also may have been sacrificing himself to placate the Lady of the Serpent Skirt goddess, Coatlicue, because the world and the soil need to be replenished with regular sacrifice.

One Aztec form of sacrifice involved flaying. Priests sometimes donned the skins that had been stripped from their victims, perhaps in homage to the "flayed lord."

And you thought *The Silence of the Lambs* was creepy.

Xochiquetzal The goddess of flowers and fruits, Xochiquetzal, or Feather Flower, is the mother of Quetzalcoatl. With her twin brother, **Xochipilli**, the flower prince, she rules over beauty, love, female sexuality, happiness, and youth. When Quetzalcoatl departs the empire, she takes less interest in the affairs of humans. Very much akin to the Mesopotamian Inanna and other Near Eastern love goddesses, she protects lovers and prostitutes in her role as moon goddess. Symbolized by flowers, Xochiquetzal also protects marriage and is a fertility goddess who may have committed incest with her brother, Xochipilli, the flower prince and god of lust. In Aztec myth, Xochipilli is the guardian of the spirits of brave warriors who die and become richly plumed birds.

THE MYTHS OF THE INCAS

Rulers of one of the largest and richest empires in the Americas, the Incas began their rise about 1200 CE and began to expand into an empire in the 1400s, until they dominated a vast region that centered on the capital, Cuzco. The empire extended more than 2,500 miles along the western coast of South America until the Incas—reeling from an epidemic that led to civil war—fell to Spanish forces soon after their arrival in 1532. But their cultural heritage is still evident today in the highlands of Peru, where descendants of the Incas still speak Quechua, the Incan language, and perform traditional healing ceremonies.

Was the "lost city" of Machu Picchu really a "sacred place"?

Sure, "lost city" sounds a lot more intriguing than "summer house" or "weekend getaway." But, contrary to conventional wisdom, Machu Picchu may not have been a sacred place. New archaeological evidence shows that when the Incas went to Machu Picchu, they probably kicked back, drank some *chicha* (fermented corn or berry beer), and enjoyed themselves.

For most of the nearly hundred years since Hiram Bingham, an explorer with no archaeological training, stumbled upon the ruins of Machu Picchu in 1911, the idea of a secret, sacred "lost city" has captivated imaginations. Elevated in the snow-clad Andean peaks, Machu Picchu ("old peak") has been the impetus for many a New Age odyssey to Peru in hopes of attaining enlightenment at this high, Andean "energy vortex." On the Richter scale of the world's "mystical places," Machu Picchu ranks right up there with Stonehenge and the pyramids. A walled compound large enough to accommodate upwards of 1,000 people, it is divided into two sections: an agricultural area with terraced fields, canals to bring water, and massive stone retaining walls; and an "urban" area that included more than a hundred residences, warehouses, baths, fountains, and two temples, one of which had a window that allowed the sun to shine through on the summer solstice.

Machu Picchu and the Incas who built it are fascinating, but a lot less exotic than the stories and theories that grew up around this fabled "lost city." As *New York Times* science correspondent John Noble Wilford recently wrote, "Bingham, a historian at Yale, advanced three hypotheses—all of them dead wrong. . . . The spectacular site was not, as Bingham supposed, the traditional birthplace of the Inca people or the final stronghold of the Incas in their losing struggle against Spanish conquest in the 16th century. Nor was it a sacred spiritual center occupied by chosen women, the 'virgins of the sun,' and presided over by priests who worshiped the sun god. Instead, Machu Picchu was one of many private estates of the emperor and, in particular, the favored country retreat for the royal family and Inca nobility. It was, archaeologists say, the Inca equivalent of Camp David, albeit on a much grander scale."

But nobody's is making pilgrimages to the presidential getaway to absorb its psychic energies.

The people who constructed the architectural marvel of Machu Picchu also built what was the greatest and largest civilization in the Americas before Columbus arrived (or "pre-Columbian," as the textbooks like to call it). Based in the capital of Cuzco (also spelled Cusco),* in the 12th century CE, the Incas began to expand their land holdings until they occupied a vast region. With a brilliantly engineered system of terraced agriculture and linked by a magnificent road system, this empire stretched more than 2,500 miles (4,020 kilometers) along the west coast of South America, from present-day Colombia, through parts of Ecuador, Peru, Bolivia, Chile, and Argentina. With an estimated 10 million subjects, according to *National Geographic*, this empire was really a loose confederation of tribes ruled by a single group, the Incas. It was a theocracy to a degree greater than any other American civilization—very much in line with the empires of the ancient Near East. The ruler or the "Inca" was considered divine and a direct descendant of the sun god. Below him were his family, a large ruling aristocracy, and an elaborate priesthood that practiced both human sacrifice and mummification. A great deal of recent archaeology has added considerably to the understanding of the Incas, especially in the discovery of numbers of mummified children who had been sacrificed.

*Reflecting the "omphalos" concept in which a place is viewed as the center of Creation, Cuzco means "navel" in the Incan language called Quechua.

But with their highly centralized government, the Incas at their height were easy pickings for the Spanish. When Francisco Pizarro landed on the South American coast, the Incas were already divided by an internal war and a leadership crisis, and weakened by a smallpox epidemic that had arrived from the north. In 1532, Pizarro—described by historians as "an illiterate pig breeder"—marched about 160 men into the mountains, kidnapped the Inca ruler, Atahualpa, briefly held him hostage, then executed him despite the ransom that had been paid, said to be history's largest, a substantial roomful of gold. (Pizarro later became involved in a series of intrigues and was himself beheaded by rival Spaniards in 1541. Maybe there is some "rough justice.")

While Incan insurrections and rebellions flared for nearly thirty years afterwards, the handwriting was clearly on the wall. In the end, "guns, germs and steel" were again brutally effective. But the Spanish cannons, swords, and vicious mastiff war dogs did not kill most of those Incas. Smallpox did.*

In *Plagues and Peoples*, a compelling account of the role of disease in history, William H. McNeill points out that this onset of deadly sickness had more than just the practical effect of killing large numbers of natives throughout Central and South America. As McNeill writes, "First, Spaniards and Indians readily agreed that epidemic disease was a particularly dreadful and unambiguous form of divine punishment. . . . Secondly, the Spaniards were nearly immune from the terrible disease that raged so mercilessly among the Indians. . . . The gods of the [Indians] as much as the God of the Christians seemed to agree that the white newcomers had divine approval for all they did. . . . From the Amerindian point of view, stunned acquiescence in Spanish superiority was the only possible response."

Shrouded in legends, the Inca empire that Pizarro decimated so efficiently had begun in earnest about 1438, when Pachacuti, the ninth Inca ruler, put down an invasion by the neighboring Chanca confederacy. Called the "Alexander the Great" of the Incas, Pachacuti was a military leader and an effective administrator who conquered the regions

*The story of Pizarro's ruthless subjugation of the Incas is the subject of "The Collision at Cajamarca," a fascinating recounting found in Jared Diamond's *Guns, Germs, and Steel*.

south of Cuzco and rebuilt the city as the center of the empire and a monument to Inca power. Later looked upon as a Creator god, he began the construction of Machu Picchu around 1450.

From Spanish documents, recovered pottery, and other archaeological clues, scholars estimate that Machu Picchu was largely abandoned after only eighty years. Plague, brought by the Spaniards, had left the rest of the empire in turmoil by then. But, at an elevation of 6,750 feet, remote Machu Picchu was relatively untouched—and never even seen by the Spanish. Though called a "lost city," it was really not a city at all. Just a splendid hideaway.

Nevertheless, Bingham, who found Machu Picchu and was made famous by it, was not completely mistaken about its religious aspects. There were clearly temples there, and sun worship was part of its rituals, probably along with imbibing some *chicha*. This local beer was no doubt made at a thousand-year-old site with twenty brewing vats, which was discovered in the Andes in the summer of 2004. Described by one researcher as "a large-scale state-sponsored institutional" brewery, it could produce several hundred gallons at a time. According to scholar Gary Urton, *chicha* was also brewed in the same field where Mama Huaco, one of the Incan founding ancestral sisters, was mummified and buried.

Did the Incas have a foundation myth?

Since the Incas had no written language, most of what is known of their myths and religion comes from retellings by Spanish conquerors, or Incan accounts told to their Spanish masters. As a result, many of these myths are considered suspect.* These include a variety of Incan foundation myths and legends involving a set of siblings called the Ayars, who may be based on historical figures. Anthropologist and Inca expert

*A complex Incan recording system called *quipu*, or *khipu*, involved knotted, dyed strings that could only be interpreted by priests. The system is still not fully understood today, and some experts, such as Harvard anthropologist Gary Urton, believe *khipu* was a form of Incan writing.

Gary Urton explains in *Inca Myths* that "Ayar comes from the Quechua word *aya*, 'corpse,' establishing a link between the ancestors as mythological characters and the mummified remains of the Inca kings, which were kept in a special room in the Temple of the Sun in Cusco. In addition, this same word ayar was the name of a wild strain of the quinua plant, a high-altitude grain crop of the Andes."

In one of these foundation myths, we encounter the somewhat common mythic themes of sibling rivalry and incest as four brothers and four sisters in the Ayar family emerge from caves in the mountains and found the Incan empire. Fearing that their powerful troublemaking sibling Ayar Cachi might become dominant, three of the brothers gang up on him and wall him up. Of the remaining brothers, Ayar Oco turns himself into a sacred stone; Ayar Ayca becomes the protector of the fields; and Ayar Manco (later called Manco Capac) seizes Cuzco, the Inca capital city, and marries his sister, Mama Ocllo.

In another version of the legend, the sun god and creator Inti sends his son Manco Capac, and Manco's daughter and wife Mama Ocllo, to teach civilization to men. Inti gives them a large wedge of gold and tells them to start a city wherever this magical golden block should sink into the ground of its own accord. That proved to be at Cuzco, the Incan capital.

WHO'S WHO OF INCAN GODS

Because the Incas routinely absorbed the local deities of the people they conquered, their pantheon is wide-ranging. The gods below are generally considered the most significant of the deities that have come down through the filter of Spanish colonialism.

Inti The sun god from whom the Incas trace their descent, Inti is the divine ancestor who sends his children to earth with the arts of civilization. In an Inca foundation myth, Inti's children found Cuzco and conquer the people of the Andes. Portrayed as a solar disk with a human face, Inti was the central deity worshipped at the great sun temple at Cuzco, whose walls were lined with gold, which the Incas believed was the sweat of the sun. As Incan myth evolved, Inti was said to have three sons—the gods Viracocha, Pachacamac, and Manco Capac.

Inti's wife and sister is **Mama Kilya**, the moon goddess of fertility and a protector of women. Incan rulers married within their families, as the pharaohs of Egypt did, perhaps to consolidate power.

Manco Capac Also known as Ayar Manco, Manco Capac is the legendary founder of the Incan royal house, who marries one of his sisters, **Mama Ocllo**. All later rulers of the Inca claim to be descended from Manco Capac.

Pachacamac An ancient sun god known as "earth maker," Pachacamac is a brooding character who appears in an early Peruvian Creation myth that is believed to have originated in the coastal areas rather than in the Andes Mountains. After creating the first man and woman, Pachacamac neglects them, and the man starves to death. When the woman complains about the loss of her companion, Pachacamac impregnates her with the rays of the sun and she gives birth to a baby boy. But after four days, Pachacamac grows jealous of the infant, tears him into pieces, and then turns the dismembered body parts into food. The teeth become corn, the ribs and bones become plants, and the boy's flesh becomes fruits and vegetables. In a final act of desecration, Pachacamac uses the boy's penis and navel to create another son, but he kills his first child's mother. Finally, Pachacamac creates a new human couple, who repopulate the land. His wife, **Mama Pacha**, was a dragoness who caused earthquakes and ruled over planting and harvesting the crops.

Viracocha The "foam of the lake" (or the "lake of creation"), Viracocha is a pre-Inca Peruvian deity whom the Incas adopted as their own when they conquered the region. Although there are several versions of his story from Spanish colonial sources, Viracocha is always presented as the creator who lives in Lake Titicaca and oversees sun, water, storms, and light. He is also depicted as a sad old man weeping tears of rain over his disappointing first creation—men. When Viracocha destroys his creation by—what else?—a flood, they turn into a race of giant stones. Today the remnants of these stones are believed to stand near Lake Titicaca, the world's highest lake, which is located on the border between Peru and Bolivia. Some of the many islands in the lake have ruins of civilizations that existed before the Spanish conquest.

In a second creation, Viracocha makes the divine ancestors of the Incan rulers. While those rulers emerge from one cave, ordinary mortals come out of another.

Viracocha's wife (and sister) is **Mama Cocha**, the goddess of wind and rain.

As Incan myth and religion evolved, Viracocha is presented as the son of Inti. Because the sun temple in Cuzco contains images of Viracocha and all the other Incan gods, it is believed that these gods are all manifestations of Inti.

THE MYTHS OF NATIVE NORTH AMERICA

Canada, Mexico, Massachusetts, Utah. It is difficult to go anywhere in continental North America without seeing these place names and realizing that this place is Indian country. As Alvin M. Josephy Jr. writes in *500 Nations*, "What is little understood even today . . . is that almost every community in Canada, the United States and Mexico was once an Indian community and those communities before the arrival of the whites were part of unique Indian nations that blanketed the entire continent." As many as six hundred different languages were spoken, and there were probably just as many sacred traditions. The first Europeans were fascinated by these people and where they had come from. And the speculation, debate, and controversy over that question continues today.

But each tribe knew exactly where they were from. And like every other culture throughout history, they had sacred stories to explain their beginnings. Many of the tribes of the Southwest told a story of how the first men had emerged from a sacred hole in the ground. Other traditions tell of races of great animals that lived before man, tricksters that created people, or mother goddesses who brought forth humanity and made earth fertile. But above all, the native people of North America had reverence for the sacredness of earth and everything in it, a primal idea that is found in almost all their myths.

MYTHIC VOICES

Before the creation of man, the Great Spirit (whose tracks are yet to be seen on the stones, at the Red Pipe, in the form of a large bird) used to slay buffaloes and eat them on the ledge of the Red Rocks . . . and their blood running on the rocks turned them red. One day when a large snake had crawled into the nest of the bird to eat his eggs, one of the eggs hatched out in a clap of thunder, and the Great Spirit, catching hold of a piece of the pipestone* to throw at the snake, moulded it into a man. This man's feet grew fast in the ground where he stood for many ages, like a great tree, and therefore he grew very old; he was older than a hundred men at the present day; and at last another tree grew up by the side of him, when a large snake ate them both off at the roots, and they wandered off together; from these have sprung all the people that now inhabit the earth.

—Sioux Creation account,
from Letters and Notes on the Manner, Customs and Conditions of the North American Indians
by George Catlin, cited in Parallel Myths *by J. F. Bierlein*

All the Earth was flooded with water. Iktomi sent animals to dive for dirt at the bottom of the sea. No animal was able to get any. At last, he sent the Muskrat. It came up dead, but with dirt in its paws. Iktomi saw the dirt, took it, and made the earth out of it. . . . Iktomi then created men and horses out of dirt. Some of the Assiniboine and other northern tribes had no horses. Iktomi told the Assiniboine that they were always to steal horses from other tribes.

—Assiniboine Creation account,
cited in Primal Myths *by Barbara Sproul*

*Pipestone is a soft red stone that is sacred to many native groups. It is used in making the sacred pipes central to belief, ritual, and ceremony across Native North America.

God Isn't Dead. She's Red.

—*popular bumper sticker*

Is there a "North American" mythology?

Whether it is an elegiac Edward Curtis photo of a lone chief on horseback, an action-filled Frederic Remington painting of mounted hunters, or a John Wayne movie in which massed braves on war ponies appear threateningly on a ridge, American Indians and their horses are indelible icons.

But as late as 1700, many North American tribes had no horses. The animal that transformed much of the Native American world arrived in the 1500s with the Spanish, who guarded them carefully, not wanting to surrender the great military advantage they possessed. As historian Jake Page writes in *In the Hands of the Great Spirit*, "It would not be until decades into the next (18th) century that the horses would almost totally transform the cultures of the Plains and . . . Southwest, producing some of the finest light cavalry ever known on Earth."

Native Americans connected in a state of "oneness" with their horses is one of the persistent stereotypes of the American native past. There are others, including Hollywood's stock images of "Indians" as savage, dangerous, elusive, and untrustworthy characters—"Indian givers"—unless they happen to be Lone Ranger's reliable sidekick, Tonto. The stereotypes were underscored in pidgin-English dialogue, like "You speak-um with forked tongue" and "We smoke-um peace pipe." While the worst of this nonsense has been eliminated, there was still Disney's 1995 *Pocahontas*. Playing fast and loose with history, Disney turned the story of the mercenary soldier John Smith, who ran Jamestown with an iron fist, and the ten-year-old native girl Pocahontas into a colonial version of *Romeo and Juliet*. After seeing the film, a Native American school principal noted that *Pocahontas* was the equivalent of teaching the Holocaust by having Anne Frank fall in love with a German soldier.

Once past these stereotypes, the challenge comes in getting a handle on who the North American natives were and what they believed. One of the problems is sheer numbers. There are almost as many traditions and deities as there are tribes—and there are hundreds of tribes. Even so, as Native North American myth authority David Leeming points out, "As various as Indian cultures had become by the time they moved into North America, they had and have in common an identifiable collective mythological tradition. . . . These common themes, in many cases, can be traced back probably not only to Asian roots, but also to the process by which the various people migrated across the continents. . . ."

A fundamental connection between many of these tribes is the idea that everything in life has a spiritual component. Not only was there a supernatural power or spirit present throughout the Creation, this power is also present in daily life—in the preparation for planting or hunting, constructing a home, or settling a dispute. Another common idea is that the creation of the earth and its people involves a supreme god, usually a male sky god, sky father, or "all father," but often a Mother Earth or great goddess as well. When the supreme deity is male, the messy details of Creation are often left to a helper, such as the "earth divers"—usually animals who create the earth by bringing up pieces of dry land from beneath a primal ocean. In other traditions, the helper deity is a goddess. A number of scholars point out that this scenario is a lot like "the men bring home the bacon and the women run the household." In what tribe have you heard that?

Other common threads that run through the North American traditions include shamanism, drumming, chanting, sweat lodges, and pipe smoking. All of these traditions are believed to stem from the prehistoric roots that are widely shared by the people of North America. But perhaps the most familiar "public" face of the North American native traditions is sacred dancing, a form of communal prayer that brought spirituality to life in pulsing, rhythmic performances—some outdoor, some indoor; some very secret, others public—that connected the people with the "mysteries" surrounding them. The vivid image of Native Americans dancing in a circle in a parched field as they look to the sky and implore the Great Spirit to send rain to nurture the wilting corn crop is almost iconic. But the rain dance is only one of a wide range of sacred and secular dances that exist throughout many tribes. One of many others is the feather dance, a rite held whenever an eagle feather from a ceremonial dress accidentally fell to the floor. Because the eagle was considered a sacred bird, its feather would be retrieved and "reconsecrated" in a group dance. For the Iroquois, the feather dance was also a sacred expression of thanksgiving.

Perhaps the most famous "dance" from Native American tradition is the sun dance. Typically a four-day rite, the sun dance usually took place to welcome the revival of nature after winter. In preparation for the dance, a tree was cut down and erected as a sacred pole. After two or three days of feasting, sweat lodge purification rites, and fasting, the dancers attached themselves to the pole by piercing themselves with pegs secured by long grass ropes. They then danced, straining against these tethers until their skin broke or they collapsed to the ground from exhaustion and hunger. When the ordeal ended, it was believed that the dancers had absorbed the pain and suffering of the tribe for the year to come. Missionaries and government agents eventually banned the dance in the nineteenth century.

Dancing was also typically linked to fertility rites. Among many of the southwestern tribes, there was a corn dance, which was eventually merged with Catholic feast days. And the basket dance of the Pueblos took its name from the baskets symbolizing fertility and was originally held in spring. After Catholicism was introduced, the basket dance was moved to winter, because the missionaries did not want any ritual dancing during their holy days of Lent.

The ghost dance, whose name referred to the spirits of departed ancestors and the nearly depleted buffalo, was a reaction to the coming of whites and the destruction of native ways. Appearing in 1870, the dance grew out of a religious movement initiated among the Northern Paiute in Nevada by a tribal leader named Wodziwob. A new type of religious leader, Wodziwob was regarded as one of a number of "prophets" who appeared among several tribes to restore conditions to the way they were before the white man arrived. Among his many reforms, which included a prohibition on alcohol, he proposed performing a ghost dance to the ancestors to help make this happen. This communal prayer in the form of a continuous circular dance culminated when the dancers achieved a state of ecstasy.

But in 1890, another messianic movement grew among the Paiutes. This time it was led by another prophet, Wovoka, who also wanted to return to a time before the white man's coming. Wovoka's relatively benign message included a call to perform a five-day ghost dance to bring about that change. His message spread to the Plains, which sent a delegation, including Sitting Bull, to learn more about Wovoka's vision. As the movement gained followers, it took on more militant aspects, especially among the younger Sioux, some of whom wore "ghost shirts," which, they believed, would protect them from bullets. The movement provoked hysteria among white settlers, who saw it as a dangerous conspiracy. Eventually military action was called for, resulting in the arrest and death of Sitting Bull and the massacre of more than 300 ghost dancers—Lakota men, women, and children—at Wounded Knee on December 29, 1890, marking the practical end of the ghost-dance movement.

MYTHIC VOICES

And while I stood there I saw more than I can tell and I understood more than I saw; for I was seeing in a sacred manner the shapes of all living things in the spirit, and the shapes of all shapes as they must live together in one being. And I saw the sacred hoop of my people was one of many hoops that made one circle, wide as daylight and as starlight, and in the center grew one almighty flowering tree to shelter all the children of one mother and one father. And I saw that it was holy.

—*Black Elk, from* Black Elk Speaks

WHO'S WHO OF NORTH AMERICAN NATIVE GODS

The following list includes some of the most typical and intriguing North American deities and mythic figures. These include Great Spirits—the somewhat passive, "all-father" Creator gods shared by many North American tribes—and Earth Mothers, twins, and tricksters. (Tribal origins and locations are listed.)

Coyote (many tribes and areas) The trickster god of the North American Indians of the western and southwestern United States, Coyote is the mischievous, cunning deity who causes numerous disasters to befall the world. As Richard Erdoes writes in *American Indian Trickster Tales*, "Coyote, part human, part animal, taking whichever shape he pleases, combines in his nature the sacredness and sinfulness, grand gestures and pettiness, strength and weakness, joy and misery, heroism and cowardice that together form the human character. . . . [He is] the godlike creator, the bringer of light, the monster-killer, the miserable little cheat, and of course, the lecher."

Coyote has many origins. The Maidu (California) believe that Coyote emerges from the ground and watches the creator Wonomi ("no death") make the first man and woman. When Coyote tries to do the same thing, the humans he creates are blind. So, Coyote decides it would be more interesting to make sickness, sorrow, and death to plague mankind. In short order, he accomplishes his goal.

But the joy quickly fades when Coyote's son is killed by a rattlesnake bite. Coyote tries to revive the corpse by submerging it in a lake. But the boy remains dead, so Coyote leaves the corpse to rot. Seeing what has happened, Wonomi realizes that Coyote will always be a torment, and decides to leave the earth and its affairs to his wily adversary.

In other tales, Coyote is a lecherous character with a colossal and magical penis. In a tale of the Shasta (northern California), Coyote sees two pretty maidens in a creek and desires them both. Turning himself into a salmon, Coyote/Salmon swims between the two girls and enters their bodies. As the girls ask each other if they feel something strange, Coyote emerges in his true form and laughs at them. In another Shasta tale, Coyote sees a girl digging for roots by a stream, changes his penis into the stalk of a plant, and stretches it across the stream so that it can enter her. When the girl sees the stalk, she taps at it with her digging tool. Coyote howls in pain and has to pull back his "stalk."

Glooskap (**Gluskap**) (Algonquian, Abenaki of the Northeast) A creator and trickster, Glooskap is a patriarch who makes the sun, moon, plants, animals, and people from Mother Earth's body. His troublesome brother, **Malsum**, creates insects, reptiles, and other nuisances. After Glooskap defeats his evil brother, he uses his trickster's ability to change shapes and defeat the witches, spirits, and sorcerers who threaten mankind. Glooskap performs other heroic feats, including riding on the back of a whale before leaving the world. He promises to return in times of peril.

Hahgwedhdiyu (Hodenosaunee, Northeast) Creator of the Iroquois, Hahgwedhdiyu is the son of the sky goddess **Atahensic**. His evil twin is **Hahgwehdaetgan**. After the twins' mother dies, Hahgwedhdiyu forms the sky and turns his mother's face into the sun; the moon and stars are made from her breasts, and the earth is made fertile with her body. The evil twin counters his brother by making floods, earthquakes, and other disasters. The brothers ultimately fight, and the evil sibling is defeated and banished to the murky underworld.

Hinun (Iroquois or Hodenosaunee, Northeast) The great thunder spirit and guardian of the sky, Hinun is portrayed as a powerful brave armed with a bow and arrows of fire. With help from his wife, **Rainbow**, and his friend **Gunnodyak**, Hinun fights the great serpent of the Great Lakes. When the serpent swallows Gunnodyak whole, Hinun rescues the young warrior and takes him up to the sky. After applying a magic ointment to his own eyes, Hinun is able to see the serpent in the lake and shoot it with his arrows. The great snake dies but makes a great noise as it writhes in death throes. Terrified by the noise, heaven and earth fall silent. Hinun also slays the ferocious giant stone people who dwell in the west and are planning to attack the Iroquois.

Igaluk (Inuit, Arctic regions) Igaluk is the supreme god who directs everything. He is also the moon. When Igaluk discovers that he has slept with his sister, the sun, there is great upset. His sister tears off her breasts and rises into the sky. Eventually the pair build a house in the sky that is divided in two sections. That is where they coexist.

Iktome (also **Ik-to-mi**) (Sioux, Plains) Known as Spiderman, Iktome is a trickster who does things backwards but is still a sly and cunning teacher. To the Assiniboine (Plains), he is the Creator who orders the animals to dive for bits of earth (see Mythic Voices, page 482). A man with the attributes of a spider, Iktome has a hearty sexual appetite, like his friend and frequent companion, Coyote.

In one story told by the Brule—with a slight overtone of the Little Red Riding Hood tale—Iktome tricks a beautiful young maiden he sees walking one day. Dressing himself in the clothes of an old woman, he approaches the girl and asks for permission to accompany her across a stream. She notices that his legs are very hairy, and Iktome explains that it comes with age. When he hikes up his robes, she says his backside is hairy, too, and he responds that this happens to older people. When he lifts his robe farther, the girl gasps at the sight of his penis and asks what it is. Iktome explains that it is a wart put there by a sorcerer and will only go away if he puts it between her legs. The girl complies, and the "wart" grows smaller, but Iktome suggests that if he puts it between her legs again, it may go away altogether. Despite several tries, the "wart" remains, so Iktome suggests they keep going until it disappears. The girl, who has forgotten forgot why she set out across the river, readily agrees.

Kitchi Manitou (Algonquian, Northeastern woodlands) A manifestation of the Great Spirit, Kitchi Manitou is the divine energy that lives in all things. Man tries to control the "manitou" of small things, such as fire and wood, in order to gain control over the larger forces, such as the sun, wind, and rain.

Kwatee (**Kivati**) (Puget Sound, Washington) A trickster god, Kwatee transforms the old world that is filled with giant animal people into the world that exists today. When the giant animals discover what Kwatee is doing, they try to kill him. Kwatee then rolls balls of his own flesh into human beings. After his creation is complete, he sits on a rock and leaves the world to join the setting sun.

Nayenezgani (Navajo, Southwest) "Slayer of alien gods," the translation for Nayenezgani, is the great hero and protector of the Navajo as well as the son of **Changing Woman**. Together with his twin brother **Tobadzastsini**, Slayer patrols the world, always on the lookout for evil spirits. While going to visit their father the sun god, the twins meet **Spider Woman**, who warns them of the dangers they will face on their journey. She gives them two magic feathers: one will subdue any enemy and the other will preserve life.

When they reach the sun god, he tries to kill them. First he throws sharp spikes at them. Then he tries to boil them in a great pot, but the water will not boil. The magic feathers have protected them, but now their power is used up. The brothers are about to die when **Caterpillar** gives them magical stones and they are saved. Realizing that these boys are powerful warriors, the sun god gives them weapons they can use to protect the Navajo tribe from its enemies.

Raven (Haida and others, Pacific Northwest) A trickster, Raven wants to bring fire to the world when he sees smoke coming from the village of the fire people. With his friends **Robin**, **Mole,** and **Flea**, he tries to steal the fire. But in a series of missteps, Robin's feathers are scorched and Mole burrows underground. Raven finally decides to steal the chief's baby and hold it for ransom. To get his baby back, the chief gives Raven fire and two stones with which to make sparks.

Sky Woman (Hurons, Northeast) Atahensic, or Sky Woman, is the central figure in a Creation myth of the Hurons. In the beginning, there is only water below and sky above, where the sky people live. Sky Woman is sick, and her father is afraid that she will die. A member of the tribe dreams that if they dig up the corn tree, and Sky Woman sits next to it, she will be cured. Some of the tribe object, because the tree feeds the tribe. But Sky Woman's father urges them to help his daughter. When the tree is uprooted, it falls over and opens a dark hole in the ground. A young man gets angry and kicks Sky Woman through the hole.

Falling through darkness toward the infinite sea, Sky Woman is caught by **Loon** and carried on the back of **Tortoise**. Tortoise tells the other animals to dive to the bottom of the sea and bring back a little earth from the sea floor. **Beaver** goes first, then **Otter**, then **Muskrat**, who is dead when he surfaces but has a speck of dirt in his mouth. Tortoise gives the dirt to Sky Woman, and she spreads it around his shell until it becomes a fertile island.

With land to walk on, Sky Woman gets well, and then mysteriously becomes pregnant and gives birth to a daughter, **Earth Woman**. While Earth Woman is digging potatoes, she faces east and the wind impregnates her. She gives birth to twins, a good twin and an evil twin. But the evil twin's entry into the world is rough—he breaks through his mother's side and kills her.

Sky Woman buries her daughter and raises her twin grandchildren, although she cannot love the evil twin. One day the good twin digs up his mother's body, forms a sphere from her face, and makes the sun. From the back of her head he makes more spheres, which become the moon and stars. That is how day and night are created. Watered by her mother's tears, Earth Woman's corpse starts to sprout vegetables. Over time, maize and beans grow from her body. The good twin and the evil twin then make the rest of Creation, with the good twin creating trees and cool water and the evil twin creating dangerous mountains. And for the Hurons, that's how the world came into existence.

Tirawa (Pawnee, Great Plains) Great Spirit and Creator god Tirawa holds a council and assigns tasks to the other gods. The sun god **Shakaru** is ordered to give light and warmth; the moon goddess **Pah** gives sleep and rest in the night; and the stars—Bright, Evening, Great, and Morning—are told to hold up the sky. The first humans are born when the sun and moon marry and have a boy called **Closed Man**. When the Evening Star and Morning Star finally couple, they produce a girl—known as "Daughter of Evening and Morning Star." The Pawnee believed that they were decended from these first children of the heavens.

But Tirawa gets angry and destroys his creation with a fire and then a great flood. The only survivors are an old man who carries a pipe, fire, and a drum; and his wife, who carries maize and pumpkin seeds. These two, who have been protected in a cave, re-create the human race.

White Buffalo Woman (Sioux, Northern Plains) A beautiful, long-haired figure in a white buckskin dress, White Buffalo (Calf) Woman is one of the most significant deities of the Plains tribes. Once, when the people are starving, two scouts go and search for food. They see a blur in the distance and, as it approaches, one of the scouts realizes it is the sacred White Buffalo Woman. The woman, who can read the bad thoughts of one of the young men, invites him to embrace her. But as he reaches toward her, a white cloud appears and lightning strikes the lusty man, who is killed instantly. His body is turned into a skeleton and then devoured by worms.

The second scout returns to the village to set up a great teepee for her. Once this is done, White Buffalo Woman instructs the tribes in all the sacred ceremonies. She explains how to use the pipe and teaches them seven sacred rites, including the sweat lodge, vision quest, the "ghost-keeping ceremony," in which the soul of the dead is purified, the sun dance, the *hunka* ceremony (designed to establish binding relationships among fellow human beings), girls' puberty rites, and the "throwing of the ball," a ceremony celebrating knowledge, in which a buffalo-hide ball is tossed to people standing in the four compass directions.

As she talks to the chiefs, White Buffalo Woman is a woman. But when she leaves, the people see her roll in the dust four times, bow to each corner of the universe, and then become a white buffalo before vanishing, perhaps to return again one day.

To the Plains people, no animal was more sacred than the buffalo, which completely sustained their way of life.

Which goddess gets her own "planet"?

If the Roman goddess Venus represents everything that is beautiful and good, the Inuit goddess Sedna may be her complete opposite. Queen of the underworld, Sedna gets mixed up in acts of trickery, kidnapping, murder, dismemberment, cannibalism, and revenge. The only thing that she shares with Venus is the fact that each has a heavenly body named in her honor.

While Venus was first observed by the earliest "astronomers" in prehistoric times, the Inuit goddess Sedna joined the celestial charts when the discovery of a small object orbiting the sun was announced in March 2004. Too small to qualify as a planet in the view of most astronomers, Sedna is essentially a large chunk of rock caught in a regular orbit of the sun and now thought to be the most distant object from the sun. The scientists who found this piece of the flotsam and jetsam in a very distant reach of space called the Kuiper Belt decided to name it after the Inuit sea goddess who plays a role in many myths.

In one myth, an Arctic seabird known as a fulmar, and noted for its foul (no pun intended) smell, sees Sedna and falls in love with her. Assuming human form, the bird makes himself a parka, woos Sedna, and invites her home. When they arrive, Sedna realizes that she has been tricked by the birdman and desperately calls for her father, Anguta, to help her. But he doesn't hear her cries, and she has to spend months in this awful place. When Anguta eventually finds Sedna, he kills the mischievous bird. Discovering the murder, the other birds surround the father's kayak, flap their wings in what seems like a Hitchcockian scene from *The Birds*, and create a storm that tosses the kayak in the waves.

Afraid the boat will capsize, Sedna's father decides to look out for number one—himself! To lighten the boat's load, he tosses his daughter into the sea. When she clings to the boat, Anguta takes out his knife and hacks off her fingers, one by one. In one version of the tale, each of Sedna's fingers turns into a sea animal.

Angry at her father, Sedna seeks revenge. She calls on a team of dogs to attack him and gnaw on his hands and feet. He curses and screams until the earth opens up and they all tumble into the underworld. That is where Sedna lives and reigns as queen, blessing hunters with animals and creating terrible storms. The only thing she cannot do is comb her hair, since she has no fingers.

In another myth, Sedna begins life as a beautiful young woman but later becomes a one-eyed giantess who populates the sea with ocean life while Anguta, her father, makes the earth, sea, and heavens. But Sedna's appearance is so hideous that only medicine men can bear to look at her. And some of her personal habits are pretty awful, too. On one occasion, which mirrors a scene from *Night of the Living Dead*, she feels the urge to eat some human flesh and begins to nibble on her mother and father. They wake up and discover what is going on, take Sedna far out to sea, and cast her overboard. Once again, as in other myths about her, Sedna desperately clings to the side of the boat, prompting her father to chop off her fingers. In this myth, Sedna's severed fingers turn into whales, seals, and fish as they touch the water. Sedna then sinks to the bottom of the sea, where she lives, ruler of the underworld, keeping guard over the ungrateful dead. These include her own parents, who have been devoured by sea animals.

So why name a celestial object after a gruesome, murderous child? The not-quite-a-planet is thought to be very dark and very cold, so the goddess of the Arctic underworld seems perfectly appropriate. That's the same reason Pluto was named for the Roman god of the underworld.

Sedna was actually the second of two recently discovered celestial objects to get a Native American name. In 2002, another large piece of orbiting rock was picked out of the very distant Kuiper Belt, the band of ice-and-rock objects at the very far reaches of our solar system. This rock was named Quaoar. The word comes from the Creation myth of the Tongva people, who are sometimes called the San Gabrielino Indians. The Tongva people lived in the Los Angeles area before the arrival of the Spanish and other Europeans.

Not exactly a god in the traditional sense, Quaoar is seen as the great force of Creation, who literally performs a "song and dance."

When Quaoar dances and sings, the first sky father is born. This pair continues to sing and dance, and then the Earth Mother comes into existence. Now a trio, they all sing together, and grandfather sun comes to life. As each emerging deity joins the festivities, the song becomes more complex and the dance more complicated. Grandmother moon, the goddess of the sea, the lord of dreams and visions, the bringer of food and harvests, the goddess of the underworld all eventually join in the singing, dancing, and creating, which is completed when the "earth diver" Frog brings up soil, and the other animals dance on it until it becomes the flat, wide earth. Such are the musical and mythical adventures of Quaoar.

MYTHIC VOICES

By the shore of Gitche Gumee,
By the shining Big-Sea-Water,
At the doorway of his wigwam,
In the pleasant Summer morning,
Hiawatha stood and waited.
All the air was full of freshness,
All the earth was bright and joyous. . . .

—HENRY WADSWORTH LONGFELLOW,
The Song of Hiawatha *(1855)*

What famous poem contributed to the "myth" of the Native Americans?

There was a time in the not-too-distant past when most American schoolchildren were forced to memorize at least some part of a piece of Americana that shaped their views about the Native Americans. Though its popularity has long since declined, Henry Wadsworth Longfellow's epic poem, *The Song of Hiawatha*, is still found in the pantheon of American verse. It also inspired a 1952 Disney cartoon that did little to broaden our understanding of Native American traditions.

Written in 1855, *The Song of Hiawatha* employs twenty-two long sections to tell the story of an Ojibway Indian called Hiawatha, whose life is full of triumphs and tragedies. The poem recounts the somewhat miraculous birth of Hiawatha in a time of turmoil between tribes, how he grows to become a great hunter and woos and weds the beautiful but doomed Minnehaha, commencing a golden age that will carry him toward further trials and adventures. The poem ends with the coming of the white men called "Black Robes," who bring the Christian gospel, and Hiawatha's own symbolic departure into the sunset in his canoe. As he leaves his people, to whom he has brought peace, he tells them to listen to the wisdom of the Black Robes:

But my guests I leave behind me;
Listen to their words of wisdom,

Listen to the truth they tell you,
For the Master of Life has sent them
From the land of light and morning!

Though it may sound to modern ears like Christian-mission propaganda, Longfellow (1807–1882) meant well. Writing his melodic paean in the heroic style of old sagas, he was trying to capture a sense of the humanity and nobility he saw in the Native American experience. His poetic sentiments were based on the anthropological writings of the first "experts" of his day, who were certainly not Native Americans. Most were people of European descent, who may have truly believed that the natives benefited from the coming of the white man. Along with Longfellow, these well-meaning "experts" helped create, during the eighteenth and nineteenth centuries, a highly romanticized myth of America as a "New Eden," and the native people as "noble savages." The latter concept was coined by the influential French philosopher Jean-Jacques Rousseau, who believed that true men of nature were proud and uncorrupted by civilization.

Longfellow's poem became standard classroom fare for more than a century. As it did, it left the impression that Hiawatha had done his people a great favor by leaving them in the hands of the "Black Robes." According to the poem, God Himself—the "Master of Life"—sent these Christian missionaries "from the land of light and morning" to speak "words of wisdom." It sounded like a good deal. But, in truth, the paternalistic missionaries weren't concerned with much besides bringing the "savages" to Jesus. Then there were the great masses of Americans who, by the mid-nineteenth century and certainly after "Custer's last stand" in 1876, agreed with the popular notion that "the only good Indian is a dead Indian." Of course, the grim testimony of history shows that last sentiment largely won out.

While Longfellow may have had good intentions in helping foster the "noble savage" myth, he also took poetic license with a few facts, beginning with the name of the poem's chief character. The name Hiawatha—which the poet apparently used because it fit his meter—comes from the Hodenosaunee. Commonly called the Iroquois, the Hodenosaunee lived in the Northeast, and their name meant "the people of the long house." Yet Longfellow sets Hiawatha among the

Chippewa, a tribe of the Great Lakes in the Midwest.

Which raises another question: Did a person called Hiawatha exist? According to Hodenosaunee history and lore, the answer is yes. Hiawatha was a leader in precolonial America, who probably lived during the 1500s and is credited with helping establish peace among the five major tribes—the Mohawks, Oneidas, Onondagas, Cayugas, and Senecas—who once dominated what is now Upstate New York.

For years, these tribes had been torn by raids and counterraids, in which captives were either tortured to death or, in some cases, adopted into the tribe to replace a lost family member. According to tribal legend, Hiawatha of the Onondaga had fallen into great grief after years of constant fighting and, in some versions, became a cannibal after his five daughters were killed. He was rescued from his grief and madness by Deganawida, a Huron elder said to be born of a virgin and on a mission to make peace and unite the Iroquois. With Hiawatha acting on what was believed to have been a sacred vision, the two men went from tribe to tribe, persuading them to make peace.

According to Alvin Josephy's *500 Nations*, "The Peace Maker, as Deganawida was becoming known, conceived of thirteen laws by which people and nations could live in peace and unity—a democracy where the needs of all would be accommodated without violence and bloodshed. To a modern American, it would suggest a society functioning under values and laws similar to those of the Ten Commandments and the U.S. Constitution combined. Each of its laws included a moral structure." When one chief balked at the plan, Hiawatha was able to persuade him to change his mind. According to tribal legend, the reluctant chief, Tadadaho, was an evil sorcerer whose hair was a Medusa-like tangle of snakes. Hiawatha—whose name means "he who combs"—smoothed out the tangle of snakes, cured Tadadaho's evil mind, and the Great Law of the Five Tribes was adopted. (A sixth tribe, the Tuscarora, later joined the league.)

In fact, as far back as 1751—a quarter of a century before the Declaration of Independence—Benjamin Franklin was inspired, or at least impressed, by the Iroquois League, when he proposed a colonial union in his Albany Plan. "It would be a very strange thing," wrote Franklin, "if six nations of ignorant savages should be capable of forming a scheme for such a union, and be able to execute it in such a fashion that it has subsisted for ages and appears indissoluble; and yet that a like union should be impracticable for ten or a dozen English colonies, to which it is more necessary, and must be more advantageous."

Franklin's blueprint for a colonial union failed. But there are many historians who believe that in 1789, the principles of the Iroquois Confederacy were studied by the delegates to the Constitutional Convention. However, the men who wrote the U.S. Constitution chose not to include the aspect of the Native American plan that gave men and women equality. That would not be a feature of the U.S. Constitution until 1920. So much for Franklin's "ignorant savages."

MYTHIC VOICES

There can be no peace as long as we wage war upon our mother, the earth. Responsible and courageous actions must be taken to realign ourselves with the great laws of nature. We must meet this crisis now, while we still have time. We offer these words as common peoples in support of peace, equity, justice, and reconciliation: As we speak, the ice continues to melt in the north.

—OREN LYONS,
faithkeeper of the Onondaga Nation (August 2000)

Do Native American myths still matter?

Remember the movie *Poltergeist*? You know. The one with the little girl who looks at the fuzzy television screen and says, "They're here." Made in 1982, the movie centers on a haunted house in a suburban development built over an American Indian burial ground.

What about *Close Encounters of the Third Kind*? This 1977 movie depicts benevolent aliens arriving on earth for a "close encounter" at Devil's Tower, the 1,200-foot-tall rock that seemingly erupts out of the earth in northeastern Wyoming. A popular tourist destination, especially for rock climbers, Devil's Tower is known to some Plains tribes as Mato Tipila, or Bear's Lodge, and it is sacred land to at least twenty-three native groups. Both *Poltergeist* and *Close Encounters*, which are the products of Steven Spielberg's fertile imagination, touch upon an issue of great importance to many Native Americans—what modern society is doing to their sacred spaces and religious traditions.

The Devil's Tower controversy is a case in point. On one side of the standoff are Wyoming state officials, the National Parks Service—and rock climbers—who stand for tourism and recreation. On the other side of the argument are the Native Americans who revere Bear's Lodge as a sacred place and want to restore its native name. Writing about this landmark in *Sacred Lands of Indian America*, historian Jake Page points out that "in its presence it is easy to understand why climbers are drawn to it. Easy enough to understand if you are not an Indian. For Indians, climbing the tower is an invasion of the sacred. One has to wonder what it would feel like to Christians if the steeples of churches and cathedrals suddenly became climbing destinations."

The fight over Devil's Tower, like the conflict over the construction of a large telescope near Tucson, Arizona, on Mount Graham, a mountain sacred to the Apaches, pits powerful economic interests against ancient tribal traditions. It is a fight being waged in various places around America, as development projects with a variety of purposes, including ski resorts, new highways running through reservations, and mineral rights, proliferate. These enterprises often collide headfirst with Native American sacred spaces that, to the uninformed, seem like open land or wilderness, completely suitable for modern development.

In other words, the myths—the sacred stories—of the people who have been in America longest are crashing headfirst against the desires and wishes of the federal government, science, developers, and, yes, rock climbers. These controversies have embroiled U.S. courts and Congress during the past decade in a face-off between native beliefs and government control.

In 1990, for instance, the U.S. Supreme Court ruled that states could regulate Native American religions that employed the use of peyote, a natural hallucinogen. In a majority opinion, Justice Antonin Scalia wrote, "It may be fairly said that leaving accommodation to the political process will place at a relative disadvantage those religious practices that are not widely engaged in; but that *unavoidable consequence of democratic government must be preferred*." (Emphasis added.) Scalia's opinion meant Congress—or other government bodies—can pass laws that regulate religious expression. The First Amendment, it would seem, goes only so far.*

Seeing the danger to religious expression posed by the decision and Scalia's opinion, many mainstream religious groups and other civil rights groups asked the court to reconsider, but their petition was denied. In response, Congress passed the Religious Freedom Restoration Act (RFRA) in 1993, and the American Indian Religious Freedom Act Amendments in 1994, which restored a measure of protection for Native American religions, including the use of peyote in traditional sacraments. In 1997, the Supreme Court declared RFRA unconstitutional. The court ruled that Congress had overstepped its power to legislate constitutional rights when it passed a law attempting to protect religious observances from government regulation. (Peyote use for religious ceremonies was unaffected by the decision.)

Congress had also stepped into controversial territory when it passed the Native American Graves Protection and Repatriation Act (NAGPRA), signed into law by President George Bush in 1990. Designed to protect American Indian grave sites from looting and archaeological investigation, NAGPRA also required museums to repatriate certain tribal objects to their tribes of origin. (The bill applies only to federal lands, not private property.) For centuries, Indian burial sites have been systematically looted of skeletons and burial objects. While many states have enacted similar legislation, removing bodies or objects from Indian graves is not a crime in many states. NAGPRA was invoked in the case of Kennewick man, the oldest known skeletal remains in North America, but in 2004, a federal court ruled that these remans were not covered, since Kennewick man was apparently unrelated to any tribe.

But there the situation stands. Religion and myths are still in the eyes of the beholder, as historian Jake Page convincingly demonstrates in his book *In the Hands of the Great Spirit*:

*This decision would seem to contradict the 1993 ruling about Santeria practices cited earlier (see p. 430). The difference is that this case involved compliance with an otherwise valid law governing conduct that the state can regulate—the use of illegal drugs. In other words, using illegal drugs is different from killing chickens for ritual purposes.

Most non-Indians do not look out upon the landscape and see spirits out there, spirits of such things as trees and rocks and lightning and wind. Indeed, such beliefs are considered by most Christians, at least, to be pagan and improper, even childish, and many conservative Christians today find such beliefs the work of the devil, just as the Puritans and the Spanish Franciscans and French Jesuits did five hundred years ago, in what one would like to think were less enlightened times. On the other hand, many traditional Indians find it peculiar, to say the least, that Christians and others can build a house for God, go there once or maybe twice a week, and whenever it seems like a good idea, proceed to tear God's house down and build another one, with say a bigger parking lot, on the other side of town. If the gods reside in a mountain, it is not so easy to relocate them. For Indians, a sacred site remains sacred under most circumstances.

THE MYTHS OF THE PACIFIC

By the eighteenth and nineteenth centuries, there were few places on earth unseen or unspoiled by Europeans and the rest of the "civilized" world. Most of these "last places" were islands in the vast Pacific Ocean, which occupies fully one-third of the earth's surface area. These islands would soon experience a replay of the same ruthless colonial story that had become the sad biography of Africa and the Americas.

There are literally tens of thousands of islands arranged in a rough triangle in the Pacific Ocean, with Hawaii in the north, New Zealand in the south, and Easter Island (so named by a Dutch explorer who found it on Easter Sunday in 1722) in the east. Inhabited by people who had moved out of southwestern Asia tens of thousands of years ago, the people of the Pacific islands and Australia may have island-hopped on foot when ocean levels were 400 to 600 feet lower, perhaps also using boats to settle these islands. Many of these early ocean voyagers developed separate mythologies often traceable to the Polynesians. Polynesia, which means "many islands," occupies the largest area in the South Pacific, stretching from Midway Island in the north to New Zealand, 5,000 miles (8,000 kilometers) to the south. While not part of Polynesia, Hawaii in the northern Pacific was first settled by Polynesians 2,000 years ago, and the island's myths reflect that tradition.

MYTHIC MILESTONES

Australia and the Pacific Islands

Before the Common Era

c. 8000–6000 Land bridge connecting Australia and Tasmania disappears; rising sea also covers New Guinea land bridge.

c. 6000 Migrations from southeastern Asia to Pacific islands.

c. 4000 Austronesians reach southwestern Pacific islands.

c. 2500 The dingo introduced to Australia from southeastern Asia.

c. 1500 Earliest evidence of colonization of Fiji.

c. 1000 Polynesian culture emerges on Fiji, Tonga, and Samoa.

Common Era

c. 300 Easter Island is settled.

c. 850 Polynesian ancestors of Maori settle New Zealand.

1000 First carvings and stone statues on Easter Island.

1606 Portuguese explorer Luis Váez de Torres sails around New Guinea and discovers Australia.

1642 Dutch explorer Abel Tasman finds Tasmania and New Zealand; over the next several years he will find and map Tonga, Fiji, New Guinea, and coasts of Australia.

1768 British captain James Cook's first of three voyages of discovery into the Pacific; in 1772, on his second voyage, Cook reaches Botany Bay, Australia, and claims it for Britain; in 1779, on his third voyage, Cook is killed in the Hawaiian Islands.

1788 First British settlement at Botany Bay, Australia.

First penal settlement established at Port Jackson (future Sydney), and the "first fleet" of convicts lands in New South Wales.

1789 Smallpox ravages the Aborigines of New South Wales in Australia.

Mutiny on the HMS *Bounty*; mutineers settle on Pitcairn Island.

1797 First Christian missionaries reach Tahiti.

1810 Hawaiian islands united by King Kamehameha.

1851 Gold discovered in Australia; thousands of settlers flock to Victoria, Australia.

1864 The practice of transporting prisoners to Australia is abolished.

1892 The queen of Hawaii is deposed; U.S. troops move to annex the islands.

1894 Sanford Dole proclaims the Republic of Hawaii. Hawaii is annexed by the United States in 1894 and made a U.S. territory in 1900.

Which mythic character created the Pacific Islands?

Probably the most famous Polynesian demigod was the trickster Maui, for whom the Hawaiian island of Maui is named. According to some myths, the trickster Maui is born very small, so his mother throws him away in the ocean. Surviving this attempted infanticide, Maui grows up into an oversexed trickster hero, who creates the Pacific islands by fishing them up from the bottom of the sea. The possessor of a prodigious penis, as so many tricksters are, Maui is chosen to satisfy the boundless desire of the goddess Hina. Both the bringer of fire and the cause of death, he is also credited with slowing down the sun to make days longer, either by using the jawbone of his dead grandmother or by lassoing the sun with a rope made from Hina's hair.

In one Polynesian myth, Maui is challenged by the sun god to enter the body of the goddess of death and pass from her vagina to her mouth. If he succeeds, Maui will become immortal. Attempting to accomplish this feat as the goddess sleeps, Maui is foiled when a bird sees him and laughs, waking the sleeping goddess. She kills Maui, ensuring that humanity would always suffer death.

MYTHIC VOICES

> The most puzzling question for whites was . . . why these people should display such a marked sense of territory while having no apparent cult of private property. . . . Certainly they had few external signs of religious belief: no temples or altars or priests, no venerated images set up in public places, no evidence of sacrifice or of communal prayer. . . . They carried their conception of the sacred, of mythic time and ancestral origins with them as they walked. These were embodied in the landscape; every hill and valley, each kind of animal and tree, had its place in a systematic but unwritten whole. Take away this territory and they were deprived not of "property" . . . but of their embodied history, their locus of myth, their "dreaming". . . . To deprive the Aborigines of their territory . . . was to condemn them to spiritual death.
>
> —ROBERT HUGHES, *The Fatal Shore*

What is Dreamtime?

A different but very rich tradition of the Pacific world belongs to the ancestors of today's Aborigines, or indigenous people,* who first arrived in Australia from southeastern Asia perhaps as much as 65,000 years ago. Rock engravings in Australia have been dated to 45,000 years ago, and evidence of the world's first known cremation dates to 26,000 years ago

*"Aborigines" was the word first used by the British for the native Australians they found. The word comes from the Latin phrase "ab origine," meaning "from the beginning." When spelled with a small "a," the word "aborigine" now refers to any people whose ancestors were the first to live in a country. In Australia, the official term for descendants of the native Australians is now "indigenous." Many of these Australians prefer to be referred to by their specific tribal names.

in southern Australia. Presumably these people had hopscotched the land bridge that existed between the Pacific islands at times when Ice Age climate kept sea levels lower than they are today. The number of Aborigines in Australia at the time the British arrived to create a massive penal colony in 1788 range from 300,000 to 750,000 people, scattered among at least 500 tribes. As in Africa and the Americas, a number of diverse factors nearly brought about the extinction of the Aboriginal people. These factors included disease, fighting with the British colonists, and the general depredations of a colonized people losing their land and traditional ways of life.

According to a very ancient Aboriginal Creation myth, all life today is part of a connected universe that goes back to the great spirit ancestors who existed in Dreamtime. While many tribes have variations on this concept, the idea of Dreamtime, or the Dreaming, is almost universal in Australia. It goes like this: In the beginning, the earth was in darkness. Life existed below the surface, sleeping. In the Dreaming world, the ancestor beings broke through the crust of the earth, and the sun rose out of the ground. The ancestors then traveled the land and began to shape it, creating the mountains and other features of the landscape along with all the animals, plants, and other natural elements. They also created society, teaching the songs, dances, and ceremonial rituals, and leaving behind spirits of people yet to be born. Finally, tired from this activity, the mythical ancestors sank back into earth and returned to sleep. These beings never died, but merged with nature to live on in sacred beliefs and rituals. Some of their spirits were turned into rocks, trees, or other sacred places that dot the Australian landscape.

Dreamtime is more than just a period in the past—it is ever present, and reached through sacred rituals such as the walkabout, a tribal spiritual journey taken to sacred places to renew the clan's relationship with Dreaming and the sacred landscape. An individual can go on walkabout to where the tribe originally came from, or some other place of sacred "belongingness."

Other tribal variations of native Australian myth often include a rainbow serpent—a powerful spirit of creation and fertility—whose curving movement through the sands creates river beds and other natural features. When treated carefully, the snake sleeps, but if disturbed, it creates storms and flooding. One of these snakes, called Yorlunggur, lives by a water hole. When one of two sisters falls into the hole, her menstrual blood pollutes the water, angering the serpent. The snake swallows the sisters and causes a great flood. When the floodwaters recede, the snake spits out the sisters, and the place where this happened becomes the sacred spot where adolescent boys are initiated into adulthood, a central rite for native Australians.

Another great ancestral snake called Bobbi-bobbi is responsible for what may be Australia's most identifiable "icon." The serpent drives flying-fox squirrels out into the open for people to eat, but these elusive creatures are not easy to kill. From his underground hiding place, the great serpent sees the difficulty and tosses one of his ribs up to a group of men. This becomes the first boomerang, which the men use to kill the flying foxes. Later, the men throw the boomerang into the sky and make a hole, which makes Bobbi-bobbi angry, so he takes the boomerang back for a time.

There are many other myths of Australia and the Pacific islands, a legacy of ancient people who moved across vast expanses of land and open seas. One of these ancient myths seems especially salient today. It is a story told by many of the people of the Pacific islands about a mythical race of Pygmies, two feet tall. While these "little people" sometimes shot tiny arrows at careless travelers, they otherwise lived peacefully in caves.

In October 2004, scientists announced the discovery, on a tropical island midway between Asia and Australia, of the skeletons of a race of people whose adults stood three and a half feet tall. The diminutive "Floresians," as the scientists named them, lived in a cave on Flores, an island 370 miles east of Bali. This other race of humans lived there until about 13,000 years ago—a miniature version of prehistoric man.

Myths indeed are as fresh as the headlines. And perhaps, after all, Shakespeare was right:

> *There are more things in heaven and earth,*
> *Horatio,*
> *Than are dreamt of in your philosophy.*

BIBLIOGRAPHY

Anthologies, Collections, and Translations of World Myths

Abrahams, Roger D. *African Folktales*. New York: Pantheon Books, 1983. A collection of ninety-five tales from sub-Saharan Africa, offering a taste of the rich oral tradition of African myth and legend.

Alighieri, Dante, translated by Robert Hollander and Jean Hollander. *The Inferno*. New York: Anchor Books, 2000. (Many other translations available.)

Apollodorus, translated by Robin Hard. *The Library of Greek Mythology*. New York: Oxford University Press, 1997. A source book of Greek myths from the origins of the universe to the Trojan War, compiled in the second century BCE. This is the most significant source of Greek mythology after Homer and Hesiod.

Apollonius of Rhodes, translated by Richard Hunter. *Jason and the Golden Fleece*. New York: Oxford University Press, 1993. A verse translation of the third century BCE poem about the quest for the fleece, the Argonauts, Jason, and Medea.

Bierhorst, John. *The Mythology of North America*. New York: Oxford University Press, 2002. A thorough and accessible "field guide" that breaks Native North American folklore into eleven distinct regions, with discussions of the shared mythologies, stories, and gods of each.

Birrell, Anne M. *Chinese Mythology: An Introduction*. Baltimore: Johns Hopkins University Press, 1993. English translation of some three hundred ancient Chinese myth narratives, with very scholarly notes and explanatory texts.

Birrell, Anne M. *The Classic of Mountains and Seas*. New York: Penguin Books, 2000. A treasure trove of colorful stories about more than two hundred Chinese mythical figures, most of them very unfamiliar to Western readers.

Bulfinch, Thomas. *Bulfinch's Mythology: The Age of Fable*; *The Age of Chivalry*; *Legends of Charlemagne*. Available in various editions that appeared in 1855, 1858, and 1863, respectively.

Coomaraswamy, Ananda K., and Sister Nivedita. *Myths of the Hindus and Buddhists*. New York: Dover, 1967. Gathers the most important stories from Indian mythology, which is the source of the two major religions reflected in the title.

Crossley-Holland, Kevin. *The Norse Myths*. New York: Pantheon, 1980. Retelling of thirty-two classic tales from the Viking world.

Davis, F. Hadland. *Myths and Legends of Japan*. New York: Dover, 1992. Collected retellings of classical Japanese myths.

Erdoes, Richard and Alfonso Ortiz, editors. *American Indian Myths and Legends*. New York: Pantheon, 1984. A collection of 160 folk myths and tales from eighty different tribal groups.

Erdoes, Richard and Alfonso Ortiz, editors. *American Indian Trickster Tales*. New York: Penguin, 1998. A collection of more than one hundred tales from different tribes about the colorful, mischievous, and highly oversexed characters known as tricksters, from various American Indian traditions, including Coyote, Iktomi the Spider, and Rabbit. Colorful and very earthy.

Faulkner, R. O. *The Ancient Egyptian Book of the Dead*. Austin: University of Texas Press, 1985. The religious and magical texts known to the ancient Egyptians.

Foster, Benjamin R., translator and editor. *The Epic of Gilgamesh*. New York: W. W. Norton, 2001. A recent translation of the Mesopotamian epic poem with critical notes and essays.

Gantz, Jeffrey, translator. *Early Irish Myths and Legends*. London: Penguin Books, 1981. First written down around the eighth century, these are early Celtic legends from Ireland.

Gantz, Jeffrey, translator. *The Mabinogion*. New York: Penguin Books, 1976. A collection of the eleven medieval Welsh prose tales, including some of the earliest written Arthurian legends.

Hesiod, translated by M. L. West. *Theogony* and *Works and Days*. New York: Oxford University Press, 1988. Less known than Homer, the poet Hesiod wrote a systematic genealogy of the Greek gods, from the mythological beginnings of the world.

Homer, translated by Robert Fitzgerald. *The Iliad* and *The Odyssey*. New York: Vintage Books, 1961. (Many other translations and editions available.)

Husain, Shahrukh. *The Virago Book of Erotic Myths and Legends*. London: Virago, 2002. A compilation of modern retellings of some of the erotic tales from diverse cultures.

Kinsella, Thomas. *The Táin: From the Irish Epic Táin Bó Cúailnge*. New York: Oxford University Press, 1970. A mixed prose and verse translation of the cycle of Irish heroic tales.

Leeming, David A. *The World of Myth: An Anthology*. New York: Oxford University Press, 1990. A collection of myths, organized by theme (creation, heroes, etc.).

Leeming, David A. and Jake Page. *The Mythology of North America*. Norman: University of Oklahoma Press, 1988. A collection of seventy-two representative myths from a variety of tribal groups, with commentary and introductions.

Littleton, C. Scott, general editor. *Mythology: The Illustrated Anthology of World Myth and Storytelling*. London: Duncan Baird, 2002. A large, heavily illustrated compendium of more than 300 myths from around the world.

Mascaró, Juan, translator. *The Bhagavad-Gita*. London: Penguin Books, 1962. An essential document of Hinduism, the conversation between the god Krishna and the warrior Rama before a great battle.

Mason, Herbert. *Gilgamesh: A Verse Narrative*. Boston: Houghton Mifflin, 1970. A widely read modern verse version of the ancient Babylonian epic, one of the oldest pieces of literature in human history. (Many other translations of *Gilgamesh* are also available.)

Mitchell, Stephen. *Gilgamesh: A New English Version*. New York: Free Press, 2004. Another modern translation with excellent historical and introductory notes.

Neihardt, John G. *Black Elk Speaks: Being the Life Story of a Holy Man of the Oglala Sioux*. Lincoln: University of Nebraska Press/Bison Books, 1988. First published in 1932, the now-classic "as-told-to" account of a Native American *wichasha wakon* (holy man, priest) that encompasses both the myth and history of the Oglala Sioux who fought Custer and were later massacred at Wounded Knee.

Ogden, Daniel. *Magic, Witchcraft, and Ghosts in the Greek and Roman Worlds*. Oxford, England: Oxford University Press, 2002. A scholarly translation of some of the lesser-known myths specifically involving magic.

Ovid, translated by A. D. Melville. *Metamorphoses*. New York: Oxford University Press, 1986. The Roman poet's collected accounts of transformations and changes presented in the Greek and Roman myths. (Other editions available.)

Pelikan, Jaroslav, editor. *Sacred Writings, Volume 5. Hinduism: The Rig-Veda*. New York: Quality Paperback Book Club, 1992.

Pelikan, Jaroslav, editor. *Sacred Writings, Volume 6. Buddhism: The Dhammapada*. New York: Quality Paperback Book Club, 1987.

Prabhavananda, Swami and Dr. Frederick Manchester, translators and editors. *The Upanishads*. New York: Signet, 1957. A collection of the principal holy writings of Hinduism.

Rosenberg, Donna. *World Mythology: An Anthology of Great Myths and Epics* (second edition). Lincolnwood, Illinois: NTC Publishing Group, 1994. An anthology of key myths in contemporary prose, divided by geographic region.

Rouse, W. H. D. *Gods, Heroes and Men of Ancient Greece*. New York: New American Library, 1957. A popular retelling of Greek myths by a scholar-teacher who told the tales to his students at Cambridge, England.

Spence, Lewis. *The Myths of Mexico and Peru*. New York: Dover, 1994. Reprint of a 1913 classic work on the myths of the Aztec, Inca, and other South and Central American groups.

Sproul, Barbara C. *Primal Myths: Creation Myths Around the World*. New York: Harper, 1979. A collection that cross-references various Creation stories from many cultures.

Sturluson, Snorri, translated and edited by Anthony Faulkes. *Edda*. North Clarendon, Vt.: Tuttle Publishing, 1987. The standard collection of Norse poetry, compiled in the 1200s by a poet and courtier later killed in a political intrigue. (Many other editions available.)

Tatar, Maria. *The Annotated Classic Fairy Tales*. New York: W. W. Norton, 2002. Although not about myths, this collection of twenty-six of the best-known children's tales (*Cinderella, Sleeping Beauty, Rapunzel*) illuminates some of the connections between mythic stories and familiar children's tales—and they are not always about virtues!

Tedlock, Dennis, translator. *Popol Vuh: The Mayan Book of the Dawn of Life* (revised edition). New York: Touchstone/Simon & Schuster, 1996. One of the most important texts in the native languages of the Americas, often called the Mayan Bible.

Wilson, Andrew. *World Scripture: A Comparative Anthology of Sacred Texts.* New York: Paragon House, 1991. A collection of sacred writings, organized thematically, from many diverse faiths and traditions, including the major religions as well as texts from native religions of Africa and the Americas.

Reference (includes both general references and works specific to mythology)

Achtmeier, Paul J. *The HarperCollins Bible Dictionary.* San Francisco: HarperSanFrancisco, 1996. A wide-ranging, objective, and comprehensive guide that includes many mythological connections to the Bible.

Birrell, Anne. *Chinese Myths.* (The Legendary Past series.) Austin: University of Texas Press, 2000. A brief overview of Chinese myth for Western readers.

Carpenter, Thomas H. *Art and Myth in Ancient Greece.* New York: Thames & Hudson, 1991. Scholarly appraisal of how early Greek artists used mythic themes.

Comrie, Bernard, Stephen Matthews, and Maria Polinsky, editors. *The Atlas of Languages* (revised edition). New York: Facts on File, 2003. An introduction for general readers to the world of languages and how they grew. Also includes information on the development of various writing systems.

Cotterell, Arthur. *A Dictionary of World Mythology.* Oxford, England: Oxford University Press, 1986. A concise listing of major mythic figures, divided by geographic regions.

Cotterell, Arthur. *The Macmillan Illustrated Encyclopedia of Myths and Legends.* New York: Macmillan, 1989. A very comprehensive, illustrated reference guide to world myth.

Curtis, Vesta Sarkhoshi. *Persian Myths.* (The Legendary Past series.) Austin: University of Texas Press, 1993. Overview of traditional tales and stories from ancient Iran. (Part of a series of brief monographs on world myths; other titles listed below.)

Dallapiccola, Anna L. *Hindu Myths* (The Legendary Past series). Austin: University of Texas Press, 2003. One in a series of brief monographs, this volume provides a scholarly but quick overview of Hindu legends.

Farmer, David. *Oxford Dictionary of the Saints* (fifth edition). New York: Oxford University Press, 2003. Concise accounts of the lives, cults, and artistic association of Christian saints, some of whom have their own myths.

Forty, Jonathan. *Mythology: A Visual Encyclopedia.* New York: Sterling Publishers, 2001. Heavily illustrated reference to world myths, organized geographically.

Freeman, Charles. *Egypt, Greece and Rome: Civilizations of the Ancient Mediterranean* (second edition). New York: Oxford University Press, 2004. Excellent one-volume reference on the ancient Near East and the Mediterranean world.

Gardner, Jane F. *Roman Myths.* (The Legendary Past series.) Austin: University of Texas Press, 1993. Brief overview of Roman myths and the ways they reflected specific Roman history.

Green, Miranda Jane. *Celtic Myths.* (The Legendary Past series.) Austin: University of Texas Press, 1993. Exploration of the mythology and beliefs of the pagan Celts between about 600 BCE and 400 CE.

Hart, George. *Egyptian Myths.* (The Legendary Past series.) Austin: University of Texas Press, 1993. This brief overview of Egyptian mythology is a scholarly but excellent introduction.

Hayes, Michael. *The Egyptians.* New York: Rizzoli, 1996. A concise, accessible, and highly illustrated introduction to Egyptian history and civilization.

James, Vanessa. *The Genealogy of Greek Mythology: An Illustrated Family Tree of Greek Myth from the First Gods to the Founders of Rome.* New York: Gotham Books, 2003. A very useful, accordion-like, illustrated foldout guide to the major names—both immortal and human—of Greek mythology.

Leeming, David A. and Margaret Leeming. *A Dictionary of Creation Myths.* New York: Oxford University Press, 1994. Alphabetically divided by traditions, covers almost every Creation account, from Sumer and Egypt to the Big Bang.

Lewis, Jon E., editor. *The Mammoth Book of Eyewitness Ancient Egypt.* New York: Carroll & Graf, 2003. Documenting three thousand years of Egyptian history through actual documents and eyewitness accounts of mummification, temple building, and the real Cleopatra.

McCall, Henrietta. *Mesopotamian Myths.* (The Legendary Past series.) Austin: University of Texas Press, 1993. Brief overview of the myths of Mesopotamia and their influence on the Greeks and Hebrews.

Macrone, Michael. *By Jove! Brush Up Your Mythology.* New York: HarperCollins, 1992. How the Greek myths live on in the English language through such words and expressions as "Titanic" or "Wheel of Fortune."

Manguel, Alberto and Gianni Guadalupi. *The Dictionary of Imaginary Places.* New York: Harcourt, Brace, 2000. An encyclopedic guide to places that never were, including the legendary sites of mythology.

Occhiogrosso, Peter. *The Joy of Sects: A Spirited Guide to the World's Religious Traditions.* New York: Doubleday, 1994. A somewhat irreverent but very useful guide to the practices of various religions around the world.

Orchard, Andy. *Cassell's Dictionary of Norse Myth and Legend.* London: Cassell, 1997. More than one thousand entries detailing a range of Scandinavian myths, sagas, and legends.

Page, R. I. *Norse Myths* (The Legendary Past series.) Austin: University of Texas Press, 1993. The influential stories and legends of pagan Scandinavia and Germanic tribes.

Pattanaik, Devdutt. *Indian Mythology: Tales, Symbols, and Rituals From the Heart of the Subcontinent*. Rochester, Vt.: Inner Traditions, 2003. Scholarly and somewhat abstract, this introduction to the streams of Hindu thought and myths is still highly useful.

Powell, Barry B. *Classical Mythology* (fourth edition). Upper Saddle River, New Jersey:Pearson/Prentice Hall, 2004. A textbook that breaks the usual textbook mold; well written, entertaining, insightful.

Price, Simon and Emily Kearns, editors. *The Oxford Dictionary of Classical Myth and Religion*. New York: Oxford University Press, 2003. Drawn from *The Oxford Classical Dictionary*, third edition (Oxford, 1996), this is an excellent resource focusing on Greek and Roman myths and religion and their relationship to Judaism and Christianity in the Greco-Roman world.

Romann, Chris. *A World of Ideas: A Dictionary of Important Theories, Concepts, Beliefs and Thinkers*. New York: Ballantine Books, 1999. From "a priori" to "Zoroastrianism," a useful compendium of philosophy, faith, and the people behind the ideas.

Shaw, Ian, editor. *The Oxford History of Ancient Egypt*. New York: Oxford University Press, 2000. A comprehensive single-volume reference covering Egypt's history from the Stone Age to the Roman period.

Smart, Ninian. *The World's Religions* (second edition). New York: Cambridge University Press, 1998. An academic overview of the great religions, including material on their mythic origins.

Smith, Huston. *The Illustrated World's Religions: A Guide to Our Wisdom Traditions*. San Francisco: HarperSanFrancisco, 1994. An overview of world religions and how they have evolved since primal times.

Tarnas, Richard. *The Passion of the Western Mind: Understanding the Ideas That Shaped Our World View*. New York: Ballantine Books, 1991. An accessible overview of Western philosophical thought from the ancient Greeks to modern times; especially useful for its discussion of the Greek worldview.

Taube, Karl. *Aztec and Maya Myths*. (The Legendary Past series.) Austin: University of Texas Press, 1993. Overview of the two central cultures of Mexico and Central America.

Teeple, John B. *Timelines of World History*. London: DK Publishing, 2002. A lively, illustrated timeline of human history, organized by date and geographic regions. An excellent reference.

Urton, Gary. *Inca Myths*. (The Legendary Past series.) Austin: University of Texas Press, 1993. Overview of the legends of the great empire based in the Peruvian Andes, which fell to the Spanish conquistadors.

Wilkinson, Richard H. *The Complete Gods and Goddesses of Ancient Egypt*. London: Thames & Hudson, 2003. A lavishly illustrated reference to the complex pantheon of Egypt from the early days of the pharaohs to Roman times.

Williams, Dr. William F., editor. *Encyclopedia of Pseudoscience*. New York: Facts on File, 2000. A critical and skeptical scientific guide to the world's many frauds, hoaxes, superstitions, and mistaken theories, many of them related to ancient myth and civilizations.

Willis, Roy, editor. *Dictionary of World Myth: An A-Z Reference Guide to Gods, Goddesses, Heroes, Heroines and Fabulous Beasts*. London: Duncan Baird, 1995. The title says it all.

General Works of History and Criticism

Armstrong, Karen. *A History of God: The 4,000-Year Quest of Judaism, Christianity and Islam*. New York: Knopf, 1993. Best-selling account of the rise of the three dominant monotheistic religions with discussion of their pagan or mythical roots. Scholarly but very accessible.

Armstrong, Karen. *Buddha*. New York: Penguin Books, 2001. An insightful, brief historical and "philosophical" biography of the legendary founder of Buddhism.

Ballard, Robert D. and Toni Eugene. *Mystery of the Ancient Seafarers: Early Maritime Civilizations*. Washington, D.C.: National Geographic Society, 2004. Heavily illustrated account of recent discoveries in the Mediterranean and Black Seas that shed light on the Phoenicians, Egyptians, Greeks, and other early sailors; by the man who found the *Titanic*.

Benedict, Jeff. *No Bone Unturned: Inside the World of a Top Forensic Scientist and His Work on America's Most Notorious Crimes and Disasters*. New York: Perennial, 2003. Fascinating account of the Smithsonian scientist who rebuilds skeletons and is at the center of the controversy of "Kennewick man," the oldest known human remains in North America.

Bierlein, J. F. *Parallel Myths*. New York: Ballantine Wellspring, 1994. Accessible work discussing the themes common to many mythologies.

Boorstin, Daniel J. *The Discoverers: A History of Man's Search to Know His World and Himself*. New York: Random House, 1983. How humanity learned much of what it knows. Erudite and fascinating.

Boorstin, Daniel J. *The Seekers: The Story of Man's Continuing Quest to Understand His World*. New York: Random House, 1998. A survey of the history of philosophy, religion, and the sciences in the Western world.

Cahill, Thomas. *How the Irish Saved Civilization*. New York: Doubleday, 1995. The first in the Hinges of History series, this book recounts the little-known role of medieval Irish monks in preserving history through illuminated manuscripts. Entertaining and accessible history.

Cahill, Thomas. *Sailing the Wine-Dark Sea: Why the Greeks Matter*. New York: Nan A. Talese/Doubleday, 2003. The fourth volume in the best-selling Hinges of History series, an entertaining and accessible history of the legacy of the Greeks.

Campbell, Joseph. *The Hero With a Thousand Faces*. Princeton, N.J.: Princeton University Press, 1949. Campbell's first classic account of the role of the hero in myths.

Campbell, Joseph. *The Mythic Image*. Princeton, N.J.: Princeton University Press, 1974. A heavily illustrated exploration of the relation between dreams, mythology, and artistic imagery.

Campbell, Joseph. *The Power of Myth*. New York: Broadway Books, 1988. A summing-up of the televised conversations about mythology between Joseph Campbell, the great teacher of mythology, and journalist Bill Moyers.

Camus, Albert, translated by Justin O'Brien. *The Myth of Sisyphus and Other Essays*. New York: Vintage Books, 1983. One of the most famous modern uses of ancient mythology, by the Nobel Prize–winning French existentialist who explored the idea of living in a universe devoid of meaning.

Ceram, C. W. *Gods, Graves and Scholars: The Story of Archeology* (second revised edition). New York: Knopf, 1967. Although somewhat dated, this is still an excellent introduction to the history of modern archaeology, focusing on the incredible real-life adventures of the likes of Heinrich Schliemann and Howard Carter.

Chadwick, Henry. *The Early Church (The Penguin History of the Church, Volume 1*, revised edition). New York: Penguin Books, 1993. Scholarly but accessible overview of the beginnings of Christianity and its rapid expansion throughout the Roman world.

Clayre, Alasdair. *The Heart of the Dragon*. Boston: Houghton Mifflin, 1985. Companion to a twelve-part PBS television documentary, an accessible introduction to China's past, with an emphasis on philosophy and ancient religions.

Davidson, Basil. *The Search for Africa: History, Culture, Politics*. New York: Random House, 1994. A collection of essays about African history by a veteran journalist-historian, including material on the roots of Africa's ancient kingdoms.

Davidson, James. *Courtesans and Fishcakes: The Consuming Passions of Classical Athens*. New York: St. Martin's Press, 1997. Sex and seafood in classical Athens. Interesting insights into what life was really like when the ancient Greeks were in their glory. Not quite as accessible as Thomas Cahill's *Sailing the Wine-Dark Sea* (see above).

Deloria, Vine Jr. *God Is Red: A Native View of Religion* (thirtieth anniversary edition). Golden, Colo.: Fulcrum Publishing, 2003. First published in 1972, this is the third edition of a seminal work on Native American religious views. Challenging, angry, and provocative opinions on Native American history and spirituality and how they have often been mischaracterized.

Devereux, Paul. *The Sacred Place: The Ancient Origins of Holy and Mystical Sites*. New York: Sterling, 2001. Illustrated photographic guide to many of the worldwide sites, both man-made (Stonehenge, Chichén Itzá) and natural (Ayers Rock, Mount Olympus), that significantly figure in world mythology.

Diamond, Jared. *Guns, Germs, and Steel: The Fates of Human Societies*. New York: W. W. Norton, 1999. Winner of the Pulitzer Prize, a fascinating assessment of history that focuses on geography, disease, and technology, and repudiates many traditional—and often racist—views of the rise of civilizations.

Ebrey, Patricia Buckley. *The Cambridge Illustrated History of China*. New York: Cambridge University Press, 1996. A beautifully and heavily illustrated introduction to Chinese history; scholarly but very accessible.

Eliade, Mircea. *The Sacred and the Profane: The Nature of Religion*. New York: Harcourt, 1987. A classic academic work that traces the movement of spirituality from primitive to modern times.

Fage, J. D. *A History of Africa*. New York: Knopf, 1978. A volume in the *History of Humanity* series, this is highly scholarly (and dated), but offers a sound overview of early African history.

Feiler, Bruce. *Abraham: A Journey to the Heart of Three Faiths*. New York: William Morrow, 2002. A search for the legendary figure who is patriarch of three of the world's great faiths.

Feiler, Bruce. *Walking the Bible: A Journey by Land Through the Five Books of Moses*. New York: Perennial, 2002. A modern journey in search of the history behind the mythical crossing of the Red Sea and climbing of Mount Sinai.

Fox, Robin Lane. *Pagans and Christians*. New York: Knopf, 1986. A scholarly but accessible history of the transition from paganism to early Christianity in Rome.

Frazer, Sir James. *The Golden Bough* (abridged). New York: Dover, 2002. Originally published in twelve volumes in 1890, this classic study of mythology explores the universal theme of the dying-and-resurrected god, tracing its roots to the worship of Diana. (This is the author's 1902 abridged version.) Highly academic and dated, this is still a significant work in the field of mythic studies.

Galeano, Eduardo, translated by Cedric Belfrage. *Open Veins of Latin America: Five Centuries of the Pillage of a Continent* (twenty-fifth-anniversary edition). New York: Monthly Review Press, 1997. Written by an Uruguayan journalist, an exposé of the exploitation of the Latin America, beginning with the colonial period and continuing through the twentieth century. An eye-opening account for those who know little of America's largely destructive involvement in Latin American history.

Germond, Philippe. *An Egyptian Bestiary*. New York: Thames & Hudson, 2001. With magnificent illustrations of Egyptian art and architecture, this work depicts the extraordinary role played by animals in Egypt's myth and daily life.

Graves, Robert. *The White Goddess: A Historical Grammar of Poetic Myth*. New York: Farrar, Straus & Giroux, 1975. The author best known for historical novels such as *I, Claudius* takes a highly academic look at the "white goddess of birth, love, and death," who was worshipped in Europe under many names.

Green, Miranda J. *The World of the Druids*. New York: Thames & Hudson, 1997. An elaborately illustrated and accessible introduction to the Celtic world, its priests, and the myths they inspired.

Hamilton, Edith. *Mythology*. Boston: Little, Brown, 1969. The renowned classic introduction to the gods of Greece, Rome, and the Norse; still popular but somewhat dated.

Hathaway, Nancy. *The Friendly Guide to Mythology: A Mortal's Companion to the*

Fantastical Realm of Gods, Goddesses, Monsters and Heroes. New York: Penguin Books, 2001. A breezy and readable introduction to world myths, with a particular focus on goddess stories.

Herodotus, translated by Aubrey De Sélincourt. *The Histories*. New York: Penguin Books, 1996. In this masterpiece of classic literature, the "father of history" examines the Mediterranean world of the fifth century BCE. With useful notes. (Other editions available.)

Hughes, Robert. *Fatal Shore: The Epic of Australia's Founding*. New York: Vintage, 1986. Compelling history of Australia with much information on the unhappy interaction between the original inhabitants and the British.

Jung, Carl G., editor. *Man and His Symbols*. New York: Dell Laurel Books, 1964. A collection of essays by the Swiss psychologist and several associates which explores the role of myths and symbols in human psychology. Not easy reading, but still a valuable introduction to Jung's influential ideas.

King, Ross. *Michelangelo and the Pope's Ceiling*. New York: Walker, 2003. Best-selling narrative of the intrigue behind the art and architecture of the famed Sistine Chapel; includes a discussion of the introduction of mythic figures into Christian art during the Renaissance.

Klingaman, William K. *The First Century: Emperors, Gods, and Everyman*. New York: Harper Perennial, 1990. A highly readable narrative of the years 1–100 CE—in both East and West—during which both Christianity and Buddhism flourished.

Kramer, Samuel Noah. *History Begins at Sumer: Thirty-nine Firsts in Recorded History* (third revised edition). Philadelphia: University of Pennsylvania Press, 1981. The first love song, tax cut, system of law, and schools all belonged to ancient Sumer, which is illuminated in this accessible study by one of the foremost experts on ancient Mesopotamian civilization.

Lapatin, Kenneth. *Mysteries of the Snake Goddess: Art, Desire and the Forging of History*. New York: Da Capo, 2002. A fascinating archaeological detective story that casts doubt on some long-accepted notions of ancient Minoan art and society.

Leick, Gwendolyn. *Mesopotamia: The Invention of the City*. New York: Penguin Books, 2001. An overview of the rise of the first twelve cities in the first civilization. Scholarly but still accessible.

Lévi-Strauss, Claude. *Myth and Meaning: Cracking the Code of Culture*. New York: Schocken Books, 1995. A collection of five essays based on radio interviews, which serves as an introduction to the ideas of one of the most influential social anthropologists of recent times. Although highly theoretical, this slim volume is far more accessible than the author's many other works, such as *The Raw and the Cooked*, *Tristes Tropiques*, and *Structural Anthropology*.

McNeill, J. R. and William McNeill. *The Human Web: A Bird's-Eye View of World History*. New York: W. W. Norton, 2003. Father-son authors show the set of connections that link people, creating a web of interaction in human history.

McNeill, William H. *Plagues and Peoples*. New York: Anchor Books, 1998. A fascinating narrative of the impact of disease on history, including the decimation of Native Americans by Europeans and the transfer of diseases to the Americas through the slave trade.

Mithen, Steve. *After the Ice: A Global Human History, 20,000–5000 BC*. Cambridge, Mass.: Harvard University Press, 2004. Told through the eyes of a fictional world traveler, a look at the globe as the last great Ice Age was ending, and that change's impact on human development.

Morton, W. Scott, and Charlton M. Lewis. *China: Its History and Culture* (fourth edition). New York: McGraw-Hill, 2005. A concise, accessible overview of China from neolithic times to the present.

Morton, W. Scott, and J. Kenneth Olenik. *Japan: Its History and Culture* (fourth edition). New York: McGraw-Hill, 2005. A concise chronology and good overview of Japanese history from earliest known civilizations to the modern era.

Moynahan, Brian. *The Faith: A History of Christianity*. New York: Doubleday, 2002. A very accessible narrative history of two thousand years of Christianity and its impact on world history.

Mysliwiec, Karol, translated by Geoffrey L. Packer. *Eros on the Nile*. Ithaca, N.Y.: Cornell University Press, 2002. Not as sexy as the title sounds, a fascinating but scholarly work on just how "hot" the Egyptians were.

Nash, Ronald H. *The Gospel and the Greeks: Did the New Testament Borrow from Pagan Thought?* (Original title: *Christianity and the Hellenistic World*.) Phillipsburg, N.J.: P&R Publishing Company, 2003. A scholarly work that refutes the idea that Christianity was an outgrowth of Greek philosophy and religion.

Nuland, Sherwin B. *Doctors: The Biography of Medicine*. New York: Alfred A. Knopf, 1988. A history of medicine that touches on the mythical beginnings of the healing arts.

Page, Jake. *In the Hands of the Great Spirit: The 20,000-Year History of American Indians*. New York: Free Press, 2003. Drawing on the latest archaeology and other research, a comprehensive overview of American Indian history.

Page, Jake, editor. *Sacred Lands of Indian America*. New York: Harry Abrams, 2001. A photographic collection with essays that ask the very important question "What makes a place sacred?" and, even more important, "How can such places be protected?" Beautiful and provocative.

Pagels, Elaine. *Adam, Eve, and the Serpent*. New York: Vintage, 1989. Prizewinning scholar's look at how early Christians viewed sex and transformed the pagan world.

Pagels, Elaine. *The Origin of Satan*. New York: Vintage, 1996. The Christian view of good and evil and how it influenced the rise of Christianity.

Pelikan, Jaroslav. *Jesus Through the Centuries: His Place in the History of Culture*. New Haven, Conn.: Yale University Press, 1985. A very readable examination of the changing image of Jesus over the course of two hundred years, written by a leading historian of religion.

Pelikan, Jaroslav. *Mary Through the Centuries: Her Place in the History of Culture.* New Haven, Conn.: Yale University Press, 1996. An assessment of the changing views of Virgin Mary.

Perrottet, Tony. *The Naked Olympics: The True Story of the Ancient Games.* New York: Random House, 2004. A highly entertaining and revealing account of the 1,200-year history of the ancient games. Very readable.

Pinch, Geraldine. *Egyptian Mythology: A Guide to the Gods, Goddesses, and Traditions of Ancient Egypt.* New York: Oxford University Press, 2002. Very comprehensive and reflecting much recent scholarship, a brief overview of Egyptian myths.

Plato. *The Republic.* New York: Vintage, 1991. The classic Socratic dialogues. (Many other editions available.)

Porter, J. R. *The Illustrated Guide to the Bible.* New York: Oxford University Press, 1995. Book by book, a look at the "Good Book."

Porter, Roy. *Blood and Guts: A Short History of Medicine.* New York: W. W. Norton, 2002. An entertaining overview of the history of healing, including medicine in the time of legendary healers in Egypt and Greece.

Restall, Matthew. *Seven Myths of the Spanish Conquest.* New York: Oxford University Press, 2002. A revisionist approach to the popular account of the Spanish conquest of the Americas. The author persuasively argues that the widely held notion that the Native Americans mistook the Spaniards for gods is myth.

Sagan, Carl, and Ann Druyan. *Shadows of Forgotten Ancestors: A Search for Who We Are.* New York: Random House, 1992. Better known for his writings about space (*Cosmos*), Sagan examines human experience in this wide-ranging, challenging, and fascinating book.

Seznec, Jean. *The Survival of the Pagan Gods: The Mythological Tradition and Its Place in Renaissance Humanism and Art.* Princeton, N.J.: Princeton University Press, 1981. A highly academic history of the revival of the Greek gods in art and literature during the European Renaissance.

Sowerby, Robin. *The Greeks: An Introduction to Their Culture.* London: Routledge, 1995. A concise, wide-ranging introduction to ancient Greece, from the age of Homer to the end of the classical period.

Stark, Rodney. *The Rise of Christianity: How the Obscure, Marginal Jesus Movement Became the Dominant Religious Force in the Western World in a Few Centuries.* San Francisco: HarperSanFrancisco, 1997. A sociological explanation of the rise of Christianity in a pagan world.

Tuchman, Barbara W. *The March of Folly: From Troy to Vietnam.* New York: Random House, 1984. Starting with the fatal mistake made by the Trojans, the Pulitzer Prize–winning and best-selling author catalogues a series of bad decisions made by governments in time of war. A largely ignored plea for applying the lessons of history.

Vogler, Christopher. *The Writer's Journey: Mythic Structure for Writers* (second edition). Studio City, Calif.: Michael Wiese Productions, 1998. A fascinating textbook that draws heavily on the work of Carl G. Jung and Joseph Campbell in utilizing myth for modern storytellers.

Voytilla, Stuart. *Myth and the Movies: Discovering the Mythic Structure of 50 Unforgettable Films*. Studio City, Calif.: Michael Wiese Productions, 1999. An interesting critical assessment of such classic films as *The Godfather*, *Jaws*, *The African Queen*, and *Citizen Kane* from a mythical perspective. Draws heavily on the themes laid out by Christopher Vogler (see above).

Wade, Nicholas. *The New York Times Book of Archeology*. Guilford, Conn.: Lyons Press, 2001. Collected articles from the newspaper's Science Times section record some of the major recent discoveries in archaeology.

Warner, Marina. *Alone of All Her Sex: The Myth and the Cult of the Virgin Mary*. New York: Vintage, 1983. A scholarly but accessible account of the changing historical perspectives given to the mother of Jesus, including the influence of ancient mythical characters on the image of Mary, especially in the early Christian period.

ACKNOWLEDGMENTS

Sometimes it is difficult to comprehend that this Don't Know Much About series started nearly twenty years ago with the simple idea about writing a book about something I loved—American history. It has grown into a series of books for adults and children that has exceeded my wildest imaginings. That could only have happened with the hard work, support, and determination of a large supporting cast. A great many people have been part of the long journey I have been on, and I wish to thank and recognize some of them for their unique contributions to making my work possible.

I start with a teacher somewhere out there who once read Homer's *Odyssey* to a group of fifth-graders in Mount Vernon, New York. To her, and the all the other teachers who inspire young minds every day in schools around America, I say thank you for doing what you do. It is the most important job in America, but is not usually seen that way. America owes an enormous debt of gratitude to the teachers who are so dedicated to the work of challenging young minds in difficult times.

For the past few years, it has been my great pleasure to work with an excellent group of committed, dedicated colleagues at HarperCollins, starting with Jane Friedman, who has been so supportive of my work. I would also like to especially thank Carrie Kania, Christine Boyd, Shaina Gopen, David Koral, Suzie Sisoler, Roberto de Vicq de Cumptich, Will Staehle, Susan Weinberg, Diane Burrowes, Patti Kelly, Leslie Cohen, my copyeditor, Olga Galvin Gardner, and my tireless publicist, Elly Weisenberg.

Most of all, I am deeply indebted to the pushing, prodding, and vision of my editor, Gail Winston. Her assistant, Katherine Hill, was also instrumental in making this book possible.

I have been fortunate to work with some of the nicest people in the book business at my longtime literary agency, the David Black Agency. I am not only lucky to have such hardworking, dedicated people on my team, I feel fortunate to consider all of them my friends: Jessica Candlin, Leigh Ann Eliseo, Linda Loewenthal, Gary Morris, Susan Raihofer, Jason Sacher, Joy Tutela, and the maestro, David Black.

Over the years, a great many other people have provided moral support, laughs, and the encouragement that make the work of writing bearable, and I am indebted to all of them for their friendship: Star Gibbs, Ellen Giusto, Jim and Esther Gray,

Joyce Waldon, and Linda Louise Watson. I also thank the wonderful people at one of America's great independent bookstores, the Northshire, in Manchester Center, Vermont.

I would like to add a special note of gratitude to April Prince, who has been a friend and a great help in making these books, especially this one, in recent years.

My deepest and greatest gratitude must always go to my family. First, my mother, Evelyn Davis, who made those trips to the local public library such a significant part of my young life. Without anybody knowing it or predicting it, those regular visits to the temple-like Mount Vernon Public Library set me on the road to becoming a writer.

My children, Colin and Jenny, are my joy and inspiration, and they have had to put up with a father who was often distracted or preoccupied.

And, finally, I thank the young woman who once told a bookstore owner who was interviewing me for a job, "Hire the kid." That same woman later told me I should be writing books, not selling them. She was so smart, I married her. Thank you, Joann. More than I can ever say.

—DORSET, VERMONT
May 2005

INDEX

ALSO BY KENNETH C. DAVIS

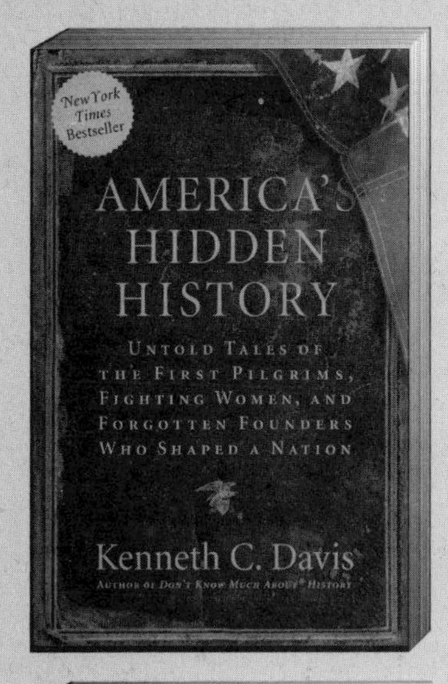

AMERICA'S HIDDEN HISTORY

UNTOLD TALES OF THE FIRST PILGRIMS, FIGHTING WOMEN, AND FORGOTTEN FOUNDERS WHO SHAPED A NATION

ISBN 978-0-06-111819-7 (paperback)

Kenneth C. Davis presents a collection of extraordinary stories, each detailing an overlooked episode that shaped the nation's destiny and character. Dramatic narratives set the record straight and bring to light fascinating facts from a time when the nation's fate hung in the balance.

"Once again Kenneth C. Davis proves that what you don't know can be shocking. Do yourself a favor. Read this book."
—Richard M. Cohen, author of *Strong at the Broken Places*

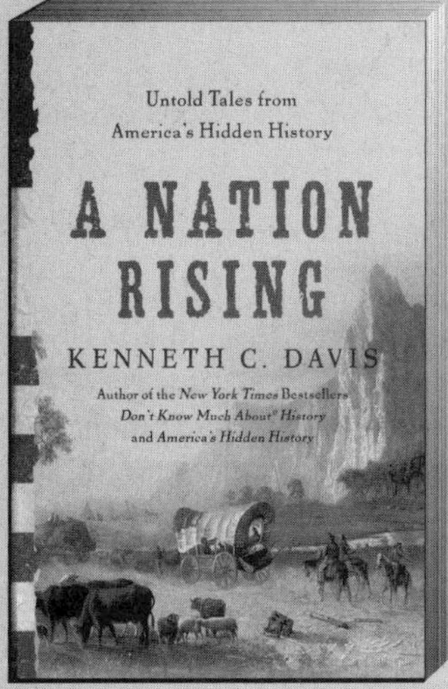

A NATION RISING

UNTOLD TALES FROM AMERICA'S HIDDEN HISTORY

ISBN 978-0-06-111821-0 (paperback)

Kenneth C. Davis explores the gritty first half of the 19th century, among the most tumultuous in the nation's short life, when the United States moved meteorically from a tiny, newborn nation, desperately struggling for survival on the Atlantic seaboard, to a near-empire that spanned the continent.

"Davis evokes the raw and violent spirit not just of an 'expanding nation,' but of an emerging and aggressive empire."
—Ray Raphael, author of *Founders*